Advances in Information Security

Volume 89

The purpose of the *Advances in Information Security* book series is to establish the state of the art and set the course for future research in information security. The scope of this series includes not only all aspects of computer, network security, and cryptography, but related areas, such as fault tolerance and software assurance. The series serves as a central source of reference for information security research and developments. The series aims to publish thorough and cohesive overviews on specific topics in Information Security, as well as works that are larger in scope than survey articles and that will contain more detailed background information. The series also provides a single point of coverage of advanced and timely topics and a forum for topics that may not have reached a level of maturity to warrant a comprehensive textbook.

Tiffany Bao • Milind Tambe • Cliff Wang
Editors

Cyber Deception

Techniques, Strategies, and Human Aspects

Editors
Tiffany Bao
Arizona State University
Tempe, AZ, USA

Milind Tambe
Harvard University
Cambridge, MA, USA

Cliff Wang
Army Research Office
Adelphi, MD, USA

ISSN 1568-2633 ISSN 2512-2193 (electronic)
Advances in Information Security
ISBN 978-3-031-16612-9 ISBN 978-3-031-16613-6 (eBook)
https://doi.org/10.1007/978-3-031-16613-6

This Springer imprint is published by the registered company Springer Nature Switzerland AG
The registered company address is: Gewerbestrasse 11, 6330 Cham, Switzerland

Preface

This book introduces cutting-edge research works in cyber deception, a research topic that has been actively studied and significantly advanced in the past decade. With the focus on cyber deception, this book spans a wide variety of areas, including game theory, artificial intelligence, cognitive science, and cybersecurity. This book will address three core cyber deception research elements as follows:

1. Understanding of human's cognitive behaviors in decoyed network scenarios
2. Development of effective deceptive strategies based on human behaviors
3. Design of deceptive techniques enforcing deceptive strategies

The research introduced in this book will identify the scientific challenges, highlight the complexity, and inspire future research of cyber deception.

This book can be used as a professional book by cybersecurity practitioners and researchers, or a supplemental textbook for educational purposes. Readers will be able to learn the state of the art in cyber deception to conduct follow-up research or translate related research outcome to practice.

Tempe, AZ, USA — Tiffany Bao
Cambridge, MA, USA — Milind Tambe
Triangle Park, NC, USA — Cliff Wang

Acknowledgments

We would like to thank all the contributors for their dedication to this book. Special thanks go to Ms. Susan Lagerstrom-Fife and Ms. Shanthini Kamaraj for their kind support of this book. Finally, we thank the Army Research Office for their financial support under the grant numbers W911NF-17-1-0370.

Contents

Diversifying Deception: Game-Theoretic Models for Two-Sided Deception and Initial Human Studies

Mohammad Sujan Miah, Palvi Aggarwal, Marcus Gutierrez, Omkar Thakoor, Yinuo Du, Oscar Veliz, Kuldeep Singh, Christopher Kiekintveld, and Cleotilde Gonzalez

1 Introduction

Both civilian and military computer networks are under increasing threat from cyberattacks, with the most significant threat posed by Advanced Persistent Threat (APT) actors. These attackers use sophisticated methods to compromise networks and remain inside, establishing greater control and staying for long periods to gather valuable data and intelligence. These attackers seek to remain undetected, and estimates from APT attacks show that they are often present in a network for months before they are detected [31].

Cyber deception methods use deceptive decoy objects like fake hosts (honeypots), network traffic, files, and even user accounts to counter attackers in a variety of ways [1, 13, 28]. They can create confusion for attackers, make them more hesitant and less effective in executing further attacks, and can help to gather

An earlier version of some parts of the work was published in the proceedings of the 53rd Hawaii International Conference on System Sciences (HICSS), 2020, pp. 01–20.

M. S. Miah · M. Gutierrez · O. Veliz · C. Kiekintveld (✉)
Department of Computer Science, University of Texas at El Paso, El Paso, TX, USA
e-mail: msmiah@miners.utep.edu; mgutierrez22@miners.utep.edu; osveliz@utep.edu; cdkiekintveld@utep.edu

P. Aggarwal · Y. Du · K. Singh · C. Gonzalez
Carnegie Mellon University, Pittsburgh, PA, USA
e-mail: palvia@andrew.cmu.edu; yinuod@andrew.cmu.edu; kuldeep2g@andrew.cmu.edu; coty@cmu.edu

O. Thakoor
University of Southern California, Los Angeles, CA, USA
e-mail: othakoor@usc.edu

T. Bao et al. (eds.), *Cyber Deception*, Advances in Information Security 89,
https://doi.org/10.1007/978-3-031-16613-6_1

information about the behavior and tools of various attackers. They can also increase the ability of defenders to detect malicious activity and actors in the network. This deception is especially critical in the case of APT attackers, who are often cautious and skilled at evading detection [32]. Widespread and effective use of honeypots and other deceptive objects is a promising approach for combating this class of attackers.

However, the effectiveness of honeypots and other deceptive objects depends crucially on whether the honeypot creators can design them to look similar enough to real objects, to prevent honeypot detection and avoidance. This design goal especially holds for APT threats, which are likely to be aware of the use of such deception technologies and will actively seek to identify and avoid honeypots, and other deceptive objects, in their reconnaissance [32, 35]. A well-known problem with designing successful honeypots is that they often have characteristics that can be observed by an attacker that will reveal the deception [14]. Examples of such characteristics include the patterns of network traffic to a honeypot, the response times to queries, or the configuration of services which are not similar to real hosts in the network. However, with some additional effort, these characteristics can be made more effective in deception (e.g., by simulating more realistic traffic to and from honeypots).

In this chapter, we introduce a game-theoretic model of the problem of designing effective decoy objects that can fool even a sophisticated attacker. In our model, real and fake objects may naturally have different distributions of characteristic features than an attacker could use to tell them apart. However, the defender can make some (costly) modifications to *either* the real or the fake objects to make them harder to distinguish. This model captures some key aspects of cyber deception that are missing from other game-theoretic models. In particular, we focus on whether the defender can design convincing decoy objects, and what the limitations of deception are if some discriminating features of real and fake objects are not easily maskable.

We also present several analyses of fundamental questions in cyber deception based on our model. We analyze how to measure the informativeness of the signals in our model and then consider how effectively the defender can modify the features to improve the effectiveness of deception in various settings. We show how different variations in the costs of modifying the features can have a significant impact on the effects of deception. We also consider the differences between modifying only the features of deceptive objects and being able to modify both real and deceptive objects (two-sided deception). While this is not always necessary, in some cases, it is essential to enable effective deception. We also consider deception against naïve attackers, and how this compares to the case of sophisticated attackers.

Next, we present an exploratory study that looked into the effectiveness of a two-sided deception technique using a human-attackers trial. In the experiment, we used a network topology with an equal number of real machines and honeypots where we modify the features of a system using an experimental test bed (HackIT). We first categorized the adaptable features of both real machines and honeypots, then changed the features' characteristics and observed the attackers' behavior after modification. Finally, we discuss the results of our case study in no-deception, one-sided, and two-sided situations.

Later section in this chapter will discuss how our game model relates to work in adversarial learning and how this model could be applied beyond the case of honeypots to, for example, generating decoy network traffic.

2 Motivating Domain and Related Work

While the model we present may apply to many different types of deception and deceptive objects, we will focus on honeypots as a specific case to make our discussion more concrete and give an example of how this model captures essential features of real-world deception problems. Honeypots have had a considerable impact on cyber defense in the 30 years since they were first introduced [29].

Over time, honeypots have been used for many different purposes and have evolved to more sophisticated designs with more advanced abilities to mimic real hosts and to capture useful information about attackers [5, 18, 20]. The sophistication of honeypots can vary dramatically, from limited low-interaction honeypots to sophisticated high-interaction honeypots [9, 18, 22].

Here, we do not focus on the technological advancements of honeypots, but rather on the game-theoretic investigation of honeypot deception. There have been numerous works that emphasize this game-theoretic approach to cyber deception as well. Our work builds upon the Honeypot Selection Game (HSG), described by Píbil et al. [13, 21]. Much like the HSG, we model the game using an extensive form game. We extend the HSG model with the introduction of *features*, which are modifiable tokens in each host that enable more robust deceptions and allow to model more realistic settings. Several game-theoretic models have been established for other cyber defense problems [4, 17, 24, 26], specifically for deception as well [25, 33]; however, these consider attribute obfuscation as the means of deception rather than use of decoy objects.

Reference [34] notably investigate the use of honeypots in the smart grid to mitigate denial-of-service attacks through the lens of Bayesian games. Reference [16] also model honeypots mitigating denial-of-service attacks in a similar fashion but in the Internet-of-Things domain. Reference [8] tackle a similar "honeypots to protect social networks against DDoS attacks" problem with Bayesian game modeling. These works demonstrate the broad domains where honeypots can aid. This work differs in that we do not model a Bayesian incomplete information game.

A couple of works also consider the notion of two-sided deception, where the defender deploys not only *real*-looking honeypots but also *fake*-looking real hosts. Rowe et al. demonstrate that using two-sided deception offers an improved defense by scaring off attackers [23]. Caroll and Grosu introduced the signaling deception game where signals bolster a deployed honeypot's deception [6]. Our work differs in that we define specific features (signals) that can be altered and revealed to the attacker. Shi et al. introduce the mimicry honeypot framework, which combines real nodes, honeypots, and *fake*-looking honeypots to derive equilibria strategies to bolster defenses [27]. They validated their work in a simulated network. This notion of two-sided deception is quickly becoming a reality; De Gaspari et al. provided a prototype proof-of-concept system where production systems also engaged in active deception [7].

3 Feature Selection Game

Feature Selection Game (FSG) models the optimal decisions for a player (the defender) who is trying to disguise the identity of real and fake objects so that the other player (the attacker) is not able to reliably distinguish between them. Each object in the game is associated with a vector of observable features (characteristics) that provides an informative signal that the attacker can use to detect fake objects more reliably. The defender can make (limited) changes to these observable features, at a cost. Unlike many models of deception, this game model considers the possibility that the defender can make changes to both the real and fake objects; we refer to this as 2-sided deception.

The original feature vector is modeled as a move by nature in a Bayesian game. Real and fake objects have different probabilities of generating every possible feature vector. How useful the features are to the attacker depends on how similar the distributions for generating the feature vectors are; very similar distributions have little information while very different distributions may precisely reveal which objects are real or fake. The defender can observe the features and may choose to pay some cost to modify a subset of the features. The attacker observes this modified set of feature vectors and chooses which object to attack. The attacker receives a positive payoff if he selects a real object, and a negative one if he selects a honeypot.

To keep the initial model simple, we focus on binary feature vectors to represent the signals. We will also assume that the defender can modify a maximum of one feature. Both of these can be generalized in a straightforward way, at the cost of a larger and more complex model.

3.1 Formal Definition of Feature Selection Game

We now define the Feature Selection Game (FSG) formally by the tuple $G = (K^{\mathrm{r}}, K^{\mathrm{h}}, N, v^{\mathrm{r}}, v^{\mathrm{h}}, C^{\mathrm{r}}, C^{\mathrm{h}}, P^{\mathrm{r}}, P^{\mathrm{h}}, \tau, \chi)$.

- K^{r} denotes the set of real hosts and K^{h} denotes the set of honeypots. Altogether, we have the complete set of hosts $K = K^{\mathrm{r}} \cup K^{\mathrm{h}}$. We denote the cardinalities of these by $k = |K|$, $r = |K^{\mathrm{r}}|$, $h = |K^{\mathrm{h}}|$.
- $[n]$ is the set of features that describe any given host. The sequence of feature values of a host is referred to as its *configuration*. Thus, the set of different possible configurations is $\{0, 1\}^n$.
- v^{r}, v^{h} denote the importance values of the real hosts and honeypots, respectively.

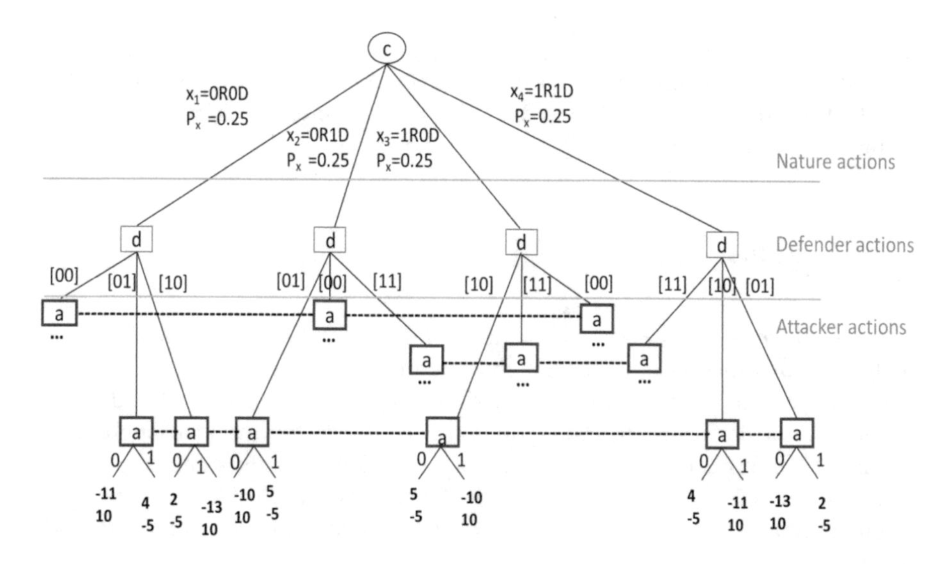

Fig. 1 The extensive form game tree with one real host, one honeypot and 1 feature in each host. The importance value of real host is 10, whereas the modification cost of a feature is 3. The same values for the honeypot are 5, 1, respectively

- C^{r}, C^{h} denote the cost vectors associated with modifying a single feature of a real host and a honeypot, respectively, and are indexed by the set of features N. Thus, C_i^{r} is the cost of modifying the ith feature of a real host.
- $P^{\mathrm{r}} : \{0, 1\}^n \rightarrow [0, 1]$ is probability distribution over feature vectors for real hosts.
- $P^{\mathrm{h}} : \{0, 1\}^n \rightarrow [0, 1]$ is the probability distribution over feature vectors for honeypots.
- The collection of all possible information sets is denoted by τ.
- $\chi : \{0, 1\}^{kn} \times D \rightarrow \tau$ is a function that given the initial network and a defender action, outputs the attacker's resultant information set $I \in \tau$. Here, D is the set of defender actions.

An example of a small FSG with 1 real host, 1 honeypot, and 1 feature for each host is shown in Fig. 1. The probability distributions $P^{\mathrm{r}}(0) = P^{\mathrm{r}}(1) = 0.5$ and $P^{\mathrm{h}}(0) = P^{\mathrm{h}}(1) = 0.5$ are randomly generated for each feature combination.

3.2 Nature Player Actions

We assume that both players know the probability distributions P^r and P^h that define how the feature vectors are selected by nature for real and honeypot hosts, respectively. Nature generates the network configurations as per the distributions P^{r} and P^{h}. Thus, the network state $x = (x_1, \ldots, x_k)$ is generated as per the joint distribution P^x where $P^x(x) = \prod_{i=1}^{r} P^r(x_i) \times \prod_{i=r+1}^{k} P^h(x_i)$. Both players can compute the distribution P^x. For example, in Fig. 1 $P^x = 0.25$ for network 0R1D is calculated from $P^r(0) = 0.5$ and $P^h(1) = 0.5$. Here, 0R1D refers to the 1st feature's status of real host and decoy object (honeypot).

3.3 Defender Actions

The defender observes the network configuration $x \in X$, selected by nature as per probability distribution P^X. Then he chooses an appropriate action $d \in D$, which is to change at most one feature of any single host. Thus, D has $nk + 1$ different actions. This action results in a configuration $x' \in \{0, 1\}^{nk}$ that the attacker observes, defining his information set $I \in \tau$ as described previously. In the example of Fig. 1, given the initial network configuration 0R0D, the defender can alter a feature which results into 0R1D or 1R0D, or make no change leading to 0R0D as the attacker's observation.

3.4 Attacker Actions

The attacker observes the set of feature vectors for each network but does not directly know which ones are real and which are honeypot. Thus, any permutation of the host configurations is perceived identically by the attacker. Hence, the attacker's information set is merely characterized by the combination of the host configurations and thus represented as a multiset on the set of host configurations as the Universe. For example, in Fig. 1, the networks 0R1D and 1R0D belong to the same information set. Given the attacker's information set, he decides which host to attack. When indexing the attack options, we write the information set as an enumeration of the k host configurations, and we assume a lexicographically sorted order as a convention. Given this order, we use a binary variable a_i^I to indicate that when he is in the information set I, the attacker's action is to attack host $i \in K$.

3.5 Utility Functions

A terminal state t in the extensive form game tree is characterized by the sequence of actions that the players (nature, defender, attacker) take. The utilities of the players can be identified based on the terminal state that the game reaches. Thus, given a terminal state t as a tuple (x, j, a) of the player actions, we define a function $U(t) = U(x, j, a)$ such that the attacker gains this value while the defender loses as much. That is, this function serves as the zero-sum component of the player rewards. In particular, if the action a in the information set $\chi(x, j)$ corresponds to a real host, then $U(x, j, a) = v^{\mathrm{r}}$, whereas if it corresponds to a honeypot, then $U(x, j, a) = -v^{\mathrm{h}}$. Intuitively, the successful identification of a real host gives a positive reward to the attacker otherwise gives a negative reward that is equal to the importance value of a honeypot. The expected rewards are computed by summing over the terminal states and considering the probabilities of reaching them. Finally, the defender additionally also incurs the feature modification cost C_i^{r} or C_j^{h} if his action involved modifying ith feature of a real host or jth feature of a honeypot, respectively.

3.6 Solution Approach

We solve this extensive form game with imperfect information using a linear program. For solving this game in sequence form [15], we create a path from the root node to the terminal node that is a valid sequence and consists of a list of actions for all players. Then we compute defender's behavioral strategies on all valid sequences using a formulated LP as follows, where U_d and U_a are the utilities of the defender and the attacker. To solve the program, we construct a matrix $X[0 : 2^{kn}]$ of all possible network configurations, and then the defender chooses a network $x \in X$ to modify. In network x, any action d of the defender leads to an information set I for the attacker. Different defender's actions in different networks can lead to the same information set $I \in \tau$. Then, in every information set I, the attacker chooses a best response action to maximize his expected utility.

$$max \sum_{x \in X} \sum_{j \in D} \sum_{i \in K} U_d(x, j, i)\, d_j^x\, P^x a_i^{\chi(x,j)} \tag{1}$$

$$s.t. \quad \sum_{(x,j):\chi(x,j)=I} U_a(x, j, i) d_j^x\, P^x\, a_i^I \geq$$

$$\sum_{(x,j):\chi(x,j)=I} U_a(x, j, i')\, d_j^x\, P^x\, a_i^I$$

$$\forall i, i' \in K \quad \forall I \in \tau \tag{2}$$

$$d_j^x \geq 0 \quad \forall x \in X \quad \forall j \in D \tag{3}$$

$$\sum_{j \in D} d_j^x = 1 \quad \forall x \in X \tag{4}$$

$$\sum_{i \in K} a_i^I = 1 \quad \forall I \in \tau \tag{5}$$

The program's objective is to maximize the defender's expected utility, assuming that the attacker will also play a best response. In the above program, the only unknown variables are the defender's actions D (the strategies of a defender in a network $x \in X$) and the attacker's actions a^I. The inequality in Eq. 2 ensures that the attacker plays his best response in this game, setting the binary variable a_i^I to 1 only for the best response i in each information set. Equation 3 ensures that the defender strategies in a network x is a valid probability distribution. Equation 4 makes sure that all probability for all network configurations sum to 1. Finally, Eq. 5 ensures that the attacker plays pure strategies.

4 Empirical Study of FSG

The FSG game model allows us to study the strategic aspects of cyber deception against a sophisticated adversary who may be able to detect the deception using additional observations and analysis. In particular, we can evaluate the effectiveness of cyber deception under several different realistic assumptions about the costs and benefits of deception, as well as the abilities of the players. We identify cases where deception is highly beneficial, as well as some cases where deception has limited or no value. We also show that in some cases, using two-sided deception is critical to the effectiveness of deception methods.

4.1 Measuring the Similarity of Features

One of the key components of our model is that real hosts and honeypots generate observable features according to different probability distributions. The similarity of these distributions has a large effect on the strategies in the game, and the outcome of the game. Intuitively, if out-of-the-box honeypot solutions look indistinguishable from existing nodes on the network the deception will be effective without any additional intervention by the defender. However, when the distributions of features are very dissimilar the defender should pay higher costs to modify the features to disguise the honeypots. In some cases this may not be possible, and the attacker will always be able to distinguish the real hosts and honeypots.

Measuring the similarity of the feature distributions is a somewhat subtle issue, since the defender can make changes to a limited number of features. Standard approaches such as Manhattan distance or Euclidean distance do not provide a good way to compare the similarity due to these constraints. We use a measure based on the Earth Mover's Distance (EMD) [19], which can be seen as the minimum distance required to shift one pile of earth (probability distribution) to look like another. This measure can be constrained by the legal moves, so probability is only shifted between configurations that are reachable by the defender's ability to change features.

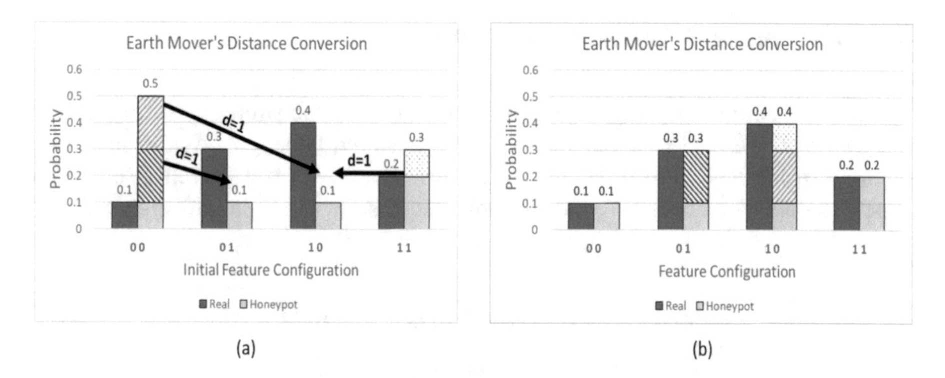

Fig. 2 Earth Mover's Distance process. (**a**) Displays the initial feature configuration probability distributions P_r and P_h and where to move slices of the distribution from P_h and (**b**) Shows the updated P_h after the conversion, resulting in a final EMD of 0.5

In the experiments, we allow the defender to modify only a single feature in the network and the EMD determines the minimum cost needed to transform a weighted set of features to another where the probability of each feature configuration is the weight. The ground dissimilarity between two distributions is calculated by the Hamming distance. This distance between two distributions of equal length is the number of positions at which the comparing features are dissimilar. In other words, it measures the minimum number of feature modification or unit change required to make two sets of feature indistinguishable. We model the distance from moving the probability of one configuration (e.g., turning [0, 0] into [0, 1]) to another by flipping of a single bit at a time with a unit cost of 1. This can be seen visually in Fig. 2 where we calculate the EMD of moving the honeypot's initial distribution into that of the real node's initial distribution.

In our experiments we will often show the impact of varying levels of similarity in the feature distributions. We generated 1000 different initial distributions for the features using uniform random sampling. We then calculated the similarities using the constrained EMD and selected 100 distributions so that we have 10 distributions in each similarity interval. We randomly select these 10 for each interval from the ones that meet this similarity constraint in the original sample. This is necessary to balance the sample because random sampling produces many more distributions that are very similar than distributions that are further apart, and we need to ensure a sufficient sample size for different levels of similarity. we present the results by aggregating over the similarity intervals of 0.1 and average ten results in each interval.

4.2 Deception with Symmetric Costs

Our first experiment investigates the impact of varying the similarity of the feature distributions. We also vary the values of real host and honeypot. As the similarity of the distributions P^r and P^h decreases, we would expect a decrease in overall expected defender utility. We can see this decrease in Figs. 3a and b as we vary

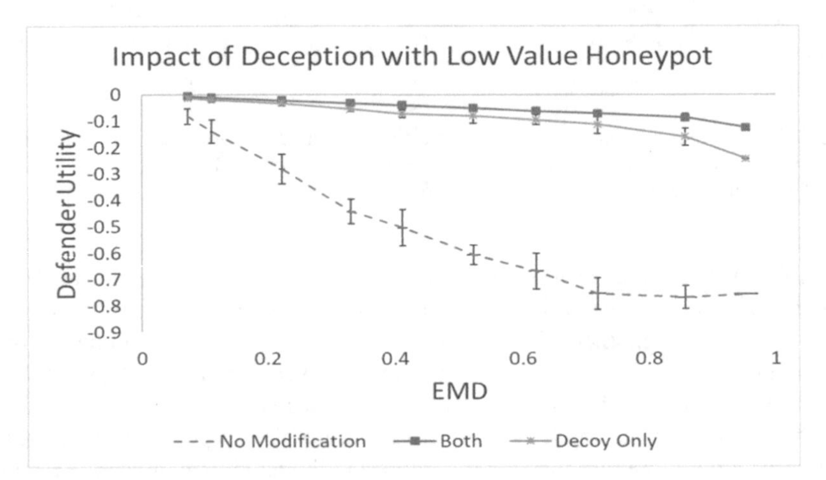

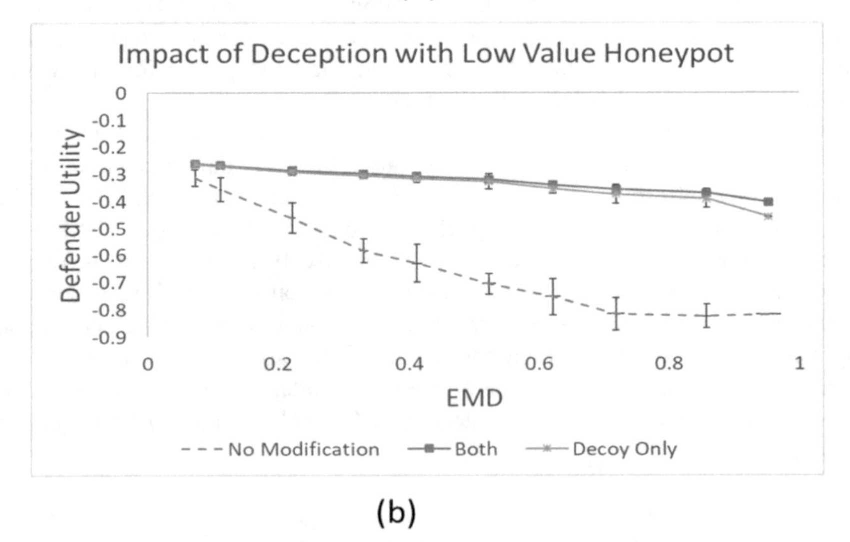

Fig. 3 Comparison of defender utility when the real host's importance value (**a**) doubles that of the honeypot and (**b**) equals that of the honeypot. Here we see one-sided deception provides a comparable defense despite a high initial dissimilarity

the similarity measured using EMD. In Figs. 3a and b, we compare the utility differences between an optimal defender that can only modify the features of the honeypot (one-sided deception), an optimal defender that can modify features of *both* the honeypot and real host (two-sided deception), and a baseline defender that cannot make any modifications against a fully rational best response attacker.

In Fig. 3a, the honeypot has the same importance value as the real host, while in Fig. 3b, the honeypot value is half of the real host. The first observation is that in both cases the value of deception is high relative to the baseline with no deception, and this value grows dramatically as the feature distributions become more informative (higher EMD). In general, the defender does worse in cases where the hosts have different values. Two-sided deception does have a small advantage in cases with highly informative features, but the effect is small. Here, the costs of modifying the features are symmetric, so there is little advantage in being able to modify the feature on either the honeypot or the real host, since the defender can choose between these options without any penalty.

To further investigate the issue of one-sided and two-sided deception, we fix the honeypot features modification costs and increased real host modification costs as reflected in Table 1. Here, we compare how increasing the real host's feature modification negatively affects the defender's expected utility. As the cost for modifying the real hosts increases relative to the cost of modifying honeypots, the defender must make more changes on honeypots in order to maximize his utility. Altering the real system in this case is not feasible and does not provide a good return on investment.

Traditionally network administrators avoid altering features in their real hosts on the network and simply employ one-sided deception, attempting to alter the honeypot to look like a real host. In the case where modifying a real host to look *less believable* might be too costly or even impossible, one-sided deception is an obvious

Table 1 Parameters used in HFSG experiments. RIV denotes real system's importance value, RMC denotes real system's feature modification cost, HpIV denotes importance value of honeypot, and HpMC denotes feature modification cost of honeypot. All numbers are normalized to 1

		RMC		HpMC		
Figure	RIV	F 1	F 2	F 1	F 2	HpIV
3a	1.0	0.25	0.1	0.1	0.25	0.5
3b	1.0	0.25	0.1	0.2	0.1	1.0
4 (Both (A))	1.0	0.25	0.1	0.1	0.2	0.5
4 (Both (B))	1.0	0.5	0.2	0.1	0.2	0.5
4 (Both (C))	1.0	1.0	0.5	0.1	0.2	0.5
5 (Exp-1)	1.0	0.1	∞	0.1	∞	1.0
5 (Exp-2)	1.0	0.1	∞	∞	0.1	1.0
6 (Exp-1)	1.0	0.2	0.2	0.2	0.2	1.0
6 (Exp-2)	1.0	0.15	0.25	0.25	0.15	1.0
6 (Exp-3)	1.0	0.1	0.3	0.3	0.1	1.0
6 (Exp-4)	1.0	0.05	0.35	0.35	0.05	1.0
6 (Exp-5)	1.0	0.0	0.4	0.4	0.0	1.0
8	1.0	0.25	0.1	0.2	0.1	1.0

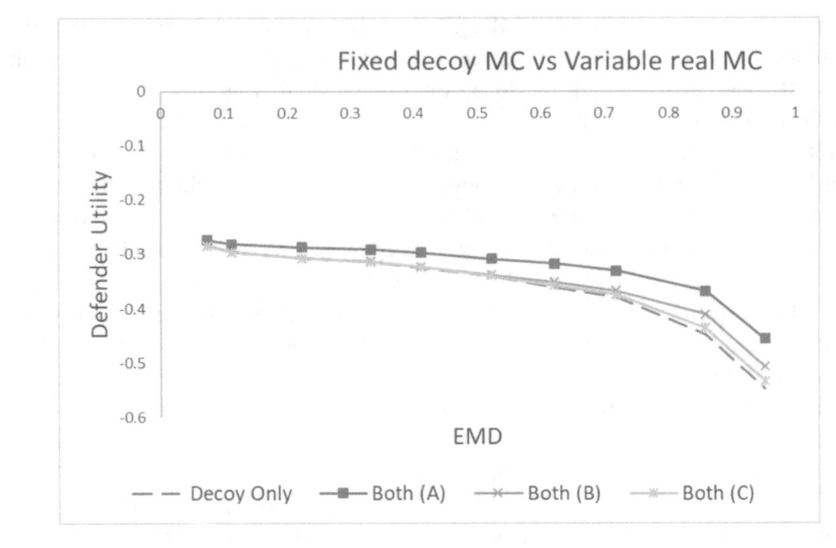

Fig. 4 Comparison of defender utility when the cost of modifying the real host features is different than modifying the honeypot features

choice as demonstrated in Fig. 4. However, when these real feature modifications are not too costly, we see that two-sided provides a noticeable increase in defenses when the feature distributions are increasingly dissimilar.

4.3 Deception with Asymmetric Costs

While the results so far have suggested that one-sided deception may be nearly as effective as two-sided deception, they have all focused on settings where the costs of modifying features are *symmetric* for real and fake hosts. We now investigate what happens when the costs of modifying different features are asymmetric. We start with the extreme case where some features may not be possible to modify at all.

In our examples with two features, we can set the unmodifiable features for the real and honeypot hosts to be the same or to be opposite. In Fig. 5, we show the results of the game when we set the modification costs of some features to infinity. If the same feature for the real host and honeypot are unmodifiable, then there is little the defender can do to deceive an intelligent attacker when they are highly dissimilar. However, when the features that cannot be modified are different for the real and honeypot hosts, we see a very different situation. In this case the defender benefits greatly from being able to use two-sided deception, since he can avoid the constraints by modifying either the real or fake hosts as needed.

In our next experiment, we investigate less extreme differences in the costs of modifying features. We set the costs so that they are increasingly different for

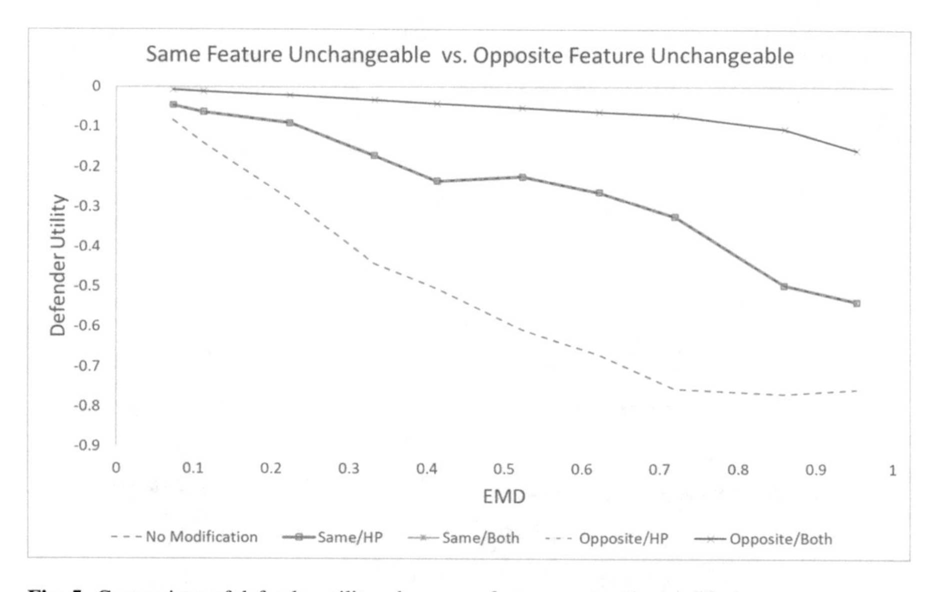

Fig. 5 Comparison of defender utility when some features cannot be modified

real and honeypot hosts, so modifying one feature is cheap for one but expensive for the other, but not impossible. We show the results of using either one or two-sided deception for varying levels of initial feature distribution similarity in Fig. 6. The specific costs are given in Table 1. We see that there is very little difference when the initial distributions are similar; this is intuitive since the attacker has little information and deception is not very valuable in these cases. However, we see a large difference when the initial distributions are informative. As the difference in the feature modification costs increases, the value of two-sided deception increases, indicating that this asymmetry is crucial to understanding when two-sided deception is necessary to employ effective deception tactics.

We also expect that the number of features available to the players will have a significant impact on the value of deception. While the current optimal solution algorithm does not scale well, we can evaluate the differences between small numbers of features, holding all else equal. Figure 7 presents the results of the modeling HFSG with variable number of features. We found that when the number of features is increased two-sided deception becomes more effective than one-sided deception. The defender in this case has more opportunity to alter the network by changing the features and make it the more confusing network to the attacker. However, the defender payoff decreases with more features due to the constraint on how many features he can modify and the total cost of modifying these features.

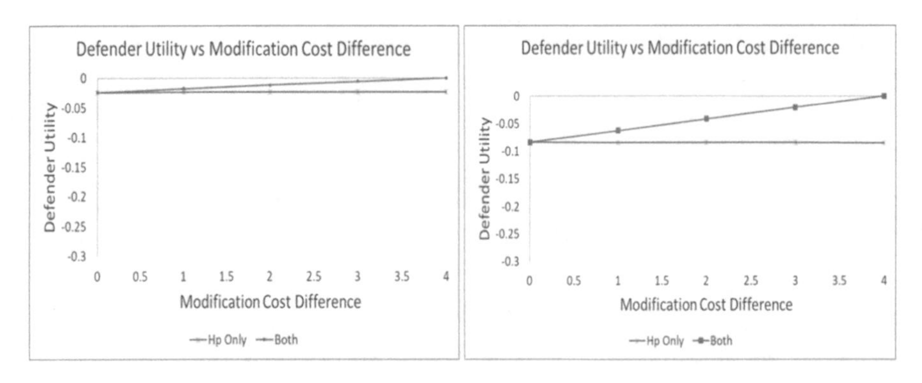

(a) EMD - 0.11 (b) EMD - 0.41

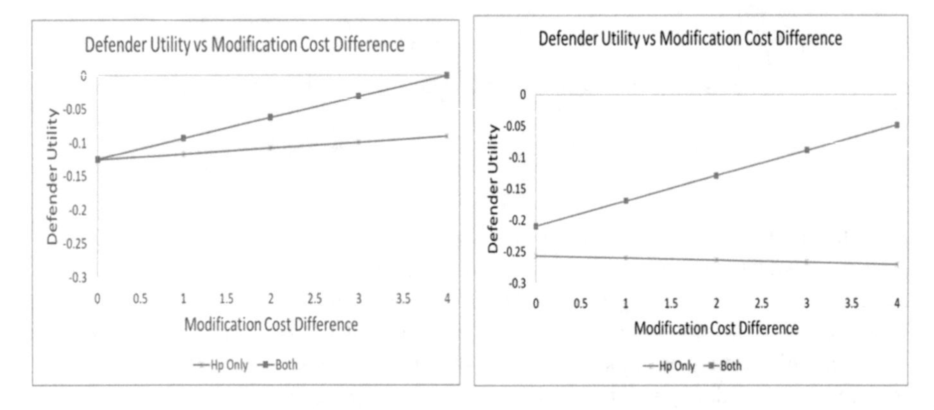

(c) EMD - 0.63 (d) EMD - 0.95

Fig. 6 Impact of modification cost over various initial similarity parameters

4.4 Deception with Naïve Attackers

The previous empirical results all assumed a cautiously rational attacker who actively avoided attacking honeypots. This is a common practice, because fully rational actors present the highest threat. In cybersecurity, these fully rational attackers might be an experienced hacker or APT. However, these are not the only threats faced in cybersecurity and we cannot assume that these attacking agents are always cautious and stealthy. For example, many attacks on networks may be conducted by worms or automated scripts that are much simpler and may be much more easily fooled by deceptive strategies.

We now consider a more naïve attacker that does not consider the defender's deception. He observes the hosts on the network and assumes no modifications were made. Based on all observations for a particular network he calculates his

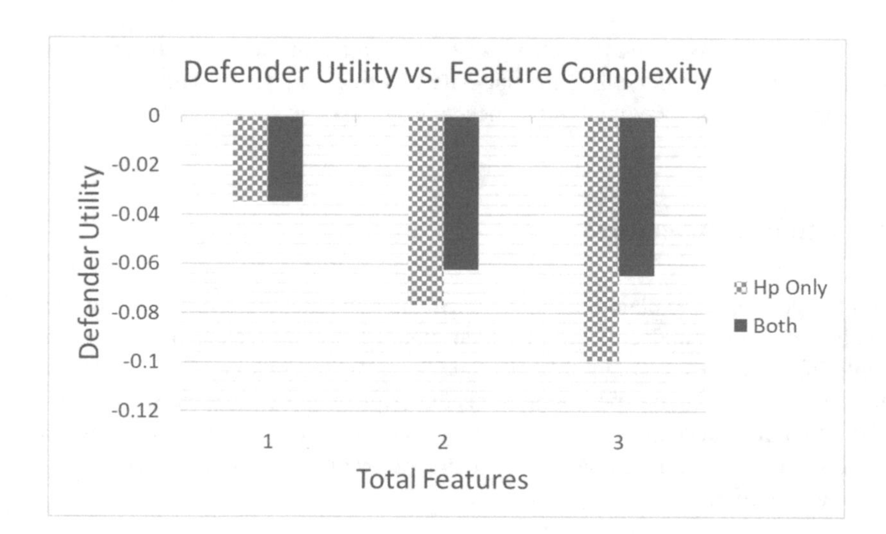

Fig. 7 Comparison of defender utility when increasing the number of features

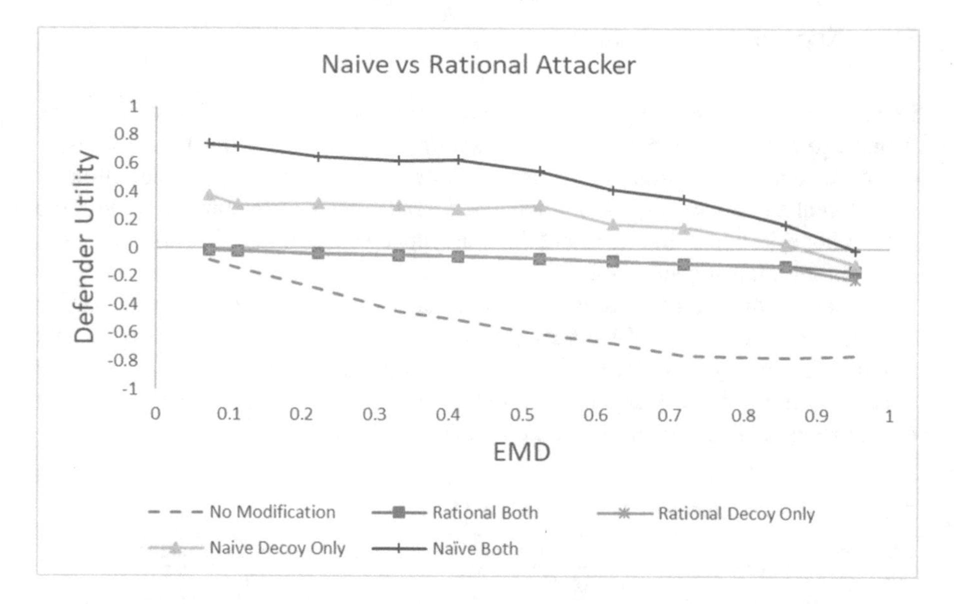

Fig. 8 Comparison of defender utility of a naïve attacker versus a fully rational attacker. Here, the naïve attacker does not consider the defender's utility or strategy at all

best response but does predict the defender's optimal strategy. The results of the experiment are shown in Fig. 8 and the costs given in Table 1.

The best case is when the defender can perform two-sided deception against a naïve attacker and the worst case is when the defender performs no deceptive actions against a fully rational attacker. These two cases form an upper- and lower-bound as seen in Fig. 8. Two-sided deception is more effective in this case when the feature distributions are similar, while the opposite was true for a rational attacker. Overall, deception strategies are much more effective against naïve attackers.

5 Human Experiment

In this section, we explain the human experiment conducted using an experimental testbed HackIT [2]. HackIT provides various capabilities such as network creation with different sizes, configurable machine features, and manipulation of topologies for conducting human experiments. This experiment was conducted with a network size of 40 machines, including 20 honeypots and 20 real machines. These machines were used with different configurations to achieve deception based on the experimental conditions.

5.1 Experimental Design

In this experiment, we study the effectiveness of a two-sided deception strategy. We define two between-subject conditions: *default*, where both real machines and honeypots have default configurations; and *2-sided deception*, where the configuration of both real and honeypot is modified. The real machines were configured with the default honeypot configuration, and the honeypots were configured with the default real configuration shown in Table 2.

In each condition, there are two identical rounds of 25 min each. The proportion of honeypots (20 out of 40 machines) remains the same across conditions and rounds. The honeypots were allocated randomly in the network. The real and honeypots used in the study have multiple features as shown in Table 2. These features were configured based on the experimental conditions.

Table 2 The configuration of Real machines and Honeypots

Features	Default honeypot	Default real machine
OS	Obsolete OS	Up to date
Ports & services	Normal ports, honeypot ports	Normal ports
Exploit information	Obsolete	Up to date
Exploit success rate	100%	40%
Link latency	1 ms	0.2 ms
Virtual/physical machine	VM	Physical
Running processes	2 processes	10 processes
File system (1 user folder with 5~6 sub-folders)	4~5 empty or access-deny folders	1 empty or access-deny folder

5.2 Experiment Task

In this experiment, the participant played the role of an attacker. The goal of an attacker was to steal a file "pin.txt" containing confidential credit card pin numbers stored in different systems on the network. The network contains 40 systems, and participants can explore and exploit as many systems as they can within a given time frame. For the exploration and exploitation of a system on the network, the participants were provided tools which were accessible through commands, as shown in Table 3.

The first step in the experimental task is information gathering by accessing available tools using the given commands nmap, checkHS, and info_ exploit commands. There are some commands which have a cost associated with it as these commands provide some extra information to the participants. The participants have limited time in the experiment, when participants use these particular commands, the cost (i.e., time) is detected from their remaining time. The *"checkHS"* command is to detect the existence of deceptive strategies. The reliability of *"checkHS"* command is only 50% and it cost approximately 10 s to participants. It generate a score from 0 to 1 representing the likelihood of the system being deceptive (0 = "not deceptive" to 1 = "deceptive"). To find the most promising attack path, participants can collect more detailed information about exploits, including its disclosure date, using *"info_exploit exploit"*. Attackers could decide to attack the system they explored or gather more information about other systems.

Table 3 Available commands for participants

Execution place	Command name	Command use
Pre-Exploitation	nmap -sL all	Scan the network and find which system are available in network
	nmap System	Scan a system to know open ports, application vulnerabilities, and OS
	nmap System [-rtt]	Provides benchmark RTT and RTT of the current system
	checkHS	Check Honeypot score, i.e., detect the existence of deceptive strategies
	info_ exploit exploit	Gather information about specific exploit
	use_exploit	Use exploit to target a system
Post-Exploitation	ls	List all files present in the current directory
	cd	Traverse through the directories
	ps -A	Lists all running processes on the system
	checkVM	To check machine is a physical system or a virtual environment
	scp	To transfer "pin.txt" file from the system to attacker system

When the participants are ready to exploit a machine, they need to select an exploit for that system. To exploit a vulnerability on a machine, they need to execute the *"use_ exploit"* command. For example, the participant can run the following command to exploit SQL injection vulnerability on System1: *"use_exploit sql_injection System1."* The "use_exploit" command might require multiple tries to gain success. The participants can also switch to another vulnerability on the same system.

Once the participants gain access to the server, they may do further reconnaissance with the following commands- "ls," "cd," "ps -A," and "checkVM." After collecting enough information and traversing through various directories, participants may decide whether to steal the file using *scp* command or leave the system without stealing the file. The scp command takes two parameter filename and destination as follow: *scp pin.txt address*. The participant gets feedback about the success or failure of the operation. If the system is a real machine and participants steal "pin.txt" file, they earn 100 points and if they logout without stealing the file, then they earn 30 points. Similarly, if the system is a honeypot and they steal "pin.txt" file, they lose 100 points and if they do not steal and logout, then they lose 30 points. Participants can continue exploration and exploitation until they run out their allocated time.

5.3 Participants

We recruited participants through advertisements on different social media platforms, targeted cybersecurity groups, university email groups, and Amazon Mechanical Turk. We used a cybersecurity screening test to select the participants for the study. The screening test was based on various practical and theoretical knowledge of cybersecurity such as- information gathering, network scanning, network exploitation, and tools used in these tasks. A total of 10 questions from a pool of 18 were presented to each participant, and participants had to score 70% to get selected. A total of 71 participants passed, out of 95 participated in the screening test. Qualified participants were scheduled for an online study of 60 min. A total of 35 participants appeared in to the main study. Participants were randomly assigned to one of the two conditions, default ($n = 14$) and 2-sided deception condition (n=21). Participants aged between 18 and 62 years (Mean: 33, SD: 9.9). After the successful completion of the experiment, all participants were given a base payment of $12 and could earn a bonus up to $10.

5.4 Results

We measured the performance of participants in different experimental conditions. We measured the proportion of honeypot exploitation and the proportion of data

Table 4 Proportion of Attacks in two experimental conditions

Honeypots	Default	2-sided
Exploitation	0.36 (0.04)	0.62 (0.04)
Data exfiltration	0.36 (0.05)	0.52 (0.04)

exfiltration on honeypots. Table 4 represents the mean and standard error of the dependent variables.

The proportion of honeypot attacks is higher in 2-sided deception condition compared to the default condition. There is a statistically significant difference between groups as determined by one-way ANOVA ($F(1, 33) = 21.44$, $p = 0.0001$). The statistical analysis revealed that the proportion of honeypot attacks are significantly higher in 2-sided condition compared to the default condition. Our results suggest that modifying the features of both real and honeypot machines produces more attacks than using the default features.

After successful exploitation of any system, data exfiltration is the next step in the task. The proportion of data exfiltration is significantly different between groups as determined by one-way ANOVA ($F(1,33) = 4.84$, $p = 0.035$). The statistical analysis revealed that the proportion of data exfiltration is significantly higher in 2-sided condition compared to the default condition. Our results suggest that attackers steal more data, i.e., use SCP commands more often when features are modified compared to the default condition. The detailed results of human experiment are presented in another paper [3].

6 Discussion and Further Applications

Our model gives a new and more nuanced way to think about the quality of different deception strategies, and how robust they are to an adversary being able to see through the deception. We can identify which features the defender should focus on modifying to make the deception more effective, including features of the real objects. In addition, we can correctly identify cases where deception is not the best solution because the costs of creating a believable deception may be higher than the value they create. We conclude by discussing some connections to adversarial machine learning and an additional case where our model could be applied beyond honeypots.

6.1 Adversarial Learning

Recently, adversarial machine learning models have shown great promise in generating deceptive objects, focusing mostly on images and video applications [11, 12, 30], though they have the potential to generalize to many other types of deceptive objects. The most well-known approach is Generative Adversarial Networks (GAN) [10], which rely on a pair of neural networks, one to generate deceptive inputs, and the other to detect differences between real and fake inputs. The intuition for these is often that the networks are playing a zero-sum game, though the interpretation is vague and there is no formal game presented. Our model can be viewed as a formalization of the game these types of machine learning algorithms are playing, though there are some differences. We specifically consider the costs of modifying different features of the objects, as well as the possibility of modifying the real distribution in addition to the fake one. On the other hand, GANs typically are used in much larger problems with vast numbers of complex features, and they do not find optimal solutions. Also, they use abstracted representations of the feature space in the learning process, and it is not clear exactly how this works or what the implications are.

We believe that further developing and scaling this model to address more complex feature deception problems will help to understand the theoretical qualities of GANs and related methods better. In particular, we can better understand the limits that these AML methods may have based on the costs and infeasibility of modifying features in some cases, as well as giving optimal or bounded approximations of the solutions to small feature deception games, which can then be used to provide clear quality comparisons for machine learning methods that may scale to much more complex problems but without specific quality guarantees.

6.2 Disguising Network Traffic

While we presented our model using honeypots as a motivating domain, there are many other possible applications. We briefly discuss another example here to make this point. There are many reasons to disguise network traffic to look like other traffic; defenders may wish to do this to generate fake traffic to support honeypots or to conceal the properties of real traffic on their networks otherwise. Attackers also may want to make their network traffic appear similar to real traffic to avoid detection.

While network traffic, in general, has a very large number of possible features, an increasing fraction of traffic is encrypted, which hides many of the deep features of the data. However, it is still possible to do an analysis of encrypted traffic based on the source, destination, routing, timing and quantity characteristics, etc. Our model can be used to analyze how to optimize the properties of real and decoy traffic to improve the effectiveness of the decoy traffic, based on the costs of modifying different features. For example, modifying traffic to be sent more frequently will clearly have costs in increased network congestion, while modifying some features of the real traffic may not be feasible at all (e.g., the source and destination). Even simple versions of our model with relatively few features could be used to optimize decoy network traffic in encrypted settings, where there is limited observable information about the traffic. The unencrypted case allows for many more possible features, so it would require larger and more complex versions of our model to analyze, which would require more scalable algorithms to solve exactly using our model, or the application of approximation methods and adversarial learning techniques.

6.3 Limitations

The time and memory complexities of the game model depend on n, k, feature modification options, and the amount of sampling; which makes the model grow exponentially. To avoid computational complexity, we tested our model with two machines, one each of type (*real* and *honeypot*) with two features in each. Extending the model to include more machine types and features is straightforward, although the optimization problem will become much more difficult to solve. A scalable algorithm will need to be developed to solve larger size games.

7 Conclusions

Deception is becoming an increasingly crucial tool for both attackers and defenders in cybersecurity domains. However, existing formal models provide little guidance on the effectiveness of deception, the amount of effort needed to sufficiently disguise deceptive objects against motivated attackers, of the limits of deception based on the costs of modifying the features of the deceptive objects. Also, most analyses only consider how to make deceptive objects look real, and not how real objects can be modified to look more like deceptive ones to make the task of deception easier. In this chapter, we present a formal game-theoretic model of this problem, capturing the key problem of disguising deceptive objects among real objects when an attacker may observe external features/characteristics. We also demonstrate the effectiveness of this technique using a human experiment.

Our model of FSG allows us to investigate many aspects of how a defender should optimize efforts to conceal deceptive objects, which can be applied to honeypots, disguising network traffic, and other domains. This also gives a more theoretical foundation to understand the benefits and limitations of adversarial learning methods for generating deceptive objects. We show that the symmetry or asymmetry of the costs of modifying features is critical to whether we need to consider 2-sided deception as part of the strategy, and we also show that in some cases deception is either unnecessary or too costly to be effective. Also, the sophistication of the attackers makes a great difference; in cases with naïve attackers deception is even more effective, even when considering a low-cost strategy.

Our results from human experiments demonstrate that it is possible to make honeypots more effective when we manipulate the features of the honeypot design compared to using a default configuration of honeypots. The empirical data show that both the rate of honeypot exploitation and data exfiltration increase when features are manipulated. Particularly, the effectiveness of deception increases by making honeypots look like real machines and additionally, by making real machines look like honeypots (2-sided deception).

Acknowledgments This research was sponsored by the Combat Capabilities Development Command, Army Research Laboratory and was accomplished under Cooperative Agreement Number W911NF-13-2-0045 (ARL Cyber Security CRA) and by the Army Research Office and accomplished under grant number W911NF-17-1-0370 (MURI Cyberdeception). The views and conclusions contained in this document are those of the authors and should not be interpreted as representing the official policies, either expressed or implied, of the Army Research Laboratory or the U.S. Government. The U.S. Government is authorized to reproduce and distribute reprints for Government purposes without standing any copyright notation.

References

1. Achleitner, S., La Porta, T., McDaniel, P., Sugrim, S., Krishnamurthy, S.V., Chadha, R.: Cyber deception: Virtual networks to defend insider reconnaissance. In: Proceedings of the 8th ACM CCS International Workshop on Managing Insider Security Threats, pp. 57–68. ACM (2016)
2. Aggarwal, P., Gautam, A., Agarwal, V., Gonzalez, C., Dutt, V.: Hackit: a human-in-the-loop simulation tool for realistic cyber deception experiments. In: International Conference on Applied Human Factors and Ergonomics, pp. 109–121. Springer (2019)
3. Aggarwal, P., Du, Y., Singh, K., Gonzalez, C.: Decoys in cybersecurity: An exploratory study to test the effectiveness of 2-sided deception. Preprint (2021). arXiv:2108.11037
4. Alpcan, T., Başar, T.: Network Security: A Decision and Game-Theoretic Approach. Cambridge University Press (2010)
5. Bringer, M.L., Chelmecki, C.A., Fujinoki, H.: A survey: Recent advances and future trends in honeypot research. Int. J. Comput. Network Inf. Secur. **4**(10), 63 (2012)
6. Carroll, T.E., Grosu, D.: A game theoretic investigation of deception in network security. Secur. Commun. Networks **4**(10), 1162–1172 (2011)
7. De Gaspari, F., Jajodia, S., Mancini, L.V., Panico, A.: Ahead: A new architecture for active defense. In: Proceedings of the 2016 ACM Workshop on Automated Decision Making for Active Cyber Defense, pp. 11–16. ACM (2016)
8. Du, M., Li, Y., Lu, Q., Wang, K.: Bayesian game based pseudo honeypot model in social networks. In: International Conference on Cloud Computing and Security, pp. 62–71. Springer (2017)
9. Garg, N., Grosu, D.: Deception in honeynets: A game-theoretic analysis. In: 2007 IEEE SMC Information Assurance and Security Workshop, pp. 107–113. IEEE (2007)
10. Goodfellow, I.: Nips 2016 tutorial: Generative adversarial networks. Preprint (2016). arXiv:1701.00160
11. Grosse, K., Papernot, N., Manoharan, P., Backes, M., McDaniel, P.: Adversarial perturbations against deep neural networks for malware classification. Preprint (2016). arXiv:1606.04435
12. Huang, L., Joseph, A.D., Nelson, B., Rubinstein, B.I., Tygar, J.: Adversarial machine learning. In: Proceedings of the 4th ACM Workshop on Security and Artificial Intelligence, pp. 43–58. ACM (2011)
13. Kiekintveld, C., Lisy, V., Pibil, R.: Game-theoretic foundations for the strategic use of honeypots in network security. Adv. Inf. Secur. **56**, 81–101 (2015)
14. Krawetz, N.: Anti-honeypot technology. IEEE Secur. Privacy **2**(1), 76–79 (2004)

15. Kroer, C., Sandholm, T.: Extensive-form game abstraction with bounds. In: Proceedings of the Fifteenth ACM Conference on Economics and Computation, pp. 621–638. ACM (2014)
16. La, Q.D., Quek, T.Q., Lee, J., Jin, S., Zhu, H.: Deceptive attack and defense game in honeypot-enabled networks for the internet of things. IEEE Internet Things J. **3**(6), 1025–1035 (2016)
17. Laszka, A., Vorobeychik, Y., Koutsoukos, X.D.: Optimal personalized filtering against spear-phishing attacks. In: AAAI (2015)
18. Mairh, A., Barik, D., Verma, K., Jena, D.: Honeypot in network security: a survey. In: Proceedings of the 2011 International Conference on Communication, Computing & Security, pp. 600–605. ACM (2011)
19. Monge, G.: Mémoire sur la théorie des déblais et des remblais. Histoire de l'Académie Royale des Sciences de Paris (1781)
20. Nawrocki, M., Wählisch, M., Schmidt, T.C., Keil, C., Schönfelder, J.: A survey on honeypot software and data analysis. Preprint (2016). arXiv:1608.06249
21. Píbil, R., Lisỳ, V., Kiekintveld, C., Bošanskỳ, B., Pěchouček, M.: Game theoretic model of strategic honeypot selection in computer networks. In: International Conference on Decision and Game Theory for Security, pp. 201–220. Springer (2012)
22. Provos, N.: Honeyd-a virtual honeypot daemon. In: 10th DFN-CERT Workshop, Hamburg, Germany, vol. 2, p. 4 (2003)
23. Rowe, N.C., Custy, E.J., Duong, B.T.: Defending cyberspace with fake honeypots. J. Comput. **2**(2), 25 (2007)
24. Schlenker, A., Xu, H., Guirguis, M., Kiekintveld, C., Sinha, A., Tambe, M., Sonya, S., Balderas, D., Dunstatter, N.: Don't bury your head in warnings: A game-theoretic approach for intelligent allocation of cyber-security alerts. In: IJCAI (2017)
25. Schlenker, A., Thakoor, O., Xu, H., Fang, F., Tambe, M., Tran-Thanh, L., Vayanos, P., Vorobeychik, Y.: Deceiving cyber adversaries: A game theoretic approach. In: AAMAS (2018). http://dl.acm.org/citation.cfm?id=3237383.3237833
26. Serra, E., Jajodia, S., Pugliese, A., Rullo, A., Subrahmanian, V.: Pareto-optimal adversarial defense of enterprise systems. ACM Trans. Inf. Syst. Secur. (TISSEC) **17**(3), 11 (2015)
27. Shi, L., Zhao, J., Jiang, L., Xing, W., Gong, J., Liu, X.: Game theoretic simulation on the mimicry honeypot. Wuhan Univ. J. Nat. Sci. **21**(1), 69–74 (2016)
28. Spitzner, L.: Honeypots: Tracking Hackers, vol. 1. Addison-Wesley Boston (2002)
29. Stoll, C.: The Cuckoo's Egg: Tracking a Spy Through the Maze of Computer Espionage. Doubleday (1989)
30. Szegedy, C., Zaremba, W., Sutskever, I., Bruna, J., Erhan, D., Goodfellow, I., Fergus, R.: Intriguing properties of neural networks. Preprint (2013). arXiv:1312.6199
31. The Mandiant®Intelligence Center™: Apt1: Exposing one of China's cyber espionage units. Mandiant, Tech. Rep (2013). https://www.fireeye.com/content/dam/fireeye-www/services/pdfs/mandiant-apt1-report.pdf
32. Virvilis, N., Vanautgaerden, B., Serrano, O.S.: Changing the game: The art of deceiving sophisticated attackers. In: 2014 6th International Conference On Cyber Conflict (CyCon 2014), pp. 87–97. IEEE (2014)
33. Wang, W., Zeng, B.: A two-stage deception game for network defense. In: Decision and Game Theory for Security (2018)
34. Wang, K., Du, M., Maharjan, S., Sun, Y.: Strategic honeypot game model for distributed denial of service attacks in the smart grid. IEEE Trans. Smart Grid **8**(5), 2474–2482 (2017)
35. Zou, C.C., Cunningham, R.: Honeypot-aware advanced botnet construction and maintenance. In: International Conference on Dependable Systems and Networks (DSN'06). pp. 199–208. IEEE (2006)

Human-Subject Experiments on Risk-Based Cyber Camouflage Games

Palvi Aggarwal, Shahin Jabbari, Omkar Thakoor, Edward A. Cranford, Phebe Vayanos, Christian Lebiere, Milind Tambe, and Cleotilde Gonzalez

1 Introduction

Rapidly growing cybercrime [13, 15, 25] has elicited effective defense against adept attackers. Many recent works have proposed *Cyber deception* techniques to thwart the reconnaissance—typically a crucial phase prior to attacking [17, 22]. One deception approach is to camouflage the network by attribute obfuscation [7, 10, 40] to render an attacker's information incomplete or incorrect, creating indecision over their infiltration plan [5, 10, 11, 30]. Optimizing such a deceptive strategy is challenging due to many practical constraints on feasibility and costs of deploying,

Section 1-4 are based on Thakoor et al. [38] and section 7-8 are based on Aggarwal et al. [2]. Additional details can be found in those papers.

P. Aggarwal (✉)
Department of Computer Science, University of Texas, El Paso, TX, USA
e-mail: paggarwal@utep.edu

S. Jabbari (✉)
Drexel University, Philadelphia, PA, USA
e-mail: shahin@drexel.edu

O. Thakoor · P. Vayanos
University of Southern California, Los Angeles, CA, USA
e-mail: othakoor@usc.edu; phebe.vayanos@usc.edu

E. A. Cranford · C. Lebiere · C. Gonzalez
Carnegie Mellon University, Pittsburgh, PA, USA
e-mail: cranford@cmu.edu; cl@cmu.edu; coty@cmu.edu

M. Tambe
Harvard University, Allston, MA, USA
e-mail: milind_tambe@harvard.edu

T. Bao et al. (eds.), *Cyber Deception*, Advances in Information Security 89,
https://doi.org/10.1007/978-3-031-16613-6_2

as well as critically dependent on the attacker's decision-making governed by his behavioral profile, and attacking motives and capabilities. Game theory offers an effective framework for tackling both these aspects and has been successfully adopted in security problems [3, 21, 31, 33].

Attacking a machine amounts to launching an exploit for a particular system configuration—information that is concealed or distorted due to the deceptive defense, thus, an attempted attack may not succeed. Recent game-theoretic models for deception via attribute obfuscation [32, 37] have a major flaw in ignoring this risk of attack failure as they assume that an attempted attack is guaranteed to provide utility to the attacker. Furthermore, assuming that humans will act and choose the best option available, in terms of expected values, is problematic, as psychologists have known for decades that humans can only be *boundedly rational* [18, 34] and act according to simple heuristics [12]. This was demonstrated recently in a human-subject experiments that aimed at evaluating an optimal defense strategy (proposed by Schlenker et al. [32]) against a random strategy [1]. Several previous works attempt to address this issue. For example, the Quantal response theory [23] asserts that humans exhibit bounded rationality. However, such strategies severely affect the performance of a deployed strategy, which has not been considered by previous work.

In this paper, we present Risk-based Cyber Camouflage Games (RCCG) — a crucial refinement over previous models via redefined strategy space and rewards to explicitly capture the uncertainty in attack success. As foundation, we first consider rational attackers and show analytical results including NP-hardness of optimal strategy computation and its mixed-integer linear program (MILP) formulation which, while akin to previous models, largely requires independent reasoning. Furthermore, we consider risk-averse attackers modeled using Prospect theory [41] and present a solution (*PT*) that estimates model parameters from data to compute optimal defense. Finally, we demonstrate the effectiveness of our approach by examining the human attacker behavior in human-in-the-loop behavioral experiments.

1.1 Related Work

Cyber Deception Games [32] and Cyber Camouflage Games (CCG) [37] are game-theoretic models for Cyber deception via attribute obfuscation. In these, the defender can mask the *true configuration* of a machine, creating uncertainty in the associated reward the attacker receives for attacking the machine. These have a fundamental limitation, namely the assumption that the attacked machine is guaranteed to provide utility to the attacker. Furthermore, they do not consider that human agents tend to deviate from rationality, particularly when making decisions under risk. Our refined model handles both of these crucial issues.

A model using Prospect theory is proposed in [43] for boundedly rational attackers in Stackelberg security games (SSG) [36]. However, it relies on using model parameters from previous literature, discounting the fact that they can largely vary for the specific experimental setups. We provide a solution that learns the parameters from data, as well as a robust solution to deal with uncertainty in the degree of risk-aversion and broadly the parametrization hypothesis. A robust solution for unknown risk-averse attackers has been proposed for SSGs in [29], however, it aims to minimize the worst-case utility, whereas we take the less conservative approach of minimizing worst-case regret. Previous works on uncertainty in security games consider Bayesian [19], interval-based [20], and regret-based approaches [24], however, these do not directly apply due to fundamental differences between RCCGs and SSGs as explained in [37].

Another approach in [43] is based on the Quantal Response model [23]. However, the attack probabilities therein involve terms that are exponential in rewards, which in turn are non-linear functions of integer variables in our model, leading to an intractable formulation. However, we show effectiveness of our model-free solution for this behavior model as well.

Machine learning models such as Decision Tree and Neural Networks have been used for estimating human behavior [9]. However, the predictive power of such models typically comes with an indispensable complexity (non-linear kernels, functions and deep hidden layers of neural nets, sizeable depth and branching factor of decision trees, etc.). This does not allow the predicted human response to be written as a simple closed-form expression of the instance features, viz, the strategy decision variables, preventing a concise optimization problem formulation. This is particularly problematic since the alternative of searching for an optimal solution via strategy enumeration is also non-viable — due to the compact input representation via a *polytopal* strategy space [16] in our model.

MATCH [26] and COBRA [27] aim to tackle human attackers in SSGs that avoid the complex task of modeling human decision-making and provide robustness against deviations from rationality. However, their applicability is limited in *Strictly Competitive* games where deviation from rationality always benefits the defender, they reduce to the standard minimax solution.

2 Risk-Based Cyber Camouflage Games

Here, we describe the components of the RCCG model, explicitly highlighting the key differences with respect to the CCG model [37].

The network is a set of k machines $\mathcal{K} := \{1, \ldots, k\}$. Each machine has a *true configuration* (TC), which is an exhaustive tuple of attributes so that machines having the same TC are identical. We use $\mathcal{S} := \{1, \ldots, s\}$ to denote the set of all TCs. The *true state of the network* (TSN) is a vector $\boldsymbol{n} = (n_i)_{i \in \mathcal{S}}$ with n_i denoting the number of machines with TC i. Note that $\sum_{i \in \mathcal{S}} n_i = k$.

The defender can disguise the TCs using deception techniques. More concretely, we assume each machine is "masked" with an *observed configuration* (OC). The set of OCs is denoted by $\mathcal{T}$. Similar to a TC, an OC corresponds to an attribute tuple that fully comprises the attacker view, so that machines with the same OC are indistinguishable from each other.

We represent the defender strategy as an integer matrix Φ, where Φ_{ij} is the number of machines with TC i, masked with OC j. The *observed state of the network* (OSN) is a function of Φ, denoted as $\boldsymbol{m}(\Phi) := (m_j(\Phi))_{j\in\mathcal{T}}$, where $m_j(\Phi) = \sum_i \Phi_{ij}$ denotes the number of machines under OC j for strategy Φ.

Deception is often costly and not any arbitrary deception strategy is feasible. To model this, we use *feasibility* constraints given by a (0,1)-matrix Π, where $\Pi_{ij} = 1$ if a machine with TC i can be masked with OC j. Next, we assume that masking a TC i with an OC j (if so feasible) has a cost of c_{ij} incurred by the defender, denoting the aggregated cost from deployment, maintenance, degraded functionality, etc. We assume the total cost is to be bounded by a *budget* B.

These translate to linear constraints to define the valid defender strategy set:

$$\mathcal{F} = \left\{ \Phi \,\middle|\, \begin{array}{ll} \Phi_{ij} \in \mathbb{Z}_{\geq 0}, & \Phi_{ij} \leq \Pi_{ij} n_i \;\forall (i,j) \in \mathcal{S} \times \mathcal{T}, \\ \sum\limits_{j\in\mathcal{T}} \Phi_{ij} = n_i \;\; \forall i \in \mathcal{S}, & \sum\limits_{i\in\mathcal{S}} \sum\limits_{j\in\mathcal{T}} \Phi_{ij}\, c_{ij} \leq B \end{array} \right\}.$$

The first and the third constraints follow from the definitions of Φ and $\boldsymbol{n}$. The second and fourth impose the feasibility and budget constraints, respectively.

A machine with TC i gets successfully attacked if the attacker uncovers the disguised OC and uses the correct exploit corresponding to TC i. In this case, the attacker receives a utility v_i—his *valuation* of TC i. Collectively, these are represented as a vector $\boldsymbol{v}$. Analogously, we define valuations $\boldsymbol{u}$ representing the defender's loss.

For ease of interpretation, we assign a 0 utility to the players when the attack is unsuccessful, which sets a constant reference point. Hence, unlike CCGs, valuations cannot be freely shifted. Furthermore, a successful attack typically is undesirable for the defender, and to let the valuations be typically positive values, they represent the defender's loss; its minimization is the defender objective unlike maximization in CCGs.

3 Rational Attackers

We first consider rational attackers which have been extensively studied in previous work. The attacker having to choose a TC-OC pair (i.e., the true configuration and observed configuration pair) as an attack here rather than just an OC as in the CCG model [37] requires entirely new techniques for our analytical results, despite a close resemblance to the optimization problem as below.

Previous work on general-sum Stackelberg games typically uses *Strong Stackelberg equilibria* (SSE), that is, in case of multiple best responses, it is assumed that the follower breaks ties in favor of the leader (i.e., minimizing defender loss). The leader can *induce* this with mixed strategies, which is not possible in RCCGs since the defender is restricted to pure strategies [14].

Hence, we consider the worst-case assumption that the attacker breaks the ties against the defender, leading to *Weak Stackelberg Equilibria* (WSE) [6]. WSE does not always exist [42], but it does when the leader can only play a finite set of pure strategies, as in CCG. Hence, we assume that the attacker chooses a best response to the defender strategy Φ, maximizing the defender loss in case of a tie. This defender utility is denoted as $U^{\text{wse}}(\Phi)$, defined as the optimal value of the inner Optimization Problem (OP) in the following, while the defender aims to compute a strategy to minimize $U^{\text{wse}}(\Phi)$ as given by the outer objective.

$$\begin{aligned} \underset{\Phi}{\operatorname{argmin}} \quad & \max_{i,j} U^{\text{d}}(\Phi, i, j) \\ & \text{s.t. } U^{\text{a}}(\Phi, i, j) \geq U^{\text{a}}(\Phi, i', j') \ \ \forall i' \in \mathcal{S}, \ \forall j' \in \mathcal{T}. \end{aligned} \tag{1}$$

First, we show results on optimal strategy computation shown for the important special cases — the zero-sum and *unconstrained* settings. While similar results have been shown for CCG, independent proof techniques are needed herein due to the difference in our model structure. We then focus our attention on general-sum games.

In the zero-sum setting, the defender loss equals the attacker reward, i.e., $\boldsymbol{v} = \boldsymbol{u}$.

Theorem 1 *Zero-sum RCCG is NP-hard.* □

For the special unconstrained setting (i.e., with no feasibility or budget constraints), we show the following.

Proposition 1 *Unconstrained zero-sum RCCG always has an optimal strategy that uses just one OC, thus computable in $O(1)$ time.* □

Although both of these results also hold for CCG, they require independent derivation. We next focus on the unconstrained RCCG.

Proposition 2 *Unconstrained RCCG always has an optimal strategy that uses just two OCs.* □

This result is crucial for an efficient algorithm to compute an optimal strategy (Algorithm 1), named Strategy Optimization by Best Response Enumeration (SOBRE). SOBRE constructs an optimal strategy with two OCs, due to Proposition 2, with attacker best response being OC 1 without loss of generality. It classifies the candidate strategies by triplets (i, n, m) (Line 2) where the attacker best response is $(i, 1)$, and OC 1 masks n machines of TC i, and m machines in total. It uses a subroutine DPBRF (Dynamic Programming for Best Response Feasibility) to construct a strategy yielding the desired best response (Line 6) if it exists, and then compares the defender utility from all such feasible candidates, to compute the optimal (Lines 7,8).

Algorithm 1: SOBRE

1 **Initialize** $minUtil \leftarrow \infty$
2 **for** $i = 1, \ldots, s; n = 0, \ldots n_i; m = n, \ldots, k$ **do**
3 **if** $(n/m < (n_i - n)/(|K| - m))$ **continue**
4 $util \leftarrow (n/m)u_i$
5 **if** $(util \geq minUtil)$ **continue**
6 **if** $DPBRF(i, n, m)$
7 **Update** $minUtil \leftarrow util$
8 **Return** $minUtil$

Theorem 2 *The optimal strategy in an unconstrained RCCG can be computed in time $O(k)^4$.* □

Since the input can be expressed in $O(st)$ bits, SOBRE is pseudo-polynomial time algorithm. However, it becomes a poly-time algorithm under the practical assumption of constant-bounded no. of machines per TC, (so that, $k = O(s)$, or more generally, if k in terms of s is polynomially bounded). In contrast, unconstrained CCG is NP-hard even under this restriction. This distinction arises since in RCCG, the best response utility given the attack strategy and the no. of machines masked by the corresponding OC depends on only the count of attacked TC as opposed to all the TCs in CCG.

We next focus on the constrained RCCG. For this setting, $U^{\text{wse}}(\Phi)$ is given by OP (1), and thus, computing its minimum is a bilevel OP. Reducing to a single-level MILP is typically hard [35]. In particular, computing an SSE allows such a reduction due to attacker's tiebreaking favoring the defender's objective therein, however, the worst-case tiebreaking of WSE does not. Notwithstanding the redefined attack strategies, a single-level OP can be formulated analogous to CCGs by assuming an ϵ-rational attacker instead of fully rational. It can be shown that for sufficiently small ϵ, it gives the optimal solution for rationality.

$$\min_{\Phi, \boldsymbol{q}, \gamma, \alpha} \gamma \tag{2}$$

$$\text{s.t.} \quad \alpha, \gamma \in \mathbb{R}, \ \Phi \in \mathcal{F}, \ \boldsymbol{q} \in \{0, 1\}^{|\mathcal{I}| \times |\mathcal{J}|}$$

$$q_{11} + \ldots + q_{st} \geq 1 \tag{2a}$$

$$\epsilon(1 - q_{ij}) \leq \alpha - U^{\text{a}}(\Phi, j, i) \qquad \forall i \in \mathcal{S} \, \forall j \in \mathcal{T} \tag{2b}$$

$$M(1 - q_{ij}) \geq \alpha - U^{\text{a}}(\Phi, j, i) \qquad \forall i \in \mathcal{S} \, \forall j \in \mathcal{T} \tag{2c}$$

$$U^{\text{d}}(\Phi, j, i) \leq \gamma + M(1 - q_{ij}) \qquad \forall i \in \mathcal{S} \, \forall j \in \mathcal{T} \tag{2d}$$

$$q_{ij} \leq \Phi_{ij} \qquad \forall i \in \mathcal{S} \, \forall j \in \mathcal{T}. \tag{2e}$$

The defender aims to minimize the objective γ which captures the defender's optimal utility. The binary variables q_{ij} indicate if attacking (i, j) is an optimal attacker strategy, and as specified by (2a), there must be at least one. As per (2b) and (2c), α is the optimal attacker utility, and this enforces $q_{ij} = 1$ for all the ϵ-optimal attacker strategies (using a big-M constant). Equation (2e) ensures that only the OCs which actually mask a machine are considered as valid attacker responses. Finally, (2d) captures the worst-case tiebreaking by requiring that γ is the highest defender loss from a possible ϵ-optimal attacker response. Using an alternate strategy representation with binary decision variables enables linearization to an MILP that can be sped up with *symmetry-breaking* cuts [37].

4 Boundedly Rational Attackers and Prospect Theory

A well-studied model for the risk-behavior of humans is prospect theory [41]. In this model, humans make decisions to maximize the *prospect*, which differs from the utilitarian approach in that the reward value and the probability of any event are transformed as follows. The theory assumes a value transformation function R that is monotone increasing and concave such that the outcome reward v (value of the machine attacked) gets perceived as $R(v)$ by the attacker. A parameterization of the form $R_\lambda(v) = c(v/c)^\lambda$ is commonly considered in the literature, with $\lambda < 1$ capturing the risk-aversion of the attacker. We use $c = \max_i v_i$ so that the perceived values are normalized to the same range as true values. Prospect theory also proposes a probability weighting function Π, such that the probability p of an event is perceived as $\Pi(p)$. A function of the form $\Pi_\delta(p) = p^\delta/(p^\delta + (1-p)^\delta)^{1/\delta}$ has been previously proposed in literature, parametrized by δ.In our problem, the attack success probability p is a non-linear non-convex function of the decision variables Φ_{ij} and applying a function as above loses tractability. For simplicity, we omit the probability weighting from our solution which shows effective results regardless. Future work could explore the benefits of incorporating this additional complexity.

Thus, each of the attacker's strategies has a prospect

$$f_\lambda(\Phi, i, j) = \frac{\Phi_{ij}}{m_\Phi(j)} R_\lambda(v_i) \tag{3}$$

as a function of the player strategies, parametrized by λ. This value transformation makes the problem inherently harder (even in the simpler zero-sum setting).

The main challenge arises from learning λ. Once λ is estimated, the defender computes an optimal strategy for the prospect-theoretic attacker, by simply modifying (2), replacing the valuations v_i with the transformed values $R_\lambda(v_i)$ as in (3). More generally, with this replacement, all results from Sect. 3 for rational attackers apply here too.

4.1 Learning Model Parameters from Data

Suppose we have data consisting of a set of instances $\mathcal{N}$ from a study such as Aggarwal et al. [1]. A particular instance $n \in \mathcal{N}$ corresponds to a particular human subject that plays against a particular defense strategy Φ_n and decides to attack (i_n, j_n) having the maximum prospect. The instances come from different subjects who may have a different parameter λ values. However, at the time of deployment, the defender cannot estimate the risk-averseness of an individual in advance and play a different strategy accordingly. Hence, we aim to compute a strategy against a fixed λ that works well for the whole population. Due to different subjects, different instances may have different attack responses for the same defender strategy, and requiring a strict prospect-maximization may not yield any feasible λ. Hence, we define the likelihood of an instance, by considering a soft-max function instead, so that the probability of attacking (i_n, j_n) is

$$P_n(\lambda) = \frac{\exp(f_\lambda(\Phi_n, i_n, j_n))}{\sum_{i,j} \exp(f_\lambda(\Phi_n, i, j))}.$$

Using the Maximum Likelihood Estimation approach, we choose λ which maximizes the likelihood $\prod_n P_n(\lambda)$, or, log likelihood $\sum_n \log P_n(\lambda)$. Finding such a solution via the standard approach of *gradient descent* does not have the convergence guarantee due to the non-convexity of the likelihood function and, hence, we resort to *grid search* instead.

5 Human-Subject Experiments

We conduct human experiment to evaluate two masking strategies, WSE and PT. The WSE strategy develops the OC to TC mapping (Φ matrix) by minimizing the expected losses for defenders against a rational attacker. In contrast, the PT strategy generates the Φ matrix by minimizing the defender's expected loss against risk-averse attackers. Table 1 presents the utility for attack and loss for the defender after a successful attack on TCs. In the WSE strategy, the rewards are considered as presented in Table 1. However, in the PT strategy, the rewards are transformed replacing the valuation v_i^a with the transformed values $R_\lambda(v_i^a)$. The value of $\lambda = 0.75$ parameter is estimated from human decisions in a random masking strategy. The data for random masking strategy was collected using the same procedure mentioned in Sect. 5.3. Using the WSE and PT strategies, we generated 10 Φ for each condition. Both strategies considered feasibility constraints and the number of TCs and OCs. In an experimental testbed called CyberVAN [8], we deployed these matrices and evaluated the performance of masking strategies against human attackers. The details of this experiment can be found in Aggarwal et al. [2].

Table 1 Attacker's rewards and defender's losses per true configuration

TC	Attacker's rewards	Defender's losses
Slackware	15	9
Xbox	11	10
ubuntu8	2	6
WinXPEmb	13	4
avayagw	14	3
FreeBSD	11	10
WinXP	2	14
win2008	11	2
win2k	7	8
win7pro	10	5
win7ent	9	8
OpenWrt	7	12
OpenBSD	15	15
Linux	6	15
cisco2500	13	12

5.1 Experimental Setup in CyberVAN

In the CyberVAN testbed, the true configuration (TC) of the machines was masked to the observable configuration (OC) using the Honeyd service. Masking of virtual machine TCs to OCs using the strategy matrix generated by the WSE and PT algorithms. The Honeyd configuration file masks the operating systems and ports of TCs with OCs to trick the network scanning tools [28].

Participants were assigned a virtual machine running kali operating system to perform the task. Each machine was configured with a scanning tool (i.e., Zenmap) and attack scripts. The task consists of 10 rounds (preceded by 1 practice round). Participants were provided a different pre-generated Φ matrix in each round that provides TC to OC mappings of 15 virtual machines.

After logging in to the virtual machine, participants were asked to start the task using the start script, as shown in Fig. 1. The start script provides the IP address range and the Φ matrix for the practice round. Similar information is also provided for the main rounds. The Φ matrix describes the type and number of machines present in the network (TC) and their corresponding masked configuration (OC). The Φ matrices were randomly selected for each participant and the virtual machine configuration was different in each round. Figure 2 presents an example of a Φ matrix used in one of the conditions. To help interpret the matrix, participants were provided with information about the way the TCs were mapped into the OCs. For example, in the sample matrix, there are 6 TCs (avayagw, Ubuntu8, Win7pro, Win7ent, WinXP, Slackware) that are mapped to 3 OCs (FreeBSD, Win7pro, and Ubuntu8). In the given matrix, for example, 5 machines are shown as FreeBSD, of which 3 are actually avayagw and 2 are Ubuntu8. In addition to mapping information, we provide the utility of each TC along with the matrix. Participants

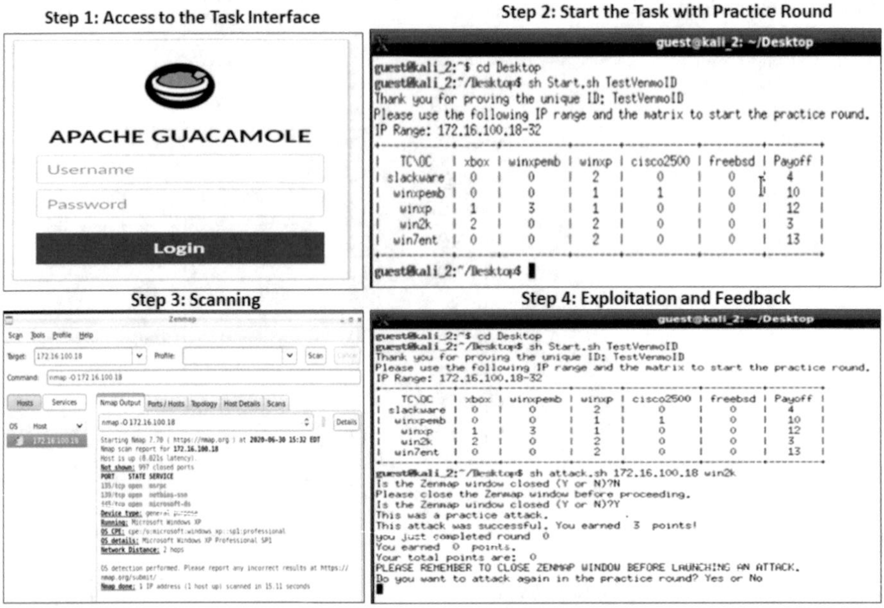

Step 1: Access to the Task Interface

APACHE GUACAMOLE

Username

Password

Login

Step 2: Start the Task with Practice Round

```
guest@kali_2: ~/Desktop
guest@kali_2:~$ cd Desktop
guest@kali_2:~/Desktop$ sh Start.sh TestVenmoID
Thank you for proving the unique ID: TestVenmoID
Please use the following IP range and the matrix to start the practice round.
IP Range: 172.16.100.18-32
+----------+------+---------+-------+-----------+---------+--------+
| TC\OC    | xbox | winxpemb | winxp | cisco2500 | freebsd | Payoff |
| slackware | 0   | 0       | 2     | 0         | 0       | 4      |
| winxpemb | 0    | 0       | 1     | 1         | 0       | 10     |
| winxp    | 1    | 3       | 1     | 0         | 0       | 12     |
| win2k    | 2    | 0       | 2     | 0         | 0       | 3      |
| win7ent  | 0    | 0       | 2     | 0         | 0       | 13     |
+----------+------+---------+-------+-----------+---------+--------+
guest@kali_2:~/Desktop$
```

Step 3: Scanning

Step 4: Exploitation and Feedback

Fig. 1 Steps involved in the CyberVAN Task for Human participants

Fig. 2 Sample Φ Matrix: columns represent the observable configuration and rows represent the true configuration.

$$\begin{bmatrix} TC\backslash OC & freeBSD & win7pro & Ubuntu8 \\ \text{avayagw} & 3 & 0 & 0 \\ \text{Ubuntu8} & 2 & 0 & 0 \\ win7pro & 0 & 2 & 0 \\ win7ent & 0 & 2 & 0 \\ winXP & 0 & 2 & 0 \\ Slackware & 0 & 0 & 1 \end{bmatrix}$$

could use this information to calculate their probability of success and the expected utility of attacking a particular machine.

In each round, participants perform two phases: exploration and attack. In the exploration phase, participants used Zenmap utility to probe the machines using the *Nmap* command and obtain information of the open ports, operating systems, and running services (according to the OC). Participants could scan machines in any order. The output of *Nmap* command provided the observable features of network to the participants. After the exploration phase, participants go through the attack phase, where they decide which machine to attack and what type of exploit to use to conduct the attack. To decide what machine to attack after exploration, participants could use the information collected in exploration phase, the payoff for each machine and the mapping matrix. Participants were provided the *rewards* that they would obtain if they were successful in their attack. Note that the utility of the attacker is different than the loss of the defender for each TC, and the information about the defender's losses is not available to attackers.

The attacker's utilities are randomly allocated between a range of 2 and 15 to represent the low, medium, and high valued machines in the network. The corresponding defender's losses were assigned with the assumption that the value of a TC may or may not be the same for the attacker and defender. Thus, some TCs have equal defender's losses, and others are either lower or higher than the attacker's gain. The attacker's rewards and the defender's losses for the TCs remained the same across all 10 rounds. Participants earned the sum of the points accumulated during the 10 rounds, which were directly translated into a bonus monetary earned to the participant.

5.2 Participants

Participants were recruited through advertisements via various university email groups, social media, and cybersecurity targeted groups. To be qualified to participate, participants were required to pass an online test of basic cybersecurity knowledge, which included questions on various attacks, network protocols, scanning tools for networks, etc. The questions were adopted from previously published research by Ben-Asher and Gonzalez [4]. Only qualified participants were scheduled for an online study of 90 min. The demographics of participants are presented in Table 2. After the successful completion of the experiment, all participants were paid a base payment of $18. In addition, for each successful exploit, participants received 1 point, which accumulated and were converted to a monetary bonus ($1 per 10 points). Participants could earn up to $15 in bonus based on their performance.

Table 2 Demographics

Demographic	Value	WSE (*N*=25)	PT (*N*=20)
Age	Mean	24.4	28.7
	SD	4.2	5.8
Sex	Male	90%	41%
	Female	10%	47%
	Not specified	0%	12%
Education	Master's	54%	29%
	Bachelor's	41%	47%
	PhD	5%	17%
Experience	Little	40%	47%
	Some	50%	35%
	A lot	5%	0%
	Expert	5%	11%
	No experience	0%	5%

5.3 Experimental Process

First, participants provided informed consent and completed the demographic questionnaire. Next, they were provided video and text instructions regarding the goal of the task and the general procedure. Instructions were followed with a brief instruction comprehension test. Participants received feedback if they incorrectly answered a question in the test. Participants were provided the contact details of the research assistant and they could ask any clarification questions before proceeding with the experiment.

During the instructions, the participants were informed that the experiment would take up to 90 min and would consist of 11 rounds. After finishing the instruction, participants were provided with login and password information for the virtual machine. Once logged on their machine, participants could see a cheat sheet to help them throughout the task. In the terminal window, participants started the task and received information such as IP addresses, the ϕ matrix, and payoffs during each round. In each round, participants were asked to probe the machines using an Nmap command like "Nmap -O 172.16.31.31" to gather information about open ports and operating systems on this IP address. They were also allowed to scan a specified range of IP addresses in each round together using a command like "Nmap -O 172.16.31.31-61." Next, using the attack script, participants decided what IP addresses to attack by selecting an appropriate exploit. Participants received points if the exploit matched with the true configuration; otherwise, they received zero points. Once they finished all rounds, we asked for their feedback regarding the experiment.

5.4 Experiment Results

Participants generally scanned all machines before launching an attack. In the practice round, each participant exploited between a minimum of 1 and a maximum of 7 machines.

We analyzed the data collected in the WSE and PT conditions during the 10 actual rounds. We randomly allocated 10 matrices to participants during the 10 rounds. To measure the effectiveness of each matrix, we measured the average attacker's utility and their success rate. We also analyzed the algorithm (i.e., defender) loss.

5.4.1 Attacker's Success Rate

We calculated the rate with which participants used the correct exploit. Table 3 shows the average success rate of participants in the WSE and PT conditions. Participants are slightly more successful when paired with the WSE compared to

Table 3 Average success rate and defender's losses for each matrix and overall in WSE and PT algorithms for Human and IBL Model and their corresponding RMSE values

	Success rate		Attacker's utility		Defender's loss	
Matrix	WSE	PT	WSE	PT	WSE	PT
1	0.37	0.70	3.42	5.50	2.50	6.25
2	0.29	0.25	2.45	3.33	3.54	2.55
3	0.54	0.25	3.08	2.75	5.71	1.50
4	0.46	0.20	2.16	2.05	5.16	2.60
5	0.33	0.10	2.66	1.10	2.50	0.20
6	0.38	0.25	1.14	2.10	2.50	1.70
7	0.46	0.30	1.75	2.95	5.17	0.90
8	0.33	0.35	2.51	5.25	4.46	3.15
9	0.46	0.25	2.62	2.70	6.75	1.50
10	0.38	0.40	3.33	4.55	2.08	1.25
Mean	0.399	0.305	2.54	3.17	4.037	2.160

the PT algorithm, but this difference was not statistically significant ($0.40 \sim 0.30$; $F(1, 42) = 3.02, p = 0.09$). The success rates in each matrix during the 10 rounds are shown in Table 3. Although it appears that human attackers exploited the machines more successfully in the WSE than the PT condition (except matrix 1), in most of the matrices the difference between WSE and PT was not significant. In matrix 1, human attackers experience an option with 100% chances of success in the PT condition. Thus, the success rate was higher in PT compared to WSE only in matrix 1.

The attacker's utility for each condition is shown in Table 3. For each successful exploit, the attacker gained points in accordance to Table 1. We observe that the attackers gained slightly more points in the PT masking algorithm compared to WSE algorithm. However, the statistical test revealed no significant difference between the masking conditions, ($2.54 \sim 3.17$; $F(1, 42) = 1.83$, p=0.18). None of the differences within each matrix was significant ($p > 0.05$). We also compared the average attacker's utility with the best option utility. The attacker's gained less points in both WSE and PT algorithms compared to the best option utility. The matrix-wise analysis in Table 3 shows that human attackers consistently earned fewer points when WSE algorithm was deployed.

5.4.2 Defender's Losses

The losses for each of the two defense algorithms against humans are shown in Table 3. For each successful exploit, the attacker gained points and the defender lost points in accordance with Table 1. We find that the WSE defender lost more points compared to PT masking algorithm. The statistical test revealed a significant difference between the masking conditions ($4.03 > 2.16$; $F(1, 42) = 10.40$, $p<0.002$). We also found that there is a significant differences between matrices ($F(9, 378) = 2.23$, p=0.02) and interaction between conditions and matrices, (F(9,

378) = 3.19, $p<0.001$). The average defender's losses per defense strategy (Φ matrix) are shown in Table 3. The defender's losses were higher for WSE algorithm compared to the PT algorithm for all matrices except in matrix 1.

6 Summary

In the cybersecurity domain, it is difficult to gain an understanding of the attacker's decision-making due to the lack of such decision data. Defense algorithms often rely on the assumption that attackers are rational decision makers and often take the best course of action. Using human experiments, [1] provided insights that human attackers have risk-aversion bias while making cyberattack decisions. In this paper, we present Risk-based Cyber Camouflage Games (RCCG) to capture the crucial uncertainty in the attack success. First, for rational attackers, we show NP-hardness of equilibrium computation, a pseudo-polynomial time algorithm for the special *unconstrained* setting, and an MILP formulation for the general *constrained* problem. Furthermore, to tackle attackers with risk-averseness, we propose a Prospect theory-based approach (PT) that estimates the attacker's behavior from human data in random masking strategy and generate optimal masking scenarios.

Our numerical results show that PT shows a significant improvement for homogeneous populations and for a high-risk aversion compared to WSE.

To validate the numerical findings, we conducted an experiment with human attackers. We tested the effectiveness of PT and WSE algorithms against human attackers. The PT strategy was calibrated using human attacker's data collected in an experiment in which humans were pitted against random strategies. This data set helped estimate the risk-averse parameter, $\lambda = 0.75$, for the PT strategy. The results of the comparison between WSE and PT strategies showed that the strategies were not different with respect to the attackers success, but they were different with respect to the defender loss. The PT strategy resulted in lower defender losses compared to WSE. These results against human attackers are in agreement with the numerical findings in Thakoor et al. [39] which evaluated these strategies against simulated risk-averse attacker populations. In other words, these results support the idea that game theoretic and ML methods that account for human bounded rationality can produce better defense strategies than methods that assume full rationality, both in theory and in practice, against human attackers.

Through human experimentation, [1] provided insights about human's risk-aversion bias and [39] developed a masking algorithm to exploit such behavior in attacker's decisions. To accurately represent the risk-aversion, we collected human data with a random masking strategy and adapted the PT model to the risk-aversion parameter. This research validates the numerical findings of Thakoor et al. [39]'s masking algorithm in a human experiment.

Although the algorithm and the experiments in this paper have been conducted for a limited number of nodes and simple network structures, the masking algorithms are capable of including network constraints that apply in other realistic settings. Through experiments, we developed an understanding of how human attackers make decisions. Attackers are not rational; instead they act according to decision biases including certainty and risk-aversion. Human attackers shift from the expected optimal actions that some defense algorithms assume; they make suboptimal decisions. When defense algorithms are designed to exploit such biases in attacker decision-making, they could reduce the overall losses incurred from cyberattacks.

Acknowledgments This work is sponsored by the Army Research Office (grant W911NF-17-1-0370).

References

1. Aggarwal, P., Thakoor, O., Mate, A., Tambe, M., Cranford, E.A., Lebiere, C., Gonzalez, C.: An exploratory study of a masking strategy of cyberdeception using CyberVAN. In: HFES (2020)
2. Aggarwal, P., Thakoor, O., Jabbari, S., Cranford, E.A., Lebiere, C., Tambe, M., Gonzalez, C.: Designing effective masking strategies for cyberdefense through human experimentation and cognitive models. Comput. Secur. **117**, 102671 (2022)
3. Alpcan, T., Başar, T.: Network Security: A Decision and Game-Theoretic Approach (2010)
4. Ben-Asher, N., Gonzalez, C.: Effects of cyber security knowledge on attack detection. Comput. Human Behav. **48**, 51–61 (2015)
5. Berrueta, D.: A Practical Approach for Defeating Nmap OS-Fingerprinting (2003)
6. Breton, M., Alj, A., Haurie, A.: Sequential Stackelberg equilibria in two-person games. J. Optim. Theory Appl. (1988)
7. Chadha, R., Bowen, T., Chiang, C.J., Gottlieb, Y.M., Poylisher, A., Sapello, A., Serban, C., Sugrim, S., Walther, G., Marvel, L.M., Newcomb, E.A., Santos, J.: CyberVAN: A cyber security virtual assured network testbed. In: MILCOM 2016 - 2016 IEEE Military Communications Conference, Nov 2016. https://doi.org/10.1109/MILCOM.2016.7795481
8. Chadha, R., Bowen, T., Chiang, C.-Y.J., Gottlieb, Y.M., Poylisher, A., Sapello, A., Serban, C., Sugrim, S., Walther, G., Marvel, L.M., et al.: CyberVAN: A cyber security virtual assured network testbed. In: MILCOM 2016-2016 IEEE Military Communications Conference, pp. 1125–1130. IEEE (2016)
9. Cooney, S., Wang, K., Bondi, E., Nguyen, T., Vayanos, P., et al.: Learning to signal in the goldilocks zone: Improving adversary compliance in security games. In: ECML/PKDD (2019)
10. De Gaspari, F., Jajodia, S., Mancini, L.V., Panico, A.: Ahead: A new architecture for active defense. In: SafeConfig (2016)
11. Ferguson-Walter, K., LaFon, D., Shade, T.: Friend or faux: Deception for cyber defense. J. Inf. Warfare (2017)
12. Gigerenzer, G., Todd, P.M.: Simple Heuristics That Make Us Smart. Oxford University Press, USA (1999)
13. Goel, V., Perlroth, N.: Yahoo Says 1 Billion User Accounts Were Hacked, December 2016. https://www.nytimes.com/2016/12/14/technology/yahoo-hack.html
14. Guo, Q., Gan, J., Fang, F., Tran-Thanh, L., Tambe, M., An, B.: On the inducibility of Stackelberg equilibrium for security games. CoRR, abs/1811.03823 (2018)
15. Gutzmer, I.: Equifax Announces Cybersecurity Incident Involving Consumer Information (2017). https://investor.equifax.com/news-and-events/news/2017/09-07-2017-213000628
16. Jiang, A.X., Chan, H., Leyton-Brown, K.: Resource graph games: A compact representation for games with structured strategy spaces. In: AAAI (2017)

17. Joyce, R.: Disrupting Nation State Hackers. USENIX Association, San Francisco, CA (2016)
18. Kahneman, D.: A perspective on judgment and choice: mapping bounded rationality. American Psychologist **58**(9), 697 (2003)
19. Kiekintveld, C., Marecki, J., Tambe, M.: Approximation methods for infinite Bayesian Stackelberg games: Modeling distributional payoff uncertainty. In: AAMAS (2011)
20. Kiekintveld, C., Islam, T., Kreinovich, V.: Security games with interval uncertainty. In: AAMAS (2013)
21. Laszka, A., Vorobeychik, Y., Koutsoukos, X.D.: Optimal personalized filtering against spear-phishing attacks. In: AAAI (2015)
22. Mandiant: Apt1: Exposing one of China's cyber espionage units (2013)
23. McKelvey, R., Palfrey, T.: Quantal response equilibria for normal form games. Games Econ. Behav. **10**(1), 6–38 (1995)
24. Nguyen, T.H., Yadav, A., An, B., Tambe, M., Boutilier, C.: Regret-based optimization and preference elicitation for Stackelberg security games with uncertainty. In: AAAI (2014)
25. Peterson, A.: OPM says 5.6 million fingerprints stolen in cyberattack, five times as many as previously thought, September 2015. https://www.washingtonpost.com/news/the-switch/wp/2015/09/23/opm-now-says-more-than-five-million-fingerprints-compromised-in-breaches
26. Pita, J., John, R., Maheswaran, R., Tambe, M., Kraus, S.: A robust approach to addressing human adversaries in security games. In: ECAI, pp. 660–665 (2012a)
27. Pita, J., John, R., Maheswaran, R., Tambe, M., Yang, R., Kraus, S.: A robust approach to addressing human adversaries in security games. In: AAMAS, pp. 1297–1298 (2012b)
28. Provos, N.: Honeyd-a virtual honeypot daemon. In: 10th DFN-CERT Workshop, Hamburg, Germany, vol. 2, p. 4 (2003)
29. Qian, Y., Haskell, W., Tambe, M.: Robust strategy against unknown risk-averse attackers in security games. In: AAMAS (2015)
30. Rahman, M., Manshaei, M., Al-Shaer, E.: A game-theoretic approach for deceiving remote operating system fingerprinting. In: CNS, pp. 73–81 (2013)
31. Schlenker, A., Xu, H., Guirguis, M., Kiekintveld, C., Sinha, A., Tambe, M., Sonya, S., Balderas, D., Dunstatter, N.: Don't bury your head in warnings: A game-theoretic approach for intelligent allocation of cyber-security alerts (2017)
32. Schlenker, A., Thakoor, O., Xu, H., Fang, F., Tambe, M., Tran-Thanh, L., Vayanos, P., Vorobeychik, Y.: Deceiving cyber adversaries: A game theoretic approach. In: AAMAS (2018)
33. Serra, E., Jajodia, S., Pugliese, A., Rullo, A., Subrahmanian, V.S.: Pareto-optimal adversarial defense of enterprise systems. ACM Trans. Inf. Syst. Secur. (TISSEC) **17**(3), 11 (2015)
34. Simon, H.A.: Rational choice and the structure of the environment. Psychological Review **63**(2), 129 (1956)
35. Sinha, A., Malo, P., Deb, K.: A review on bilevel optimization: From classical to evolutionary approaches and applications. IEEE Trans. Evol. Comput. **22**(2), 276–295 (2018)
36. Tambe, M.: Security and Game Theory: Algorithms, Deployed Systems, Lessons Learned (2011)
37. Thakoor, O., Tambe, M., Vayanos, P., Xu, H., Kiekintveld, C., Fang, F.: Cyber camouflage games for strategic deception. In: GameSec (2019)
38. Thakoor, O., Jabbari, S., Aggarwal, P., Gonzalez, C., Tambe, M., Vayanos, P.: Exploiting bounded rationality in risk-based cyber camouflage games. In: GameSec (2020a).
39. Thakoor, O., Jabbari, S., Aggarwal, P., Cleotilde, G., Tambe, M., Vayanos, P.: Exploiting bounded rationality in risk-based cyber camouflage games. In: International Conference on Decision and Game Theory for Security (2020b)
40. Thinkst: Canary (2015). https://canary.tools/
41. Tversky, A., Kahneman, D.: Prospect theory: An analysis of decision under risk. Econometrica **47**(2), 263–291 (1979)
42. von Stengel, B., Zamir, S.: Leadership with commitment to mixed strategies. Technical report, 2004
43. Yang, R., Kiekintveld, C., Ordonez, F., Tambe, M., John, R.: Improving resource allocation strategy against human adversaries in security games. In: ICJAI (2011)

Adaptive Cyberdefense with Deception: A Human–AI Cognitive Approach

Cleotilde Gonzalez, Palvi Aggarwal, Edward A. Cranford, and Christian Lebiere

1 Introduction

The decision making process of cyber defenders who protect information networks is highly specialized and complex. Cyber defenders (i.e., analysts) constantly monitor the network for possible intrusions. The existence of multiple sensors results in a large amount of diverse network activity data, which is used in making critical decisions such as stopping potentially malicious processes and restoring systems to a secure state. Cyber security tools, such as intrusion detection systems (IDS) and Machine Learning (ML) techniques, support traffic monitoring, filter out data, and organize large amounts of network events by preprocessing and classifying data, reducing the information workload of the human analyst. However, multiple limitations in the current technologies for cyber defense remain, including that most current defense technologies are static, they generate a large number of false positives, they do not adapt according to the status of the network, they do not consider predictions of potential actions of attackers and regular users of the network, and ultimately they do not support the work of human analysts appropriately [24].

Our research program spearheaded the idea of generating dynamic, adaptive, and personalized cyber defense capabilities using deception. Our research has made it clear that current defense algorithms based on game theory and ML techniques are effective in theory, but often ineffective when paired against actual human attack

C. Gonzalez (✉) · E. A. Cranford · C. Lebiere
Carnegie Mellon University, Pittsburgh, PA, USA
e-mail: coty@cmu.edu; cranford@cmu.edu; cl@cmu.edu

P. Aggarwal
University of Texas, El Paso, TX, USA
e-mail: paggarwal@utep.edu

T. Bao et al. (eds.), *Cyber Deception*, Advances in Information Security 89,
https://doi.org/10.1007/978-3-031-16613-6_3

actions [2]. This is largely due to the assumptions that these algorithms make regarding human rationality and their lack of capabilities for real-time adaptation to human actions [2, 6]. In this research program, it has become clear that to properly design adaptive cyber defense strategies, one needs quantitative, robust models of human cognitive decision making that have been validated in the context of cyber defense [24]. Furthermore, the design and study of deceptive defense strategies (e.g., decoying, signaling, masking) is essential to this program. Deceptive techniques play a key role for the defender to learn about the adversaries and to impair the attacker's strategies and the trust they may have in their own tools and sensors. In this research program, we created a framework in which ML and game theoretic models can be informed by cognitive models of attackers and users, and we have tested its potential effectiveness in a variety of platforms from abstract to naturalistic cyber deception settings [25].

This chapter will summarize the current state of this research program. We will also outline the next steps required to achieve a complete long-term vision of dynamic, adaptive, and personalized levels of autonomy for Human–AI teaming in cyber defense.

2 A Research Framework and Summary of New Insights for Adaptive Cyber Defense

Our research program over the past years has addressed a number of challenges and advanced the concept of cyber defense towards adaptive and personalized deception strategies. A research framework was first discussed in [25] including potential deception strategies, defense algorithms, testbeds, and cognitive models used in developing and testing personalized defenses. This initial research framework has evolved in the past years with additional research on the benefits and effectiveness of other deception strategies, our demonstration of this approach in complex and naturalistic testbeds, and our consideration of end-users who are often the initial target of cyberattacks through phishing.

Figure 1 presents an updated view of our research framework. The research framework is composed of 5 steps: (1) Generate a defense strategy; (2) Deploy the defense strategy in a testbed; (3) Collect human decisions through experimentation (i.e., attacker and/or end-user); (4) Generate cognitive modeling data through simulation of human decision processes (i.e., "cognitive clone"); and (5) Use the beliefs and data generated from cognitive clones to improve defense strategies. Each of these steps is explained below, including a summary of the current state of our research.

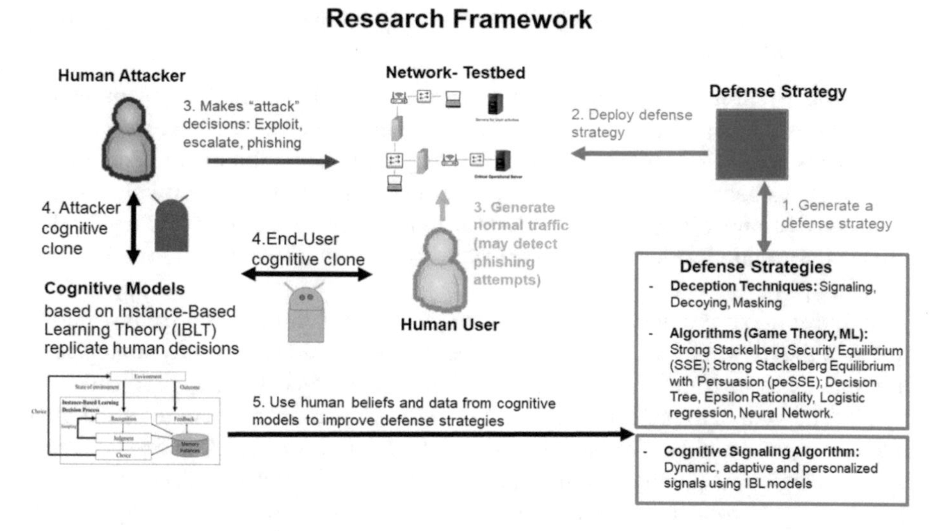

Fig. 1 Current research framework for cyber defense with deception

2.1 Generate a Defense Strategy

Defense strategies involve the selection of deception techniques (e.g., signaling, decoying, and masking) and algorithms for the distribution of defense resources (e.g., Strong Stackelberg Equilibrium with Persuasion, peSSE, and ML algorithms including Decision Tree, and Epsilon Rationality).

2.1.1 Deception Techniques

The literature regarding deception in cybersecurity proposes a taxonomy of deception techniques that correspond to the game-theoretic notions of private information, actors, actions, and duration [28]. The authors use these game-theoretic notions to describe a set of 6 techniques of deception: perturbation, moving target defense, masking, mixing, decoying, and signaling.

Perturbation refers to the application of noise in the information provided regarding machine capabilities (e.g., operating system, ports, and services). Moving target defense refers to the idea of changing attack surfaces and creating random configurations; for example, by using probabilistic strategies. Masking refers to hiding valuable information using attributes or adding noise; for example, deceptive routing of traffic. Mixing relates to a technique of hiding valuable information in an attempt to make the entry and exit nodes unlinkable; for example, masking information of real servers or honeypots. Decoying refers to using "fake" machines such as honeypots, honeynets, honeybots, etc. Finally, signaling refers to the strategic use of information to bias human actions without changing the underlying structure of the network. In our research, we have used three techniques of deception: Signaling, decoying, and masking.

Signaling Strategy Signaling theory addresses a fundamental problem in the communication between a sender (the signaler) and a receiver: whether the sender's *message* is conveying the truth or manipulating the information to her benefit [21]. Signaling has been used in Stackelberg Security Games (SSGs) in a way that makes it incentive-compatible for a sender to transmit a message that partially reveals her private information, since the receiver cannot know the underlying information with certainty [28].

In the context of cybersecurity, attackers may gather information from scanning nodes in the network (i.e., "pull information"); but also, defenders may strategically use signals to provide deceptive information to the attacker (i.e., "push information"). The goal is to prevent attackers from attacking and lead them to reveal their intentions and identity. We have investigated signaling strategies in SSGs, where these strategies essentially identify a proportion of times in which a deceptive signal could be sent from the defender to the attacker (e.g., how often to say that an unprotected node is protected or say that a protected node is unprotected). Research regarding signaling is a promising area of research in SSGs [12, 35].

Deceptive warning messages or explicit information such as network structure, number of nodes in the network, operating systems, ports, services, network traffic, round trip time information, and unpatched vulnerabilities in the network could all be used by the defender to deceive the attacker. In our research, we have considered three relevant dimensions for the investigation of signaling in SSGs: (1) The frequency of deceptive signals, (2) the level of information revealed to the attacker, and (3) the type of signal and content of the signal.

The frequency of deceptive signals is a common theme of current research in SSGs. However, current algorithms optimizing the signal frequency are less successful than expected [12, 35]. The reason is that humans are not rational, they learn from their experience, and they adapt accordingly. For example, if the defender deceives too frequently, the attacker will have a chance to learn this tendency, leading the defense strategy astray. Generally, any non-adaptive algorithm of defense will tend to be ineffective against human attackers, and SSG researchers believe that there is a *Goldilocks Zone*, an optimal level of deception that could be more effective to improve the attacker's compliance in cybersecurity games [13].

Our current work on signaling strategies for deception is largely summarized in a companion chapter in this volume. A major conclusion from that work is the key role that cognitive models of attackers' behavior have played in informing the development of adaptive and personalized signaling algorithms in Cyberdefense [15, 18]. We elaborate on these adaptive and personalized signaling algorithms in the companion chapter and in the sections below.

Masking Strategy Masking has been used to hide facts about reality (e.g., a defender can mask vulnerabilities to showcase that the computer is secure). For example, before implementing the real defense, a defender could mask the *Server Message Block* service version to showcase it is a patched and secure network.

Defenders also use mimicking software and services to imitate the ground truth [30]. The intention is to manipulate the features of a system to make it appear more or less valuable to the attacker, and to increase the attacker's time spent in planning and compromising the network. Masking is done during reconnaissance, which is the first step in the *cyber kill chain* cycle (i.e., involving reconnaissance, lateral movement, and exploitation). During reconnaissance, attackers gather information about a target using different scanning tools (e.g., Nmap, Nessus, Nikto, etc.) to learn about the network infrastructure, services, and vulnerabilities. These scanning tools provide information such as the number of systems and their connections, operating systems, and ports and services in the network.

The major research challenge is to determine how to accomplish masking to minimize the expected losses from an attack. Past research used game-theoretic solutions to design efficient masking algorithms [30]. Specifically, the authors developed a zero-sum SSG intended to design an optimal association of systems' true configurations with the observed configurations that minimize the utility of the adversary. This masking strategy is designed to optimize how the network will deceptively respond to the adversary's actions during reconnaissance. The optimal strategy was tested against synthetic "powerful" (i.e., who are fully aware of how the defender masks the information during reconnaissance) and naive adversaries (i.e., an adversary with a fixed set of preferences over the observed information).

In our work, we have investigated the effect of masking during the network reconnaissance phase [2, 34]. Specifically, we have evaluated "optimal" masking strategies against *human* adversaries and compared human performance in the optimal strategy against a random masking strategy [2, 6]. We observe that, contrary to what is observed in simulation studies with optimal making algorithms, in experimental studies with human attackers, the optimal masking algorithms were unsuccessful and similar to random masking strategies. Our analyses suggest that this is due to a general effect of risk aversion in humans. Human attackers often try to attack machines where the probability of success is high, even when the potential reward is low.

In our most recent work [6], we relaxed the assumption of rationality of the attackers made by Game Theory/Machine Learning defense algorithms, and we provide a cognitive model of human attackers that can inform these defense algorithms (more of this will be discussed below). We generated two masking strategies of defense, *risk averse* and *rational*, and the effectiveness of these two masking strategies were compared in an experiment with human attackers. The results indicate that the risk-averse strategy, which accounts for human bounded rationality, can reduce the defense losses compared to the rational masking strategy.

Decoy Strategy A decoy tactic is another popular concept used by defenders to identify attackers and gather information about their techniques [7]. Honeypots, honeynets, and honeytokens are classic examples of decoy deception. In our research, we have used decoys in abstract games to evaluate models of defense [5, 26]. Specifically, we investigate the effectiveness of various algorithms for defensive cyber deception in an adversarial decision making task using human experiments. A combinatorial Multi-Armed Bandit task represents an abstract version of a realistic problem in cybersecurity: allocating limited resources for defense in a way that an adversary can be most successfully deceived to attack "fake" nodes (i.e., honeypots) instead of real ones. We proposed six algorithms with different degrees of determinism, adaptivity, and customization to the human adversary's actions and tested those algorithms in six separate behavioral studies. We found that humans learned and took advantage of defense algorithms that are deterministic, non-adaptive, and not customized. At the same time, not all dynamic algorithms were effective, but our results suggested that adaptivity is an important feature of defense algorithms.

In recent work, we have investigated the design of honeypots. Specifically, the effectiveness of honeypots depends on their configuration, which would influence whether attackers perceive honeypots as "real" machines or not. We investigated the design of honeypots and real machines used in a simulated network and manipulated the features of the machines to test the effectiveness of the decoying strategies against humans attackers. We found that any type of deception (on honeypots and on honeypots and real machines) is better than no deception at all, and our study provides defenders with information on how to manipulate the observable features of honeypots and real machines to create more uncertainty for attackers and improve Cyberdefense [3].

2.1.2 Game Theory and Machine Learning Algorithms for Allocation of Defense Resources

Our collaborators (see Chaps. 1 and 2 in this volume) have developed innovative combinations of SSG algorithms for the distribution of limited defense resources and optimization methods from game theory, signaling theory, and ML. We have used their algorithms in combination with the deception strategies explained above in human-in-the-loop experiments. We contributed to this line of SSG research by (1) providing insights from human experiments regarding human trust to signals that are deceptive and truthful and (2) creating cognitive models that represent the decisions made by would-be human attackers that can inform the algorithms for the allocation of defense resources. The results from our experiments have informed the ML and game theory algorithms to make them more effective against human attackers [34].

Table 1 summarizes the algorithms that we have used in experimental settings, particularly in the insider attack game (see next section). The **Insider Attack Game** (IAG) was deployed under various experimental conditions to assess the effectiveness of deceptive signals on the decision making of attackers. The table highlights the high-level feature differences between the defense algorithms. The level of signaling was varied at three levels: no signal, signal uncovered nodes (1-sided signaling), and signals on both covered and uncovered nodes (2-sided signaling). Our research approach also considers the attacker type, i.e., signaling algorithms for both rational attacks and boundedly rational attackers. Finally, the

Table 1 Signaling algorithms used in experiments with the Insider Attack Game (IAG)

Deception	Algorithm	Signal on	Attacker type	Adaptive
No signal	peSSE	None	Rational	Non-adaptive
No signal	Epsilon rationality	None	Rational	Non-adaptive
1-sided	peSSE	Uncovered nodes	Rational	Non-adaptive
1-sided	peSSE-FI	Uncovered nodes	Rational	Non-adaptive
1-sided	Epsilon rationality	Uncovered nodes	Rational	Non-adaptive
1-sided	Decision Tree	Uncovered nodes	Boundedly rational	Adaptive
1-sided	*Cognitive Signaling*	Uncovered	Boundedly rational	Personalized
2-sided	peSSE	Both	Rational	Non-adaptive
2-sided	Decision tree	Both	Boundedly rational	Adaptive
2-sided	Neural network	Both	Boundedly rational	Adaptive
2-sided	Epsilon rationality	Both	Rational	Non-adaptive

algorithms vary between different levels of adaptability: non-adaptive, i.e., the signaling algorithm does not consider the past actions of the attacker, adaptive, i.e., learn the distribution of the attacker's actions and personalized, i.e., adapt the signal based on the individual attacker. Below is a brief summary of these algorithms. Expanded explanations and results from experimental work with humans against each of these algorithms are summarized in a companion chapter of this volume (Aggarwal et al.).

No-signaling Algorithm is a baseline condition where no signal is used. A signal is never presented to the attacker, regardless of whether a defender is present or absent (i.e., no deception was used). The No-signaling algorithm uses Stackelberg Security Games and calculates *Strong Stackelberg Equilibrium (SSE)* [35] to allocate defenders in the network.

1-sided Deception uses the Strong Stackelberg Equilibrium with Persuasion (peSSE) algorithm [35]. This algorithm improves defense against a perfectly rational attackers compared to strategies that do not use signaling. For a given target, the peSSE finds the optimal combination of bluffing (sending a deceptive message that the target is monitored when it is not) and truth-telling (sending a truthful message that the target is covered) so that a rational attacker would not attack in the presence of a signal. The peSSE algorithm exploits the information asymmetry between defender and attacker. Defenders have more and accurate information about the network whereas attackers could only observe the mixed strategy. Xu et al. [35] exploited this asymmetry by strategically injecting information to attackers via signaling. In this technique, defender (sender) strategically reveals information about their strategy to the attacker (receiver) to influence the attacker's decision making. The peSSE signaling scheme presents signals with probabilities calculated according to the peSSE algorithm, as described above. The **peSSE-FI** (Full-Information) signaling scheme extends the assumption of perfect rationality by ensuring that attackers have full knowledge of the probabilities of deception available to them, in addition to monitoring probabilities. In another version of peSSE, **Epsilon Rationality** defenders consider an epsilon rational model for resource allocation. All the algorithms mentioned above are *non-adaptive*, as they do not consider the actions of the attacker for generating signals. A **Decision Tree** algorithm predicts the attacker's actions to generate signals. In the 1-sided version of this algorithm, the decision tree is considered adaptive, given the algorithm relies on attack prediction for generating defense.

2-sided Deception was first introduced by Cooney et al. [13]. They extended the peSSE by considering deceptive signals on both covered as well as on uncovered nodes. Cooney et al. [13] developed **2-way peSSE** algorithm which lowers the overall frequency of showing a signal and introduces uncertainty for the rational attacker when no signal is shown. Cooney et al. [13] also focused on increasing the compliance of boundedly rational attackers by manipulating the frequency of signals. Two additional ML models, **decision tree (DT)** and a **neural network (NN)**, were used for identifying the *Goldilocks zone* and generating signals against a boundedly rational attacker.

A **Cognitive Signaling** algorithm is different type of algorithm from the other algorithms in this list. This was developed by Cranford et al. [18] using the attacker's "cognitive clone" in the insider attack game. As we will explain later, a cognitive clone is a cognitive model that aims at emulating the decisions a human makes in a task. This model generates human attack predictions and these predictions are used to modify the signaling strategy dynamically and in a personalized way (i.e., based on the particular actions of an individual attacker) [18, 19].

2.2 Deploy Defense Strategies in Testbeds that Vary in Realism and Complexity

One of the strengths of our research program is that we have been able to demonstrate our approach to cyber deception in a large variety of testbeds and interactive security games, where we have experimented with several deception techniques. Figure 2 classifies the testbeds and games we have used into two dimensions: the complexity of the task and the realism of the environment.

The **Box Game** is an abstract, 2-stage, 2-alternative SSG. In stage 1, a defender allocates resources to one of two boxes with 0.5 probability according to the optimal resource allocation SSE algorithm (i.e., pure strategy) [35]. In stage 2, the defender sends a signal to influence the attacker's decision making to her benefit by exploiting the fact that the attacker is unaware of the pure strategy at any given time. Using this game, we have investigated how humans acting as "attackers" (i.e., treasure hunters) behave under various frequencies of deceptive signals. In other words, the question is how often should a defender send a deceptive signal to gain the most benefit?

The **Honey Game** is an abstract representation of a common cybersecurity problem. A defender (i.e., the defense algorithm) assigns decoys to protect network resources and an adversary (i.e., a human) aims to capture those resources. In this

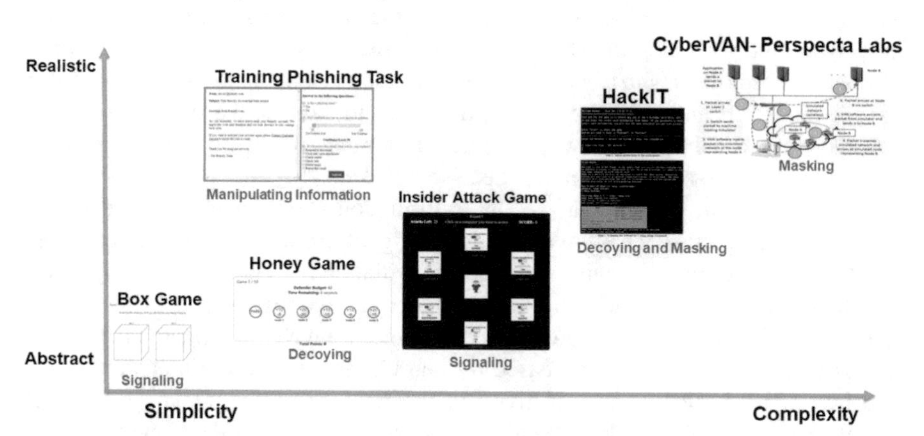

Fig. 2 Cybersecurity games and testbeds used in our research program vary in two dimensions: Complexity and realism

game, the network has 5 nodes. Each node in the network is assigned a value of the node, a cost of attacking the node, and a cost of defending nodes. The reward of attacking a non-honeypot appears as a positive number on the top of each node. The cost of attacking a honeypot appears as a negative number at the bottom of the node.

At the beginning of each round, the defender spends a budget to turn some subset of the nodes into honeypots, such that the total cost is lower than the budget. Once the defender deploys honeypots the adversary selects a node to attack. If the adversary's chosen node is not a honeypot, the adversary receives the reward and the defender receives a reward of zero. If the adversary's chosen node was a honeypot, the adversary receives the negative reward and the defender receives the positive reward. However, the adversary may also decide to "pass", which is a "safe" choice, where an adversary and the defender receive a reward of 0. Each game is composed of a set of 50 rounds. At the end of each game, the game resets, so that at the beginning of each game a new action from the defense algorithm is drawn (i.e., a combination of nodes are selected to be honeypots). The adversaries are only informed of the rewards they receive after each action, and they do not directly observe all selected honeypots in the defense action.

This game has been reported in studies [5, 26], where we investigate the effectiveness of various defense algorithms, from which we have demonstrated the importance of adaptivity in a defense algorithm: compared to other non-adaptive algorithms, adaptivity leads to more successful defense.

The IAG is an escalation of the box game, as it increases the number of nodes to six, and adds more contextual information to the nodes (e.g., gains and losses, and probabilities of coverage) [14, 19]. The allocation of defense resources is more complicated as only two out of the six nodes can be protected at a time. The attacker has access to the abstract information about the node, i.e., the value of each node, losses if the node is protected, and the probability of each node being protected. As an example, the IAG has been used to test the signaling strategy for deception in a cybersecurity scenario where the participants play a role of an attacker (a company's employee) who tries to attack the computers to gain points. The company has six computers and only two security defenders to monitor these computers. The defenders could only protect two computers at a time. To secure more computers, we use signaling to send warnings to deceive attackers [13].

In each trial, the player analyzes the information about each node and selects a computer to attack. They may receive a signal from the defender and then decide whether to proceed with the attack or withdraw it. The signaling algorithm determines whether to send a truthful signal or a deceptive signal [13]. As it is summarized in companion chapters in this volume, we have investigated the human response to deceptive signaling based on many defense algorithms, and we have also investigated the human sensitivity to the framing of the signal [17]. Using cognitive models that emulate the attacker's actions, we have developed adaptive and personalized signaling schemes that learn about attacker's actions and adjust signals accordingly [18]. An adaptive signaling scheme starts by sending truthful signals to the attacker (i.e., to gain trust), then, according to the attacker's level of trust, the algorithm adapts to whether the next signal should be a truthful or deceptive by balancing the benefits of exploit trust in the signal against the costs of rebuilding trust in the signal if it is found to be deceptive. We demonstrate how this scheme reduces the probability of attack, although at the expense of giving up more attacks in the first few trials [18, 19]. Most recent results on this game are reported in the next two chapters.

Many cyberattacks start by taking advantage of end-user cognitive and social vulnerabilities through phishing [29]. In our research program, we have also considered this important aspect of cyber defense by experimenting and building cognitive models of users' decisions to classify emails as phishing or benign.

The **Phishing Training Task** consists of an interface in which an email is presented and three responses are requested: an identification decision of whether an email was phishing or not; confidence in the classification decision; and potential responses to such an email. The task may present feedback regarding the accuracy of the classification decision after each trial. A player in this task earns points according to the accuracy of the classification decisions made, which are accumulated throughout the game. The emails in this task are a set of 186 phishing samples from a phishing email corpus ($N = 680$) collected in a past study [29]. Emails were classified based on the performance with which participants detected these as phishing emails in the study [29], as complex or simple phishing emails. This task has been used to demonstrate ways in which email phishing detection can be improved through training [31, 32], and to illustrate cognitive models of end-users that emulate the accuracy of phishing detection has been created [16]. Our current research is exploring the use of the cognitive model of the user to guide the selection of training messages, to demonstrate the benefit of adaptive training.

The **HackIT Game** is a generic web-based framework for cybersecurity to study human learning and decision making of attackers and defenders [1]. HackIT includes more semantic information such as network nodes, representing the characteristics of real nodes; deception tactics: masking, decoying; and commands, which are used for communication with the network. The defender protects the real nodes using deception tactics and the attacker's goal is to identify the real network nodes and exploit them. In HackIT, the attacker gathers information (pull information) such as operating systems, open and closed ports, services on the network nodes, and vulnerabilities from the network using probing action. Attackers could communicate with the network in HackIT using tools such as Nmap and gain information about network nodes, topologies, and configurations. However, attackers are not aware of the strategies used by defenders and they must learn those strategies overtime by playing multiple rounds. HackIT has the potential to simulate many real-world dynamic situations in the laboratory: manipulating deception tactic (e.g., decoying and masking); frequency of deceptive signals (e.g., using different proportion of honeypots in the network or testing optimal placement of honeypots); and manipulating the content of the signal (e.g., use different configurations of honeypots).

HackIT has been used recently to investigate the effects of honeypots in various network topologies and the design of the features involved in deception. In human-in-the-loop experiments using HackIT we have studied the effectiveness of deploying honeypots in different network topologies [4]: Layered and Star topologies, in which honeypots are randomly allocated. Results indicate that Layered topologies result in more exploits on honeypots compared to the Star topology.

HackIT has also been used to investigate the 2-way deception strategy with honeypots [3]. We advance past research by experimentally testing the effectiveness of 2-way deception in a game that presents attackers with a simulated network with confidential assets potentially stored on the machines. Participants who play the role of attackers are provided with commands to exploit the security of the network. The machines in the network are protected by honeypots. Our experimental conditions present participants with traditional honeypots (i.e., Default), or honeypots that look like real machines (i.e., 1-sided deception), or both, i.e., honeypots that look like real machines and real machines that look like honeypots (i.e., 2-sided deception). We find it is possible to make honeypots more effective when we manipulate the features of the honeypot design compared to using a default configuration of honeypots. Particularly, the higher the confusion level created by manipulating the design of honeypot features and real nodes, the more honeypots are attacked and the more data exfiltration is performed on honeypots. **CyberVAN** is a security testbed built on top of Virtual Ad hoc Network (VAN) for cybersecurity research [11]. CyberVAN is capable of speedy creation of high-fidelity strategic and tactical network scenarios using virtual machines, simulated networks, physical nodes and physical networks. These scenarios could be controlled by either GUI or commands on a console. CyberVAN is capable of generating realistic cyber experimentation environments which include simulated cyberattacks, cyber defense, providing synthetic users for creating realistic network traffic, and creating human-in-the-loop environments for validating various defense algorithms. Specifically, for cyber deception experiments, CyberVAN can provide different deception tactics such as masking (by hiding/faking the configuration of nodes) and decoying (by using honeypots, honeynets, honeytokens, etc.). The information manipulated for creating deception includes network structure, number of nodes in the network, operating system, ports, services, vulnerabilities, network round trip time, network traffic, etc. The proportion of deception could be controlled using different defense algorithms which could be integrated in CyberVAN. Attackers could interact with virtual machines using various network scanning tools (e.g., Nmap) to gather information during the probing phase.

Recently, we have used CyberVAN to test the effectiveness of masking strategies in a realistic scenario and to help in the development of cognitive models of cyber attackers (see below) [6]. Masking strategies for Cyberdefense (i.e., disguising network attributes to hide the real state of the network) are predicted to be effective in simulated experiments. However, it is unclear how effective they are against human attackers. Two masking strategies of defense were generated using Game Theory and Machine Learning (ML) algorithms. The effectiveness of these two masking strategies of defense, *risk averse* and *rational*, were compared in an experiment with human attackers. We collected attacker's decisions against the two masking strategies. The results indicate that the risk-averse strategy can reduce the defense losses compared to the rational masking strategy.

2.3 Collect Human Decisions Through Experimentation and the Construction of Cognitive Clones

Based on the descriptions above, it is clear that our research program heavily relies on two behavioral methods: experimentation and cognitive modeling.

The studies we conduct are most commonly designed to test specific hypotheses regarding the effectiveness of a defense algorithm or a deception technique. We conduct these human experiments using an interactive testbed (see Fig. 2), to collect human decisions online. The results also help inform the development of cognitive models that emulate human behavior (i.e., attackers or end-users). Such cognitive models (i.e., "cognitive clones") are digital representations of human memory holding an individual's experience of a particular task (also called "cognitive twins," see [33]). Cognitive models are dynamic and adaptable computational representations of the cognitive structures and mechanisms involved in cognitive tasks such as processing information for decision making. Also, cognitive models are generative, in the sense that they actually make decisions in similar ways like humans do, rather than being purely data-driven approaches [24]. In this regard, cognitive models differ from purely statistical approaches, such as machine learning, that are often capable of evaluating stable, long-term sequential dependencies from existing data but fail to account for the dynamics of human cognition and human adaptation to novel situations (see Lebiere et al. chapter in this volume).

We have built cognitive clones of the attacker in many tasks, and of the end-user in the training phishing task. All cognitive models developed in this project specifically rely on Instance-Based Learning Theory (IBLT) [23], a theory of decisions from experience. IBLT's algorithm and mechanisms have been published in multiple papers in the past [14, 15, 18, 23], thus, here we provide only a general description.

The general IBLT process involves: (1) recognition and retrieval of past experiences (i.e., instances) according to their similarity to a current decision situation; (2) generation of the expected utility of various decision alternatives by using past experiences; (3) choice of the option that best generalizes past experiences to new decisions; and (4) feedback processes that update past experiences based on the observation of decision outcomes. In IBLT, an "instance" is a memory structure that results from the potential alternatives evaluated. These memory representations consist of three elements: a situation (a set of attributes that give a context to the decision); a decision (the action taken corresponding to an alternative in a state); and a utility (expected utility or experienced outcome of the action taken in a state). Each instance in memory has an *Activation* value, which represents how readily available that information is in memory, and it is determined by its history (especially recency and frequency), its similarity to the current situation, and random noise [9]. Activation of an instance is used to determine the probability of retrieval of an instance from memory as a function of its activation relative to the activation of all instances in memory. The expected utility of a choice option is calculated based on *blending* past outcomes. The blending mechanism used in decision making models is defined by the sum of all past experienced outcomes weighted by their probability of retrieval (e.g., [22, 27]). Reflecting the general idea of an *expected value* in decision making, the blended value involves the experienced probability of events, which is based on the activation equation. At each time step, the IBL algorithm recognizes a situation in the environment (based on similarity), calculates the expected utility of the option being evaluated (through blending past experiences), determines when to stop evaluating additional alternatives, and at that point decides to make a choice by selecting the option that has the maximum blended value. Feedback, which might be immediate or delayed, updates the instance(s) in memory that lead to this particular outcome in the task. This process goes on over time, as past instances determine current decisions, which lead to learning new instance(s).

The cognitive clones developed in this project have illustrated how humans behave and exhibit nominally "irrational" behaviors (e.g., confirmation bias) that reflect capacity and information limitations, and how their decisions are based on past experience. Importantly, cognitive clones can predict individualized human decisions at any particular point going forward in time. This characteristic of cognitive models is important because it provides predictions of human decisions dynamically and applicable to the particular experience of a decision maker.

The strength of our cognitive modeling approach for adaptive signaling is explained in detail in the context of the Insider Attack Game and the Signaling deceptive technique, in a separate chapter of this volume (Lebiere et al.). However,

IBL models have also been used in the current program to generate cognitive clones in all other tasks of this program (shown in Fig. 2), and to demonstrate other deception techniques (decoying and masking). For example, Aggarwal et al. [6], recently proposed an IBL cognitive model that accurately represents and predicts the attacker's decisions in the CyberVAN environment. The model is able to capture the data at the aggregate and at the individual levels of attackers making decisions in both rational and risk-averse defense algorithms. Furthermore, this model was used to generate simulated data that represents attack decisions in CyberVAN, which was used to inform ML defense algorithms to generate new defense strategies.

2.4 Improving the Adaptivity of Defense Strategies

An attacker interacts with a network to gather information about the network structure, the number of nodes in the network, their configuration, protocols, and unpatched vulnerabilities by passively or actively probing the network [8]. Active and passive probing leaves information about attackers in the network which could be used by defenders to learn about attackers and improve their defense based on the attackers activities. Similarly, an end-user interacts with the network often through handling email, which might allow an adversary access to the network.

Having created "cognitive clones" of the attacker and end-user in handling email, we can use such cognitive models to influence the defense strategy in real-time, and adapt the defense in a dynamic fashion. A cognitive signaling algorithm closes the loop (see Fig. 1), by using the predictions of the cognitive clones and modifying the defense strategy dynamically. Such an approach has been initially illustrated in [18, 19]. However, we have not yet demonstrated how cognitive clones can influence Machine Learning and Game Theory algorithms directly or how they can help select appropriate defense strategies. This is the last step in our approach and the current stage in our research project.

3 Conclusion: Towards Adaptive Human–AI Teaming for Cyber Defense

To fully "close the loop" and be able to achieve a full level of dynamic and adaptive autonomy, we need to advance the science of Human–AI *teaming*. To advance the capabilities of cyber defense to a whole new level of effectiveness, collaboration among AI, cognitive clones, and humans will be required. A Human–AI team will aim at deploying the most effective defense strategies utilizing cyber deception. Our long-term goal is to advance the analysts' capabilities for early detection of cyberattacks against constantly evolving adversaries, and to reduce the defenders' overhead by approaching defense activities in collaborative teams. Figure 3 illustrates this vision.

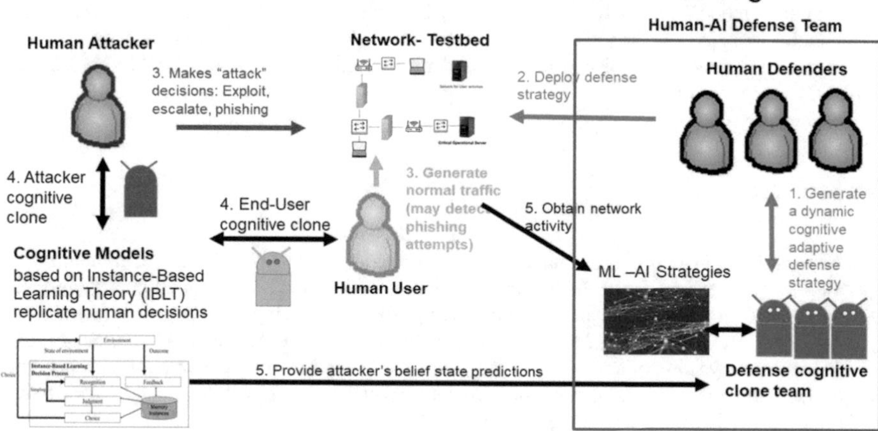

Fig. 3 A vision of Human–AI cyberdefense teams

To develop such advanced vision of cyber defense, we will need to engage in research involving collaborative and adversarial hybrid teams involving humans and machines. In such future framework, we will generate cognitive clones of the human defenders that will collaborate with the human and AI to determine the defense strategies to deploy in real-time.

In our vision of the future of adaptive cyber defense, AI defenders will have significantly larger computing capabilities than humans analysts. AI defenders will rapidly obtain activity throughout the network and be able to detect and predict the potential vulnerabilities in the network. Cognitive models will play a critical role in working with AI defenders. Cognitive models will trace human actions (attackers, defenders, and end-users) observed in the network in real-time and make predictions regarding human beliefs and potential next actions. Cognitive clones of human defenders will emerge. In the past, we have already created cognitive models of human defenders [20], but these models will need to be significantly improved, given the most recent research advances.

Furthermore, our framework for adaptive cyber defense will no longer be an individualistic approach to defense. We will need to engage in research that will help create a *team* of cognitive models that are informed by the human defender's actions, the predictions of attackers and end-users' cognitive clones, and the ML/AI capabilities. These Human–AI teaming capabilities will need to be significantly advanced by considering research regarding human behavior in teams [10]. The Human–AI Cognitive teams will determine the explicit cyber defense strategies that the human defender can decide to deploy in the network.

Our current research program has achieved significant theoretical and empirical progress towards developing the next generation of Human–AI teaming using cognitive models integrated into an adaptive cyber defense framework.

Acknowledgments This research was sponsored by the Army Research Office and accomplished under grant number W911NF-17-1-0370 (MURI Cyberdeception). Some of the work discussed in this chapter was sponsored by the Combat Capabilities Development Command Army Research Laboratory and was accomplished under Cooperative Agreement Number W911NF-13-2-0045 (ARL Cyber Security CRA).

References

1. Aggarwal, P., Gautam, A., Agarwal, V., Gonzalez, C., Dutt, V.: HackIT: a human-in-the-loop simulation tool for realistic cyber deception experiments. In: International Conference on Applied Human Factors and Ergonomics, pp. 109–121. Springer (2019)
2. Aggarwal, P., Thakoor, O., Mate, A., Tambe, M., Cranford, E.A., Lebiere, C., Gonzalez, C.: An exploratory study of a masking strategy of cyberdeception using cybervan. In: Proceedings of the Human Factors and Ergonomics Society Annual Meeting, vol. 64, pp. 446–450. SAGE Publications Sage CA, Los Angeles, CA (2020)
3. Aggarwal, P., Du, Y., Singh, K., Gonzalez, C.: Decoys in cybersecurity: An exploratory study to test the effectiveness of 2-sided deception. Preprint (2021a). arXiv:2108.11037
4. Aggarwal, P., Du, Y., Singh, K., Uttrani, S., Dutt, V., Gonzalez, C.: Effectiveness of deploying honeypots in different network topologies (2021b)
5. Aggarwal, P., Gutierrez, M., Kiekintveld, C.D., Bošanskỳ, B., Gonzalez, C.: Evaluating adaptive deception strategies for cyber defense with human adversaries. Game Theory and Machine Learning for Cyber Security, pp. 77–96 (2021c)
6. Aggarwal, P., Thakoor, O., Jabbari, S., Tambe, M., Cranford, E.A., Lebiere, C., Gonzalez, C.: Designing effective masking strategies for cyberdefense through human experimentation and cognitive models. Computers and Security (2021d)
7. Al-Shaer, E., Wei, J., Hamlen, K.W., Wang, C.: Honeypot deception tactics. In: Autonomous Cyber Deception, pp. 35–45. Springer (2019)
8. Almeshekah, M.H., Spafford, E.H.: Cyber security deception. In: Cyber Deception, pp. 23–50. Springer (2016)
9. Anderson, J.R., Lebiere, C.J.: The Atomic Components of Thought. Psychology Press (2014)
10. Buchler, N., Rajivan, P., Marusich, L.R., Lightner, L., Gonzalez, C.: Sociometrics and observational assessment of teaming and leadership in a cyber security defense competition. Comput. Secur. **73**, 114–136 (2018)
11. Chadha, R., Bowen, T., Chiang, C.Y.J., Gottlieb, Y.M., Poylisher, A., Sapello, A., Serban, C., Sugrim, S., Walther, G., Marvel, L.M., et al.: Cybervan: A cyber security virtual assured network testbed. In: MILCOM 2016-2016 IEEE Military Communications Conference, pp. 1125–1130. IEEE (2016)
12. Cooney, S., Vayanos, P., Nguyen, T.H., Gonzalez, C., Lebiere, C., Cranford, E.A., Tambe, M.: Warning time: Optimizing strategic signaling for security against boundedly rational adversaries. In: Proceedings of the 18th International Conference on Autonomous Agents and MultiAgent Systems, pp. 1892–1894. International Foundation for Autonomous Agents and Multiagent Systems (2019)
13. Cooney, S., Wang, K., Bondi, E., Nguyen, T., Vayano, P., Winetrobe, H., Cranford, E.A., Gonzalez, C., Lebiere, C., Tambe, Milind: Learning to signal in the goldilocks zone: Improving adversary compliance in security games. In: Joint European Conference on Machine Learning and Knowledge Discovery in Databases. Springer (2019)
14. Cranford, E.A., Lebiere, C., Gonzalez, C., Cooney, S., Vayanos, P., Tambe, M.: Learning about cyber deception through simulations: Predictions of human decision making with deceptive signals in Stackelberg security games. In: CogSci (2018)
15. Cranford, E.A., Gonzalez, C., Aggarwal, P., Cooney, S., Tambe, M., Lebiere, C.: Towards personalized deceptive signaling for cyber defense using cognitive models. In: 17th Annual Meeting of the International Conference on Cognitive Modelling, Montreal, CA (2019)
16. Cranford, E.A., Lebiere, C., Rajivan, P., Aggarwal, P., Gonzalez, C.: Modeling cognitive dynamics in end-user response to phishing emails. In: 17th Annual Meeting of the International

Conference on Cognitive Modelling, Montreal, CA (2019)
17. Cranford, E.A., Gonzalez, C., Aggarwal, P., Tambe, M., Lebiere, C.: What attackers know and what they have to lose: Framing effects on cyber-attacker decision making. In: Proceedings of the Human Factors and Ergonomics Society Annual Meeting, vol. 64, pp. 456–460. SAGE Publications Sage CA, Los Angeles, CA (2020)
18. Cranford, E.A., Lebiere, C., Aggarwal, P., Gonzalez, C., Tambe, M.: Adaptive cyber deception: Cognitively-informed signaling for cyber defense. In: Proceedings of the 53rd Hawaii International Conference on System Sciences (submitted). IEEE (2020)
19. Cranford, E.A., Gonzalez, C., Aggarwal, P., Tambe, M., Cooney, S., Lebiere, C.: Towards a cognitive theory of cyber deception. Cognitive Science **45**(7), e13013 (2021)
20. Dutt, V., Ahn, Y.S., Gonzalez, C.: Cyber situation awareness: modeling detection of cyber attacks with instance-based learning theory. Human Factors **55**(3), 605–618 (2013)
21. Gambetta, D.: Signaling, p. 168–194. Oxford University Press (2011). https://doi.org/10.1093/oxfordhb/9780199215362.013.8
22. Gonzalez, C., Dutt, V.: Instance-based learning: Integrating sampling and repeated decisions from experience. Psychological Review **118**(4), 523 (2011)
23. Gonzalez, C., Lerch, J.F., Lebiere, C.: Instance-based learning in dynamic decision making. Cognitive Science **27**(4), 591–635 (2003)
24. Gonzalez, C., Ben-Asher, N., Oltramari, A., Lebiere, C.: Cognition and technology. In: Cyber Defense and Situational Awareness, pp. 93–117. Springer (2014)
25. Gonzalez, C., Aggarwal, P., Lebiere, C., Cranford, E.: Design of dynamic and personalized deception: A research framework and new insights (2020)
26. Gutierrez, M., Cernỳ, J., Ben-Asher, N., Aharonov-Majar, E., Bosanskỳ, B., Kiekintveld, C., Gonzalez, C.: Evaluating models of human behavior in an adversarial multi-armed bandit problem. In: CogSci, pp. 394–400 (2019)
27. Lejarraga, T., Dutt, V., Gonzalez, C.: Instance-based learning: A general model of repeated binary choice. J. Behav. Decis. Mak. **25**(2), 143–153 (2012)
28. Pawlick, J., Colbert, E., Zhu, Q.: A game-theoretic taxonomy and survey of defensive deception for cybersecurity and privacy. Preprint (2017). arXiv:1712.05441
29. Rajivan, P., Gonzalez, C.: Creative persuasion: A study on adversarial behaviors and strategies in phishing attacks. Front. Psychol. **9**, 135 (2018)
30. Schlenker, A., Thakoor, O., Xu, H., Tambe, M., Vayanos, P., Fang, F., Tran-Thanh, L., Vorobeychik, Y.: Deceiving cyber adversaries: A game theoretic approach. In: International Conference on Autonomous Agents and Multiagent Systems (2018)
31. Singh, K., Aggarwal, P., Rajivan, P., Gonzalez, C.: Training to detect phishing emails: Effects of the frequency of experienced phishing emails. In: Proceedings of the Human Factors and Ergonomics Society Annual Meeting, vol. 63, pp. 453–457. SAGE Publications Sage CA, Los Angeles, CA (2019)
32. Singh, K., Aggarwal, P., Rajivan, P., Gonzalez, C.: What makes phishing emails hard for humans to detect? In: Proceedings of the Human Factors and Ergonomics Society Annual Meeting, vol. 64, pp. 431–435. SAGE Publications Sage CA, Los Angeles, CA (2020)
33. Somers, S., Oltramari, A., Lebiere, C.: Cognitive twin: A cognitive approach to personalized assistants (2020)
34. Thakoor, O., Jabbari, S., Aggarwal, P., Gonzalez, C., Tambe, M., Vayanos, P.: Exploiting bounded rationality in risk-based cyber camouflage games. In: International Conference on Decision and Game Theory for Security, pp. 103–124. Springer (2020)
35. Xu, H., Rabinovich, Z., Dughmi, S., Tambe, M.: Exploring information asymmetry in two-stage security games. In: Twenty-Ninth AAAI Conference on Artificial Intelligence (2015)

Cognitive Modeling for Personalized, Adaptive Signaling for Cyber Deception

Christian Lebiere, Edward A. Cranford, Palvi Aggarwal, Sarah Cooney, Milind Tambe, and Cleotilde Gonzalez

1 A Framework for Personalized Adaptive Cyber Deception

Deceptive tactics have been deployed across many cybersecurity techniques, including the strategic allocation of honeypots [22], masking the properties of systems [32], and more recently, a technique adapted from physical security systems uses deceptive signaling to expand the perceived coverage of limited defense resources [41]. Deception typically involves the strategic presentation of truthful and false information to an adversary, to mislead and gain an advantage over them, and succeeds by exploiting human processing constraints and perceptual, cognitive, and social biases [27, 30]. Despite the many advantages of traditional cybersecurity techniques for thwarting attacks, many of them are static, and adversaries succeed in their attacks as they continuously adapt to find and exploit new vulnerabilities. Therefore, adaptive techniques are required that can continuously assure effectiveness.

C. Lebiere (✉) · E. A. Cranford
Department of Psychology, Carnegie Mellon University, Pittsburgh, PA, USA
e-mail: cl@cmu.edu; cranford@cmu.edu

P. Aggarwal · C. Gonzalez
Social and Decision Sciences Department, Carnegie Mellon University, Pittsburgh, PA, USA
e-mail: palvia@andrew.cmu.edu; coty@cmu.edu

S. Cooney
USC Center for AI in Society, University of Southern California, Los Angeles, CA, USA
e-mail: cooneys@usc.edu

M. Tambe
Center for Research in Computation and Society, Harvard University, Boston, MA, USA
e-mail: milind_tambe@harvard.edu

T. Bao et al. (eds.), *Cyber Deception*, Advances in Information Security 89,
https://doi.org/10.1007/978-3-031-16613-6_4

Human–machine interactions have typically been engineered as static systems, and cybersecurity systems are no exception. However, recent technological developments have led to a push toward personalized and adaptive interactions. To meet those requirements, underlying system policies and algorithms must be tailored to individuals and therefore built upon accurate models of human behavior. Currently, personalized system policies and algorithms are often tailored to a population, or the average human behavior, or even based on erroneous assumptions/models of human behavior (e.g., that humans make perfectly rational decisions). It is well known that humans are, at best, boundedly rational [34] and that they learn through experience and can adapt accordingly. Consequently, predicting individual human behavior can be difficult. Statistical, or machine learning, models of adversaries are very good at explaining the statistics of the environment and the probability of making decisions in particular situations but often rely on large amounts of data to make accurate predictions of individual human decisions. In this chapter, we argue that behavior generative cognitive models can do so without relying on large amounts of training data and, most importantly, can help explain human behavior.

Insights from cognitive modeling not only inform how humans learn and adapt to cyber deception techniques but can also be used to personalize and adapt system algorithms to be more robust against future attacks. Therefore, we present a framework for using cognitive models to drive personalized and adaptive cybersecurity systems. As shown by the information flow depicted in Fig. 1, as an adversary interacts with a cybersecurity system (arrow 1), static policies can limit the effectiveness of automated defense algorithms (arrow 4). Cognitive models of the adversary can be inserted into the loop to predict and explain human behavior as they interact with the system. When deployed alongside the system, a cognitive

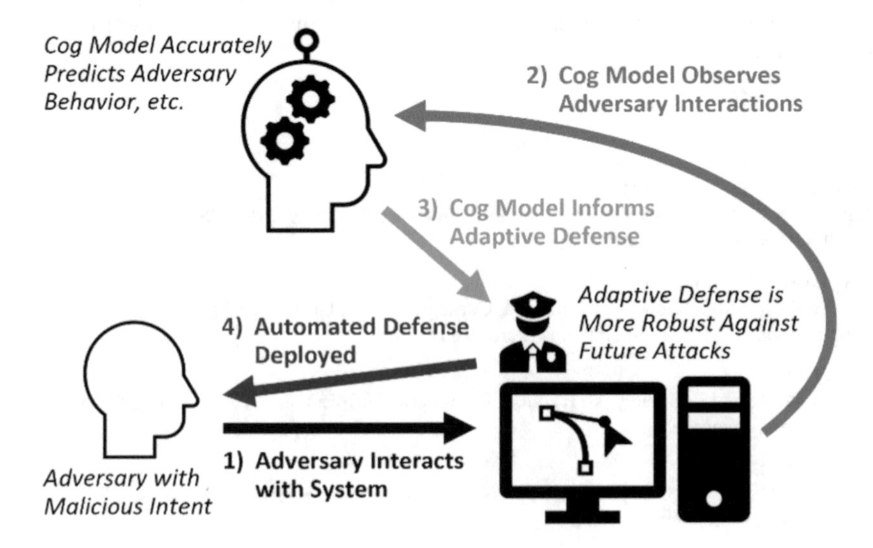

Fig. 1 Framework for using cognitive models for personalized, adaptive security systems

model can observe an adversary's interactions with the system to predict their behavior in real time (arrow 2). Combined with techniques that align the model with the individual, such as model/knowledge-tracing (discussed below in Sect. 4), a cognitive model can adapt to an individual. The cognitive model can therefore be used to inform the underlying security algorithm to drive an adaptive defense that is personalized to an individual and robust to future attacks (arrow 3).

2 Modeling the Adversary

The first step in developing personalized, adaptive cybersecurity systems is to build accurate models of the adversary. Cognitive models are particularly useful because they can offer explanations of the human reasoning and decision-making process that contribute to behavior. The predicted performance can inform how and why humans react in particular ways to particular situations.

2.1 What Is a Cognitive Model?

Cognitive models provide an introspectable, white-box abstraction of human cognitive processes. Cognitive architectures are computational implementations of the basic cognitive mechanisms that drive those processes. When cognitive models are grounded in a cognitive architecture, such as Adaptive Control of Thought—Rational (ACT-R; [2, 5]), they provide a falsifiable theory for understanding human cognition. ACT-R provides a constrained, principled framework for modeling complex human behavior that can generate predictions, and also explanations, of human reasoning and decision-making. ACT-R has been used to model complex cognition across a range of decision-making tasks, including repeated binary-choice decisions [17, 25], multi-person/multi-choice games such as Stackelberg security games (SSGs) and backgammon [1, 31, 40], dynamic environments such as social dilemmas and supply chain management [18–21, 24], and automated malware/intrusion detection systems [38, 39].

ACT-R is a hybrid architecture that represents knowledge as symbolic information while sub-symbolic computational processes operate on these structures to determine their availability and applicability in order to simulate human cognition and learning. The architecture is decomposed into multiple interacting modules that represent various cognitive faculties, such as knowledge, action selection, working memory, perception, and motor actions. Declarative knowledge is knowledge of facts and constitutes what is often referred to when talking about human memory. Meanwhile, procedural knowledge is knowledge of skills and is represented as production rules consisting of a set of conditions and actions. The procedural module uses declarative knowledge to perform tasks and make decisions. ACT-R uses buffers to hold the results of operations in each module, which can be thought of as the knowledge currently available in working memory. When the contents of the buffers match the conditions of a production rule, an action is executed, and the contents of the buffers are modified to trigger further action or knowledge elaboration.

To make decisions, declarative knowledge is retrieved from memory via a production rule. Declarative knowledge is formally represented as chunks, which consist of slot-value pairs to represent information. A chunk is retrieved from declarative memory based on its activation strength and its similarity to the content of the retrieval buffer (i.e., the requested knowledge pattern). The activation A_i of an instance i is determined by the following equation:

$$A_i = \ln \sum_{j=1}^{n} t_j^{-d} + MP * \sum_{k} \mathrm{Sim}(v_k, c_k) + \varepsilon_i \quad (1)$$

The first term reflects the power law of practice and forgetting, where t_j is the time since the jth occurrence of instance i and d is the decay rate of each occurrence, which is set to the standard ACT-R value of 0.5. The second term is a partial matching process reflecting the similarity between the current context elements (c_k) and the corresponding context elements for the instance in memory (v_k), scaled by a mismatch penalty (MP; but which was set to the ACT-R default of 1.0 in the model presented here). A variance parameter ε_i introduces stochasticity in retrieval and is a random value from a logistic distribution with a mean of zero and variance parameter s of 0.25 (ACT-R standard). Similarities between numeric slot values are computed on a linear scale from 0.0, an exact match, to -1.0. Symbolic slot values are either an exact match or maximally different, which was set to -2.5 in the presented model, a relatively large value that minimizes similarities between different types of categorical situations/actions.

A Boltzmann softmax equation determines the probability of retrieving an instance P_i based on its activation strength:

$$P_i = \frac{e^{A_i / t}}{\sum_j e^{A_j / t}} \quad (2)$$

A temperature parameter t can be used to scale probabilities according to the activation such that low temperatures result in a greater proportion assigned to the highest activated instances and high temperatures result in proportions being more randomly distributed regardless of activation strength. The model presented here sets the temperature to 1.0, which results in retrieval probabilities reflecting the original probability distribution, unbiased toward or against the most active instances.

2.2 Modeling Decisions from Experience

In many human–machine interactions, including for cybersecurity, there are ample situations where human decisions are made from experience, offering a great opportunity to leverage the powerful modeling methodology of instance-based learning (IBL) to model these decisions and adapt a system to the individual. According to IBL theory [16, 19], decisions are made by generalizing across past experiences, or instances, that are similar to the current situation. Experiences are represented by the contextual features of the decision, the action/choice made, and the outcome/utility of the decision. As humans interact with their environment, they accumulate experiences in memory. For each new situation, an expectation is generated from memory for each action/choice, based on the similarity of the current situation to past instances in memory, and their recency and frequency in memory. The choice with the best outcome, or highest expected utility, is selected. This choice, its contextual features, and the associated outcome/utility are stored in memory as a new instance that can then influence future decisions.

IBL is a domain-general, memory-based theory of experiential learning, which means that it does not require explicit engineering of strategies. Nor does it require any hand-modeling of reward functions. IBL models decision-making as learning through experiential interaction with the environment. Formally, IBL utilizes ACT-R's *blending* mechanism [23] to make aggregate retrievals from memory in order to generate expectations for the outcome of an action in a given situation. Typically, the most active chunk is retrieved from memory. However, the blending process retrieves a chunk representing the interpolation of past instances. According to blending, an expected outcome of a particular choice is the value V that best satisfies the constraints of all matching instances i weighted by their probability of retrieval, where satisficing is defined as minimizing the dissimilarity between the consensus value V and the actual answer V_i contained in instance i:

$$\underset{V}{\operatorname{argmin}} \sum_{i} P_i \times (1 - \operatorname{Sim}(V, V_i))^2 \tag{3}$$

When the values are numerical and the similarity function is linear, the process simplifies to a weighted average by the probability of retrieval, where $V_t = \sum_{i=1}^{n} P_i \times V_{it}$. Therefore, in summary, the outcomes of past instances are weighted by their recency, frequency, and similarity to the current instance (i.e., probability of memory retrieval) to produce an expected outcome via blending.

A combination of instance-based learning models, grounded in a cognitive architecture like ACT-R, can be used to drive robust adaptive and personalized systems. As an adversary interacts with a system, an IBL model can observe the human and predict their behavior. This information can then be used to inform the system to provide personalized and adaptive interventions. For the present research, we highlight the methodologies' strengths for adapting a deceptive signaling algorithm to individual attackers in an insider attack scenario.

2.3 Deceptive Signaling for Cybersecurity

Recent developments in cybersecurity have proposed using deceptive signals to deter attacks on uncovered systems beyond any capabilities of static defenses that do not use signaling or only use truthful signals. Finding the right balance of deceptive signaling so that the attacker continues to believe the signal is crucial to the success of the strategy. Recently, game-theoretic research on deceptive signaling algorithms in Stackelberg security games (SSGs) has optimized the strategic allocation of limited defenses and the rate of deception so that a rational attacker would not attack when presented with a signal [41]. SSGs model the interaction between an attacker and a defender where a defender plays a particular strategy (i.e., random patrolling of an airport terminal), the attacker observes the strategy, and then the attacker takes action. Under this framework, researchers have developed algorithms, such as the strong Stackelberg equilibrium (SSE), that optimally allocate limited defense resources across a set of targets [36]. These algorithms have been applied successfully across a number of physical security systems (e.g., protecting ports, scheduling air marshals, and mitigating poachers; [28, 33, 35, 36]). Such security practices could be applied to the cyber realm, for example, in scheduling active monitoring of security systems by network administrators (e.g., security analysts).

Xu et al. [41] extended the SSG models by incorporating elements of signaling, in which a defender (sender) strategically reveals information about their strategy to the attacker (receiver) in order to influence the attacker's decision-making [6, 8]. Sending a message that reveals the protection status of the target can influence attacker behavior. For example, a truthful message that reveals a target is monitored can deter attacks, but adversaries can attack with impunity when a message reveals the target is not monitored. However, defenders can use a combination of truthful and deceptive signals to help deter attacks on unprotected resources. Xu et al.'s solution, the strong Stackelberg equilibrium with persuasion (peSSE), improves defense against a perfectly rational attacker compared to strategies that do not use signaling. For a given target, the peSSE finds the optimal combination of bluffing (sending a deceptive message that the target is monitored when it is not) and truth-telling (sending a truthful message that the target is covered) so that a rational attacker would not attack in the presence of a signal.

In practice, the SSE allocates defenses proportionally across the set of targets so that the expected values of all targets are equal. Once defenses are scheduled, the attacker can choose a target to attack. Then, as determined by the peSSE, the defender will send a signal to the attacker revealing the protection status of the target, which may sometimes be deceptive. Based on this information, the attacker can then choose to continue the attack or withdraw. If the attacker continues the attack, then they will receive a penalty if the target is truly monitored, but a reward if the target is open. The peSSE sends deceptive signals at a rate that makes the expected value of attacking a target, given a signal, equal to the expected value of withdrawing the attack, or zero. Therefore, under the assumption of perfect rationality, when presented with a signal, an attacker will always break ties in favor of the defender and choose the safer option, to withdraw the attack.

2.3.1 Insider Attack Game (IAG)

The Insider Attack Game (IAG) is an online game designed to investigate the interaction between an attacker and defender in a cybersecurity scenario, gain a better understanding of how humans react to deceptive signals, and assess the effectiveness of various signaling schemes [11, 15]. The left side of Fig. 2 shows a screenshot of the game interface. Players take the role of the attacker at the center of the screen (i.e., a company employee) and their goal is to score points by "hacking" computers to steal proprietary data. There are six computers from which to choose to attack, but only two security analysts (i.e., defenders controlled by a computer algorithm) that can monitor one computer each. If the player attacks a computer that is monitored, they lose points denoted by the number of red stars, but if the computer is not monitored, then they win points denoted by the number of yellow stars. As shown in the top-middle of Fig. 2, each computer shows its reward for winning, penalty for losing, and the probability that the computer is being monitored (which reflects the SSE for the game).

Players make repeated attempts at attacking computers. On each turn, the player must first select a computer to attack. Then, the signaling algorithm determines whether to send a truthful signal or a deceptive signal. As depicted in the bottom-right of Fig. 2, in the IAG with six targets and two analysts, the peSSE presents a signal every time a target is monitored, or 33% of trials on average. Additionally, the peSSE sends a signal half of the time when a target is not monitored, or 33% of

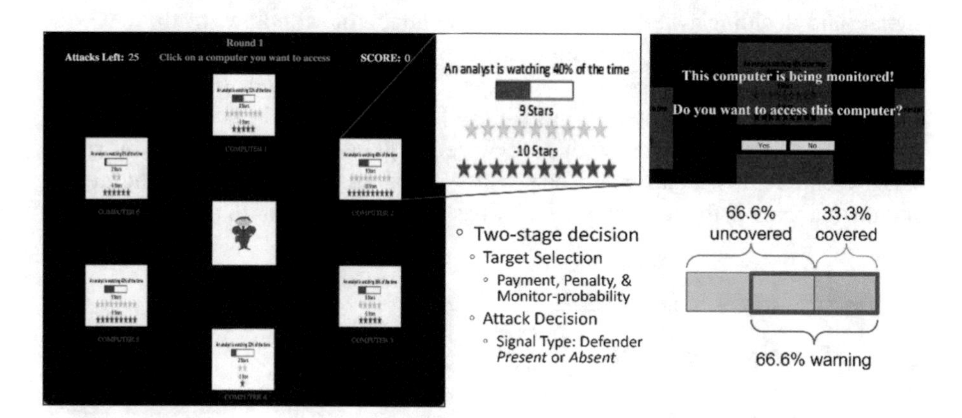

Fig. 2 Insider Attack Game interface, zoom inlay of a target, example signal message, description of the decision procedure, and the average coverage/signaling probabilities

trials on average. This means that, on average, a signal is deceptive half of the time. At this rate, the expected value of attacking given a signal is zero, the same expected value as withdrawing the attack. Therefore, a perfectly rational adversary that only attacks with a positive expected value (i.e., in the absence of a signal) is predicted to attack on 33% of trials on average (i.e., when a signal is *not* presented).

The top-right of Fig. 2 shows an example message signaling that a target is currently being monitored. If the computer is not being monitored, then the first line of the message is omitted. After reading the message, the player must decide whether to continue their attack or withdraw and earn zero points. Players play four rounds of 25 trials each (after an initial five trials of practice). The payoff structures and monitoring probabilities of the targets are different in each round. Coverage and signaling of targets were precomputed for each trial. Therefore, each individual player experiences the same coverage and signaling schedule.

2.3.2 Modeling Adversary Behavior in the IAG

Cranford et al. [15] presented the results of 98 human participants playing the IAG against the peSSE signaling scheme and a cognitive model of an attacker that accurately predicts human performance and helps explain their behavior. The left side of Fig. 3 shows the mean probability of attack across trials. The dashed line at the bottom of the graph shows the predicted probability of attack of a perfectly rational adversary (33%). The results showed that humans attacked far more often than predicted, almost 80% of trials.

It is clear that humans do not make perfectly rational decisions. Instead, human behavior can be explained as decisions from experience [19]. Our IBL model performs the task like humans by selecting targets to attack, being presented a signal, and deciding whether to continue the attack or withdraw. In the IAG, the experiences, or instances, are represented by the features of the decision. This includes the context of the selected target, the decision, and the outcome. The context includes the monitoring probability [0.0, 1.0], reward [1, 10], and penalty values [−1, −10] associated with the selected target and whether a warning signal was presented [present, absent]. The possible decisions are to attack or withdrawing, and the outcome is the reward or penalty based on the decision. In a given situation,

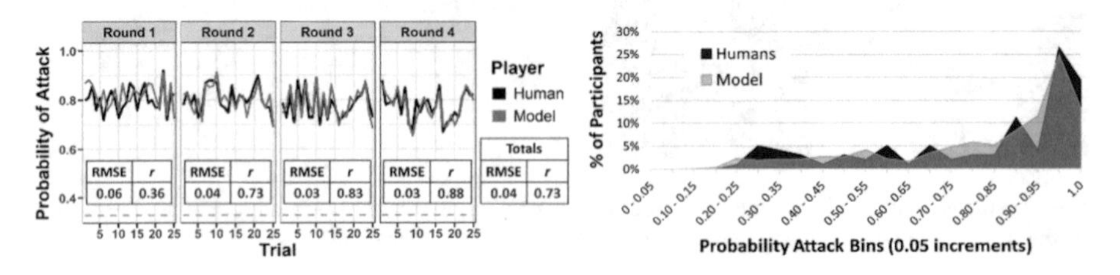

Fig. 3 Results of the IBL model compared to humans playing the IAG against the peSSE. (Figure adapted from Cranford et al. [13])

for each possible decision, an associated utility is computed through blended memory retrieval weighted by contextual similarity to past instances. The decision with the highest expected utility is made. However, withdrawing always results in zero points. Therefore, the model only needs to determine the utility of attacking in order to make a choice. If the value is greater than zero, then the model attacks, else it withdraws.

For each trial, the model first selects a target with the highest expected outcome, generated via blending, and then decides whether to continue the attack or withdraw based on whether a signal was presented. For this decision, the model uses blending to generate an expected outcome for the given target, but only on the basis of the signal and ignores the values of the target context (i.e., the target information is occluded from the participants, so it is plausible that they do not consider the target information beyond deciding which target to select initially). An instance is then saved in memory that represents the model's expected outcome. Humans tend to remember not only the actual experience but also their expectations prior to the experience [26]. This results in additional positive (or negative) instances, which in turn generates a confirmation bias whereby one's pre-conception of winning (or losing) perpetuates itself in future trials, even when it is ultimately disconfirmed. Based on the value of the expected outcome, a decision is made, and the action and outcome slots of the current instance are updated to reflect the action taken by the model and the ground-truth outcome. This final instance is saved in memory and thereby influences future decisions. Therefore, two instances are saved to memory on each trial, one reflecting the expected outcome and the other reflecting the ground truth outcome.

The model continues for four rounds of 25 trials each. The model behavior reflects its experiences. If an action results in a positive/negative outcome, then its future expectations will be increased/decreased, and the model will be more/less likely to select and attack that target in the future. Also, the impact of a particular past experience on future decisions strengthens with frequency and weakens with time.

The model was run 1000 times to simulate a population of individuals and to generate stable estimates of human performance. As shown on the left side of Fig. 3, the model is highly accurate at predicting human performance (total RMSE = 0.04), even matching the trial-to-trial variations that reflect the underlying coverage and signaling schedules (total $r = 0.73$), and that accuracy increases over time. Not only does the model match the average human performance in the IAG, but it also matches well to the individual performance. The right side of Fig. 3 shows the distribution of participants by their mean probability of attack. Like humans, some model simulations attack at a fairly low rate, while a large proportion attack 95% of the time or more.

In summary, Cranford et al. [15] show that human decision-making in the IAG is largely influenced by memory dynamics across past experiences. The peSSE suffers because human biases (e.g., recency, frequency, and confirmation) lead to overweighting of certain outcomes that often results in inflated expectations. Humans fail to fully comply with the signal because they are more likely to expect a positive outcome than a negative one as belief in the signal deteriorates. While deception is an effective tool for preventing malicious behaviors, the experience of successfully calling a bluff can reduce compliance with the signal. Regaining trust in the signal is difficult, if not impossible, to do under static signaling schemes. Therefore, an adaptive signaling scheme is needed that adjusts the rate of deception to dynamically balance (re)building trust in the signal and exploiting it, thus optimizing compliance.

3 Predicting Adversarial Behavior

Unlike a statistical, or machine learning, model of attackers that can explain the statistics of the environment and the probability of making decisions in particular situations and that rely on large amounts of data to make accurate predictions of human decisions, behavior generative cognitive models can do so without relying on large amounts of training data and, most importantly, can help explain human behavior. In fact, cognitive models need no training data at all because the goal is not to model the data but instead to model human cognition when performing the task. Therefore, this also means that cognitive models can be robust to changes in the environment, given the task procedures remain the same. Because cognitive models also have the capacity to learn quickly (i.e., at human speed), they can be used to model individual learning.

For example, our cognitive model of backgammon playing learned to play the game at a high level in about a thousand games, a scale of practice compatible with human experience [31]. That contrasts with machine learning models that can attain higher performance but at the cost of requiring hundreds of thousands or even millions of games of experience, an amount incompatible with human limitations (e.g., [37]). Cognitive models replicate fast human learning, including in the backgammon domain, through a combination of symbolic knowledge representation and sub-symbolic (statistical) generalization.

Our IBL model of the IAG can therefore predict situations (e.g., counterfactuals) for which there is no data currently available, so we used our cognitive model as synthetic subjects to explore manipulations of the environment. For example, Cooney et al. [9, 10] attempted to use game theory and machine learning algorithms to improve the deceptive signaling algorithm to account for the boundedly rational human behavior observed against the peSSE. Two of the algorithms examined included a decision tree machine learning method and an epsilon rationality game theory approach (the latter being an adaptation of the Bounded Rationality Assumption in Stackelberg Solver [BRASS] algorithm; [29]). While human experiments showed that neither of the algorithms was significantly more effective than the peSSE, the model was used to make a priori predictions of human behavior. These predictions, although discouraging for the signaling algorithms, ultimately were

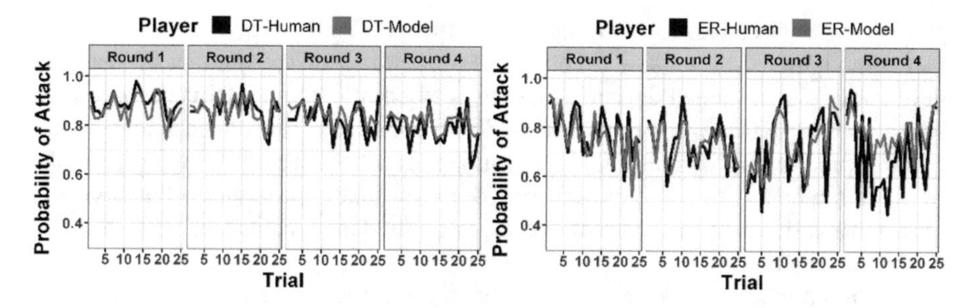

Fig. 4 Model makes accurate predictions of human behavior, generated prior to running human subjects experiments, across two alternative signaling schemes, the left based on decision trees (DT) and the right based on epsilon rationality (ER)

highly accurate when compared to human performance, as can be seen in Fig. 4. The model was able to accurately predict the mean probability of attack across trials, as well as the trial-to-trial fluctuations in performance that are attributed to the underlying, static signaling scheme. The ability of cognitive models to predict the effect of manipulations that have not yet been implemented and tested against human adversaries makes it possible to greatly speed up the development of those defensive techniques and expand the range of techniques considered.

4 Observing the Adversary: Personalizing the Model

The goal for personalized, adaptive cybersecurity is to model the individual, not the population. In the models presented above, each run of the model produces its own decision and learning experience. That is, each run represents an individual in a population. The model can reproduce the entire distribution of individual behavior due to stochasticity in the activation calculus, the IBL learning loop, and representation variants (e.g., each run begins with a randomized set of instances that represent a practice round). However, in order to drive personalized, adaptive cybersecurity systems, we must be able to model a specific individual at a specific time. It is clear that individual attackers behave differently from one another, and each may learn and adjust behavior after repeated experiences with deceptive signals. Therefore, a model of the individual must be able to reflect those differences, both initially and resulting from experience, including from the intervention itself. An adaptive signaling scheme based on cognitive principles can be used to adjust the rate of deception, tailored to an individual's behavior, so as to maintain belief in the signal.

To personalize deception, two techniques can be used. One method runs the full cognitive model alongside the human adversary, in combination with model-tracing techniques, to align the model to the human in real time and uses the model predictions to directly inform the signaling scheme. The limitation of this method is that computational processing time increases linearly with the number of instances in memory. Since human decision-making is stochastic, the model is probabilistic, and a possibly large number of Monte Carlo runs must be collected to generate reliable predictions. Therefore, such a model can become unwieldy in practice. Another method is to use a closed-form solution that approximates the cognitive model predictions. In this chapter, we focus on the latter method but first describe the former method.

To align a model's behavior with the human's experience and decisions, we use a combination of model-tracing and knowledge-tracing. Both methods are used to align the model's memory with that of the human it is tracing. Model-tracing aligns the actions and outcomes of the model with those observed of the human, while knowledge-tracing is used when we must infer the knowledge that humans have acquired to maintain accurate predictability.

4.1 Model-Tracing

As described in Cranford et al. [13], *model-tracing* is a technique commonly used to adjust feedback provided to the student in intelligent tutoring systems (see [4]). The alignment helps ensure future model predictions are adapted and optimized to the interaction with the human. For example, geometry tutors use model-tracing to keep track of where errors are made so that the learning experience can be tailored to the individual [3]. We use model-tracing to synchronize the IBL model with the human's *observed* actions and experience in the IAG task. After each trial, the instance saved in memory that represents the model's decision and outcome is changed to reflect the human's action and outcome (i.e., the action and outcome slots are changed to match the human's). Therefore, during the next trial, the model makes predictions based on the exact experience of the human and not on what it would have done based on its own past instances. With more trials, the model is expected to make more accurate predictions of a particular human's actions, as the model's memory aligns better with that of the human.

4.2 Knowledge-Tracing

Both model-tracing and knowledge-tracing are used to align the model's memory with that of the human; however, *knowledge-tracing* is used to align the model's expectations to those of the human and must therefore be inferred. When making a decision to attack or withdraw, the model produces instances that represent the expected outcome of attacking, which contributes to confirmation bias, and these instances must also be changed. Knowledge-tracing is therefore used to resolve any discrepancies in knowledge or strategy. Knowledge-tracing can be used to infer

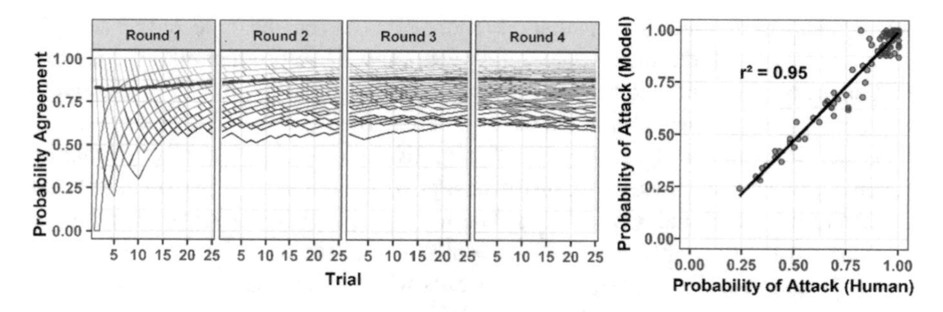

Fig. 5 Effectiveness of model/knowledge-tracing techniques for predicting human performance and learning in the IAG. (Figure adapted from Cranford et al. [13])

the expectations humans had prior to making a decision that would contribute to confirmation bias. For example, if the model and human both decided to attack (or both withdraw), then nothing needs change and the expected outcome generated by the model can be used to infer the human's expectation. However, if the model expects a positive outcome for attacking, but the human withdrew the attack, then we can infer that the human expected to lose (or vice versa). For these instances, we can modify the expected outcome slot to match the expectations of the attacker. We cannot infer this expectation precisely, so we set the expected outcome to either the reward or penalty of the selected target.

Using a combination of model- and knowledge-tracing, we tested the model's ability to predict individual behavior in the IAG. The cognitive model was run alongside the human data when paired against the peSSE. On each trial, the model makes a prediction and is then aligned to the human before making the next prediction. The left side of Fig. 5 shows the probability of agreement between each model run and the human it traced. The results show that the model is highly accurate at predicting individual behavior in the IAG, and the predictability increases the more experience the model accumulates of the individual it is tracing. After only a few instances (i.e., around trial 15), all model runs are above chance, 0.50. The right side of Fig. 5 shows the overall probability of attacking for the model and the human it traced, with an astonishing $r^2 = 0.95$. These results highlight the model's ability to adapt to an individual by aligning the model's memory with that of the human and could feasibly be used to personalize a signaling algorithm.

5 Using Cognitive Models to Inform Adaptive Defense

A traditional approach to modeling the individual attacker for purposes of adapting interventions would be to collect sufficient data on that user, training a machine learning model with that data, and then using that model to predict their behavior. That approach has several drawbacks, however. First, it is likely that the machine learning model will require considerably more data than can be collected on a single individual. Second, the data-driven model can only make predictions in situations that have been encountered before. Third, the machine learning model is effectively a black box that makes predictions without being able to explain their rationale.

Using information leveraged from cognitive models, an adaptive signaling scheme can be developed to improve deceptive techniques by personalizing the system to the individual. Our initial solution to this problem is to interleave blocks of trials with only truthful signals between blocks of trials with deceptive signals. The assumption is that experiences of rewards when a signal is present increase the probability of attacking in the future, while experiences of penalties given a signal reduce the probability of attacking in the future. Therefore, eliminating deceptive signals for a short period of time can help increase penalties and restore belief in the signal. The goal for the cognitive signaling scheme is to induce, and preserve, the belief that attacking given a signal will result in a loss.

Relying on the attacker's history of behavior, this new cognitive signaling scheme estimates the current probability of attack given a signal and judges whether the cost of issuing a truthful block outweighs the benefits of a deceptive block to effectively reduce the future probability of attack given a signal. At the beginning of each block of trials, a closed-form equation of the current probability of attack given a signal, reflecting the blending process used in generating expectations and the recency and frequency power laws in chunk activations, can be formulated based on the times t since past actual decisions made by the attacker, as

$$P_{\text{est}}^{\text{now}}\left(A|S\right) = \frac{\sum_{i}^{\text{wins}} t_i^{-d} + \sum_{j}^{\text{losses}} t_j^{-d}}{\sum_{i}^{\text{wins}} t_i^{-d} + \sum_{j}^{\text{losses}} t_j^{-d} + \sum_{k}^{draws} t_k^{-d}} \tag{4}$$

Next, we estimate the change in the probability of attack given a signal from a truthful block. Therefore, we need to make an additional assumption as to how wins and losses impact choice. We assume that the attacker will follow the same decision-making process, keeping the same format reflecting probability matching behavior:

$$P_{\text{ass}}^{\text{now}}\left(A|S\right) = \frac{\sum_{i}^{\text{wins}} t_i^{-d}}{\sum_{i}^{\text{wins}} t_i^{-d} + \sum_{j}^{\text{losses}} t_j^{-d}} \tag{5}$$

The impact of a truthful block of size b on $P_{\text{ass}}^{\text{now}}\left(A|S\right)$ $P_{\text{ass}}^{\text{now}}$ (A|S) results in a new estimate $P_{\text{ass}}^{\text{then}}\left(A|S\right)$ $P_{\text{ass}}^{\text{then}}$ (A|S) with an expected number $\frac{1}{3} * b * P_{\text{est}}^{\text{now}}\left(A|S\right)$ $P_{\text{est}}^{\text{now}}$ (A|S) of losses distributed randomly across the block, where 1/3 is the mean probability of sending a signal in a truthful block. For the present implementation, block size b is set to 10. This value was chosen as a reasonable compromise that provides enough opportunities for switching blocks while allowing for enough experience within a block to impact behavior.

The adaptive cognitive signaling scheme is as follows: the next block will use a truthful signal if the following comparison of the cost in terms of additional attacks allowed in the next block is less than its benefits (i.e., the number of attacks saved in the remaining r trials during the rest of the experiment after that block):

$$\frac{1}{3} * b * \left[1 - P_{\mathrm{est}}^{\mathrm{now}}\left(A|S\right)\right] < \alpha * r * \left[P_{\mathrm{ass}}^{\mathrm{now}}\left(A|S\right) - P_{\mathrm{ass}}^{\mathrm{then}}\left(A|S\right)\right] \tag{6}$$

where 1/3 is the difference in probability of a signal being generated between deceptive (66.6%) and truthful blocks (33.3%), and α is a discount parameter that can take any value between 0.0 and 1.0 (default is 1/3). The discount parameter is an assumption of how long the impact of the truthful block on the probability of attack given a signal will persist. If we assume that it will persist until the end and all future blocks will be deceptive blocks, then the right value would be 2/3 (i.e., the percentage of trials when a signal is generated). If it would persist indefinitely but all future blocks are truthful blocks, then that value would be 1/3. In practice, it will be somewhere between 1/3 and 2/3, depending on the mix of truthful and deceptive. The effect of the signal will dilute over time, so the minimum 1/3 is a reasonable default value.

In summary, the cognitive signaling scheme uses a closed-form version of the model decision procedure to optimize the tradeoff between the cost of building trust in the signal using blocks of truthful signals and the benefits of exploiting that trust in future blocks of deceptive signals.

5.1 Cognitive Signaling Scheme Evaluation

The effectiveness of the cognitive signaling scheme was examined through cognitive model simulations and a human behavioral experiment. The cognitive model of the attacker presented above was run through 1000 simulations against the cognitive signaling scheme, and these predictions were then compared to the performance of human participants. For the human experiment, 100 participants were recruited via Amazon Mechanical Turk (mTurk). All mTurk participants resided in the United States. For completing the experiment and submitting a completion code, participants were paid $1 plus $0.01 per point earned in the game, up to a maximum of $5.50. One participant was removed from the analysis because of incomplete data due to data recording errors, resulting in a final N of 99.

As this was an initial study, all players began with a block of truthful signals to establish baseline belief in the signal. As before, players played four rounds of 25 trials each, with a different set of targets each round. Every 10 trials overall, the algorithm determined whether to switch to a different type of block: either using only truthful signals or using deception according to the peSSE. Figure 6 shows the proportion of players that received a truthful block across each of the 10 blocks in the game. The first block is always a truthful block. From there, depending on the individual's behavior, the cognitive signaling scheme assigned more truthful or deceptive blocks. The second block was always deceptive, and the third block was

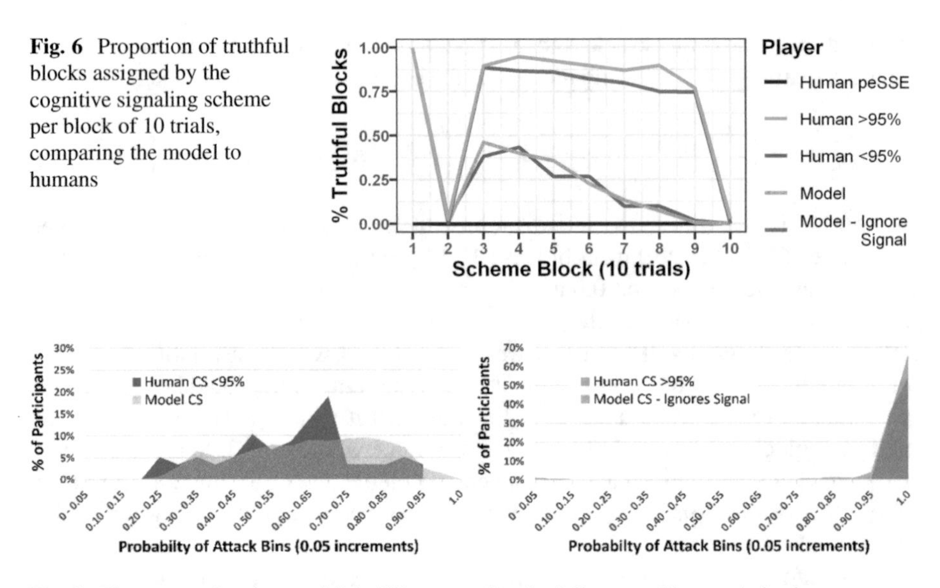

Fig. 6 Proportion of truthful blocks assigned by the cognitive signaling scheme per block of 10 trials, comparing the model to humans

Fig. 7 Histograms showing models' ability to predict the full range of human behavior

about evenly divided between truthful and deceptive. Over time, the proportion of truthful blocks declines because the estimated reduction in the future probability of attack over the remaining blocks does not outweigh the near-term term costs of issuing a truthful block.

To assess human and model performance, the data was analyzed for the probability of attack across trials. The results initially revealed that the scheme is effective at influencing human behavior beyond the peSSE, but only for some humans. As shown in the histogram in Fig. 7, the model fails to account for approximately 44% of participants that attacked at a rate of 95% or more. However, if we separate participants into two groups, the model is highly accurate at predicting the performance of the approximately 56% of participants that attack at a rate less than 95% (left side of Fig. 7). For the participants that attacked at a rate greater than 95%, the cognitive signaling scheme did not influence behavior even after giving these participants, almost exclusively, truthful blocks. Looking back, Fig. 6 shows the proportion of truthful blocks assigned per block of 10 trials for the two separate groups. The cognitive signaling scheme presented the same proportion of truthful blocks to the model as it did to those participants that attacked less than 95% of the time. However, the scheme continued to present truthful blocks to the other group of participants because they continued attacking undeterred in the face of a signal.

Figure 8 shows the probability of attack across trials for humans compared to the model when playing against the cognitive signaling scheme, which is compared to human performance when playing against the peSSE. Compared to the peSSE, the cognitive signaling scheme further reduces the probability of attack, but at the expense of giving up more attacks in the first block. Because all signals are truthful in the first block of the cognitive signaling condition, fewer signals are sent to deter

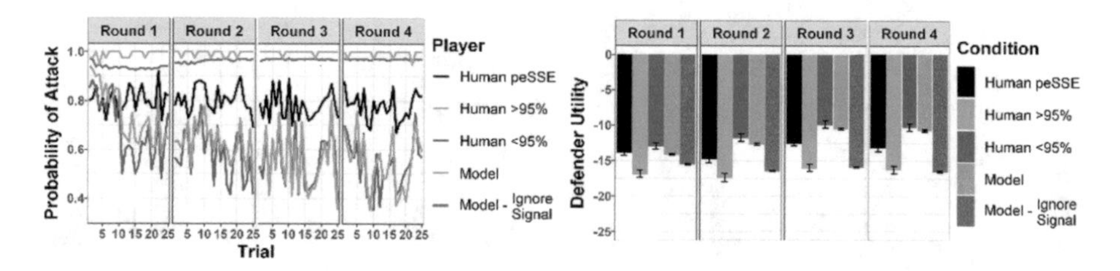

Fig. 8 Comparing the effectiveness of the cognitive signaling scheme to the peSSE for humans compared to model predictions

attacks overall. The effect of an initial truthful block is immediately observable by a relatively lower probability of attack in trials 10 through 20 (which is always a deceptive block), and this trend continues throughout the game.

To further assess the effectiveness of the signaling scheme, we examine defender utility. To compute defender utility, the defender is penalized one point every time the player attacks a target that is not monitored and zero points otherwise (e.g., if a player attacks a target that is monitored, or does not attack). This means, the more often players attack in the face of a deceptive signal, the worse will be defender utility. Since targets are not monitored on an average of 66.6% of trials, a defender utility less than −17 (i.e., >2/3 of 25 trials) means the signaling scheme is better than a purely truthful signaling scheme, while a utility greater than −9 is ideal (i.e., <1/3 of 25 trials).

As shown in Fig. 8, the cognitive signaling scheme provides better defense for a subset of humans, as indicated by low defender utility values that match what was predicted by the model. However, against some participants, the scheme performs about as poorly as would be expected given no signals. In fact, in a post-experiment survey that asked an open-ended question about what strategy participants used when faced with a signal, a majority of participants that attacked more than 95% responded explicitly that they ignored the signal.

An informal analysis was conducted with two independent coders, and the responses were categorized based on the features in which decisions were based or the reported actions taken. Discrepancies between coders were resolved through discussion. Figure 9 shows the distribution of responses for each group of participants, those that attacked greater than 95% compared to those that attacked less than 95%. For the group that attacked more than 95%, almost 23% reported that they ignored the signal while another ~10% reported that they always attacked (but did not explicitly mention whether they attended the signal or not). Approximately 10% reported that they stayed and continued attacking the same target even after suffering a loss, while ~15% switched to another target and continued attacking. The remaining participants reported a mixed strategy, weighing the risk against the reward, using intuition, some other noncategorized strategy, or no strategy, while some falsely reported that they withdrew when faced with a signal. Meanwhile, for the group that attacked less than 95%, none reported that they ignored the signal, while approximately 20% reported that they withdrew in the face of a signal, ~12%

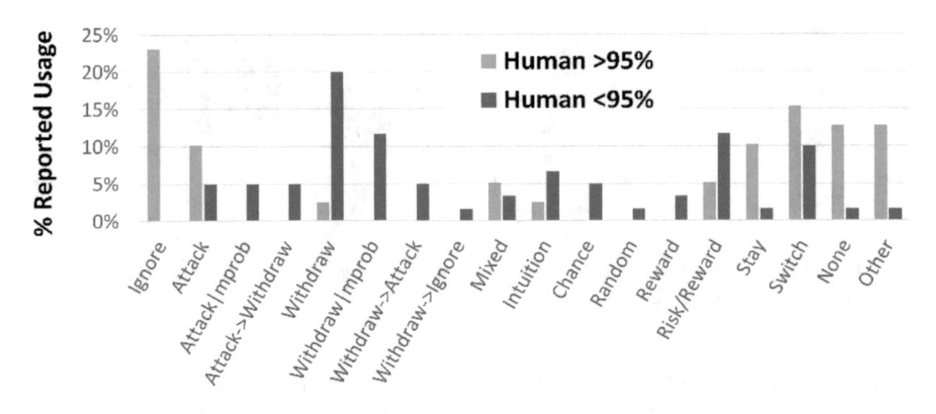

Fig. 9 Reported strategies or features used when deciding whether to continue the attack or withdraw when faced with a signal, comparing participants that attacked more than 95% to those that attacked less than 95%

withdrew if the monitoring probability was high, another ~12% weighed the risk against the reward to decide, and ~10% switched targets after seeing a signal. The remaining participants reported a mix of other strategies, decided randomly, or used no strategies. Overall, the survey results showed that some participants ignored the signal and treated all instances as if the signal was neutral. This means that the signaling scheme will not be effective against these participants because the expected value of attacking given a signal is combined with the expected value of attacking given no signal. Therefore, with only two analysts, the overall expected values would be positive, and thus deciding to always attack is a natural consequence.

Based on these findings, we created a version of the cognitive model that does not consider the signal when generating an expected outcome of attacking the selected target. For this version, blending samples equally across past instances regardless of the signal, and so only recency and frequency of past instances play a role in decisions. As shown in Fig. 8, the model attacks 96.0% of trials (SD = 15.1%), with 54% of simulations attacking 100% of trials and 35.8% attacking greater than 95%. As shown on the right side of Fig. 7, the model matches well to the distribution of humans that attack ≥95%, and the other measures such as percent of truthful blocks presented (Fig. 6) and defender utilities (right side of Fig. 8). These results stress the importance of understanding the features that individuals consider in their decisions since one's representation of the decision context strongly influences the chosen action.

5.2 *Discussion*

The results of the adaptive, personalized signaling scheme showed improvement upon traditional game-theoretic signaling schemes for cyber defense by using a computational model of human cognition. The peSSE signaling scheme offers effective defense against boundedly rational human adversaries compared to not signaling. However, the algorithm optimizes the rate of deception for perfectly rational adversaries, which results in a static scheme that is not personalized to individual attackers. Through experimentation and cognitive modeling, we learned how humans respond to deceptive signals and developed a cognitive signaling scheme that is adaptive and based on cognitive principles. Cognitive model predictions showed that the solution is promising at further influencing human behavior beyond the capabilities of the peSSE. These predictions were verified in human experiments, and the results helped shed additional light on individual differences in human behavior.

The cognitive model predicts that human decisions are made by aggregated retrieval across past experiences based on the similarity to the current situation [19]. These decisions are influenced by the frequency and recency of past experiences, cognitive biases, and representation of information in memory. These are the core assumptions for the cognitive signaling scheme.

Two key insights gleaned from the cognitive model regarding human behavior are that: (1) decisions are highly affected by confirmation bias, and (2) it is important to consider what features the individual factors in their decision. The cognitive signaling scheme leveraged this information to induce bias and influence human behavior. Specifically, by relying on observations of actual human behavior, the cognitive signaling scheme estimated the probability of attack given a signal and, if it was too high, would send only truthful signals for a period of time in an attempt to rebuild trust in the signal and ultimately increase compliance. Continued attacks given truthful signals should strengthen the expectation that attacking in the future, given a signal, will result in a loss.

5.2.1 Open Questions

An open question for the cognitive signaling scheme is how long do we need to display truthful signals to regain trust and thus compliance? Currently, the approach gives up some attacks early on with an initial truthful block, but this is done in order to increase belief in the signal for the rest of the experiment. The algorithm only determines whether to switch to a different type of signal after a block of 10 trials. Ten is a reasonable value, but the algorithm could be called as often as every trial. The implications of this are unclear at this point. It could result in too few truthful signals in a row to impact behavior, or it could help further personalize the scheme so that it is better adapted to the individual. Future research is aimed at exploring ways to optimize the proportion of truthful to deceptive signals over a period of time.

Cranford et al. [15] showed that humans seem to ignore the context of the selected target, and only consider the signal when making decisions of whether to continue to attack. This insight allowed us to simplify the cognitive signaling scheme and focus on reducing the overall probability of attack given a signal, and not need to take into account individual target values. After all, the SSE normalizes targets, so their expected values are equal [36].

An important observation from the human experiments was that the cognitive signaling scheme is only effective for some participants, while others seem to ignore the signal when making decisions. This further highlights the importance of accurately representing decision features. For participants that do not consider the signal, all targets are treated equally. Thus, trying to reduce the probability of attack given a signal by adjusting the rate of deception may prove fruitless when the overall expected values are positive for all targets. An alternative method to combat such adversaries could be to shift coverage instead of, or in addition to, adjusting the rate of deception. For example, while it might be difficult or impossible to extract attack preferences to influence behavior, it might be possible to extract selection preferences and shift coverage to induce more experiences of loss given a signal. Driving the expected value of attacking to negative values could result in attackers starting to pay attention to the signal, which in turn would raise the effectiveness of cognitive signaling. Future research is aimed at exploring the potential of this method.

5.2.2 Limitations and Extensions

A limitation of the current approach is that it relies only on deceiving when given a signal. Meanwhile, players can attack with impunity when no signal is presented. An alternative approach is to use deception in two ways: when a signal is present and when it is absent. In this way, the attacker can lose points when a signal is absent, instilling further uncertainty in their decisions. In fact, recent research explored several game-theoretic algorithms that employ two-way deception that proved better than one-way deception against human participants [9, 10]. Future research is aimed at exploring the potential of using two-way deception in the current cognitive signaling approach.

We have already used two-way deception in an alternative cognitive signal scheme, but it has not been tested against human participants [12]. In that scheme, the model- and knowledge-tracing are used to trace human behavior in real time to make predictions about the human's probability of attack given a signal and determine on a trial-to-trial basis whether to give a signal based on the underlying coverage. The scheme shows potential, and the use of two-way signaling is an enhancement over the current approach. Where the current approach stands out is in the fact that it is a closed-form solution that relies on a simplified version of the cognitive model to make predictions of individual behavior. However, there is room to refine the current cognitive signaling approach through the discount parameter, the size of the truthful block, and the assumptions concerning the likelihood of various coverage conditions. Future research will further explore the complexities of the cognitive signaling scheme.

One caveat to these approaches is that they rely on observing and tracking an individual's behavior. In the real world, it may prove difficult, if not impossible, to track all, or even some, of an adversary's actions. Luckily, the methods are robust and can be tailored to a population, sub-group, or even a time window of attacks. While not as effective as at the individual level, such a method could still reliably influence individual human behavior.

Another limitation is capturing real-world incentives. The payments offered for performance in the experiment are in line with typical practice, and by having a low base pay with a high potential for bonus tied to the points earned in the game, we ensured that participants (mTurkers who are driven by maximizing their pay per minute) were sufficiently motivated to maximize their points earned. However, the payments do not compare to the rewards and dangers present in real-world cybersecurity breaches. A version of the experiment using an initial endowment has proven effective at preventing careless exploration at the start of the experiment before rewards have been accumulated and could be lost [14]. Modeling behavior using scaled-up rewards would allow exploration of the impact of higher stakes.

5.2.3 Future Research

In conclusion, we have outlined an initial approach to deceptive signaling for cyber defense that relies on cognitive models of attacker behavior to balance the rate of deception in an attempt to keep belief in the signal high. The cognitive signaling scheme is adaptive and personalized and can therefore be used to induce biases and influence attackers to comply with the signal beyond the capabilities of any static scheme.

Future research is aimed at improving upon the current cognitive signaling scheme. Recent research has shown that, when deployed in the real world to mitigate poaching, the peSSE is less effective than predicted because it fails to account for various forms of uncertainty [7]. A potential direction includes incorporating uncertainty at a couple different levels: first in the system's ability to reliably detect an attacker, then in the system's ability to make sure that its deceptive signal is seen and correctly processed by the attacker. Another related direction is to use the cognitive model to adapt not only the deceptive signal when an attacker is detected but also the initial coverage scheme. This is particularly important since the effectiveness of the deceptive signal is partly conditioned on the adequacy of the coverage. For instance, for the portion of attackers that seems bent to always attacking, the coverage needs to be effective enough in anticipating the locations of their attacks so as to drive their expected gains into negative territory. Fortunately, the cognitive model already predicts the attacker's target, so it can be used directly to adapt coverage in a manner similar to adapting the deceptive signaling.

6 Conclusion

In this chapter, we have introduced the concept of personalized cognitive models and how they can be used to inform deceptive signaling. We provided a general introduction to cognitive modeling techniques and concepts including general goals, capabilities and limitations, cognitive architectures, and instance-based learning. We showed that cognitive models' reliance on generative mechanisms has predictive capabilities beyond those of purely data-driven techniques such as machine learning that can be used to evaluate the effectiveness of cyber defense techniques without requiring full implementation and test. Cognitive models can account for the entire range of human performance, including levels of expertise and individual differences. Techniques such as knowledge-tracing and model-tracing can align a specific cognitive model against an individual behavior trace, enabling personalized interventions. Because cognitive models are analytically tractable, they can guide, inform, and optimize the design of cyber deception techniques. We illustrated these concepts using an insider attacking game meant to abstract the dynamics and decision-making characteristics of real-world cyber defense. Future research directions in the development and application of cognitive models to cyber deception for defense offer the promise of scaling this approach to real-world problems.

Acknowledgments This research was sponsored by the Army Research Office and accomplished under MURI Grant Number W911NF-17-1-0370.

References

1. Abbasi, Y.D., Ben-Asher, N., Gonzalez, C., Kar, D., Morrison, D., Sintov, N., Tambe, M.: Know your adversary: insights for a better adversarial behavioral model. In: Proceeding of the 38th Annual Conference of Cognitive Science Society, pp. 1391–1396. Cognitive Science Society, Austin (2016)
2. Anderson, J.R., Lebiere, C.: The Atomic Components of Thought. Erlbaum, Mahwah (1998). https://doi.org/10.4324/9781315805696
3. Anderson, J.R., Boyle, C.F., Yost, G.: The geometry tutor. J. Math. Behav. **5**, 5–20 (1986)
4. Anderson, J.R., Corbett, A.T., Koedinger, K., Pelletier, R.: Cognitive tutors: lessons learned. J. Learn. Sci. **4**, 167–207 (1995). https://doi.org/10.1207/s15327809jls0402_2
5. Anderson, J.R., Bothell, D., Byrne, M.D., Douglass, S., Lebiere, C., Qin, Y.: An integrated theory of the mind. Psychol. Rev. **111**(4), 1036–1060 (2004). https://doi.org/10.1037/0033-295X.111.4.1036
6. Battigalli, P.: Rationalization in signaling games: theory and applications. Int. Game Theory Rev. **8**(01), 67–93 (2006). https://doi.org/10.2139/ssrn.635244
7. Bondi, E., Oh, H., Xu, H., Fang, F., Dilkina, B., Tambe, M.: To signal or not to signal: exploiting uncertain real-time information in signaling games for security and sustainability. Proc. AAAI Conf. Artif. Intell. **34**(02), 1369–1377 (2020). https://doi.org/10.1609/aaai.v34i02.5493
8. Cho, I.-K., Kreps, D.M.: Signaling games and stable equilibria. Q. J. Econ. **102**(2), 179–221 (1987). https://doi.org/10.2307/1885060
9. Cooney, S., Vayanos, P., Nguyen, T.H., Gonzalez, C., Lebiere, C., Cranford, E.A., Tambe, M.: Warning time: optimizing strategic signaling for security against boundedly rational

adversaries. In: Proceedings of the 18th International Conference on Autonomous Agents and Multiagent Systems (AAMAS), pp. 1892–1894. IFAAMS, Montreal (2019)
10. Cooney, S., Wang, K., Bondi, E., Nguyen, T., Vayanos, P., Winetrobe, H., Cranford, E. A., Gonzalez, C., Lebiere, C., Tambe, M.: Learning to signal in the goldilocks zone: improving adversary compliance in security games. In: Proceedings of the Joint European Conference on Machine Learning and Knowledge Discovery in Databases. Wurzburg (2019)
11. Cranford, E. A., Lebiere, C., Gonzalez, C., Cooney, S., Vayanos, P., Tambe, M.: Learning about cyber deception through simulations: Predictions of human decision making with deceptive signals in Stackelberg Security Games. In: Proceedings of the 40th Annual Conference of the Cognitive Science Society, pp. 258–263. Madison (2018)
12. Cranford, E. A., Gonzalez, C., Aggarwal, P., Cooney, S., Tambe, M., Lebiere, C.: Towards personalized deceptive signaling for cyber defense using cognitive models. In: Proceedings of the 17th Annual Meeting of the International Conference on Cognitive Modeling. Montreal (2019)
13. Cranford, E.A., Gonzalez, C., Aggarwal, P., Cooney, S., Tambe, M., Lebiere, C.: Toward personalized deceptive signaling for cyber defense using cognitive models. Top. Cogn. Sci. **12**, 992–1011. Wiley-Blackwell (2020). https://doi.org/10.1111/tops.12513
14. Cranford, E.A., Gonzalez, C., Aggarwal, P., Tambe, M., Lebiere, C.: What attackers know and what they have to lose: framing effects on cyber-attacker decision making. Proc. Hum. Factors Ergon. Soc. Annu. Meet. **64**(1), 456–460 (2020). https://doi.org/10.1177/1071181320641102
15. Cranford, E.A., Gonzalez, C., Aggarwal, P., Tambe, M., Cooney, S., Lebiere, C.: Towards a cognitive theory of cyber deception. Cogn. Sci. **45**, e13013, 1–28. Wiley-Blackwell (2021). https://doi.org/10.1111/cogs.13013
16. Gonzalez, C.: The boundaries of instance-based learning theory for explaining decisions from experience. Prog. Brain Res. **202**, 73–98 (2013). https://doi.org/10.1016/B978-0-444-62604-2.00005-8
17. Gonzalez, C., Dutt, V.: Instance-based learning: integrating decisions from experience in sampling and repeated choice paradigms. Psychol. Rev. **118**(4), 523–551 (2011). https://doi.org/10.1037/a0024558
18. Gonzalez, C., Lebiere, C.: Instance-based cognitive models of decision making. In: Zizzo, D., Courakis, A. (eds.) Transfer of Knowledge in Economic Decision-Making, pp. 148–165. Macmillan, New York (2005)
19. Gonzalez, C., Lerch, J.F., Lebiere, C.: Instance based learning in dynamic decision making. Cogn. Sci. **27**(4), 591–635 (2003). https://doi.org/10.1007/978-3-319-11391-3_6
20. Gonzalez, C., Ben-Asher, N., Martin, J.M., Dutt, V.: A cognitive model of dynamic cooperation with varied inter-dependency information. Cogn. Sci. **39**(3), 457–495 (2015). https://doi.org/10.1111/cogs.12170
21. Juvina, I., Saleem, M., Martin, J.M., Gonzalez, C., Lebiere, C.: Reciprocal trust mediates deep transfer of learning between games of strategic interaction. Organ. Behav. Hum. Decis. Process. **120**(2), 206–215 (2013). https://doi.org/10.1016/j.obhdp.2012.09.004
22. Kiekintveld, C., Lisy, V., Pibil, R.: Game-theoretic foundations for the strategic use of honeypots in network security. In: Jajodia, S., Shakarian, P., Subrahmanian, V., Swarup, V., Wang, C. (eds.) Cyber Warfare, pp. 81–101. Springer, Cham (2015). https://doi.org/10.1007/978-3-319-14039-1_5
23. Lebiere, C.: A blending process for aggregate retrievals. In: Proceedings of the 6th ACT-R Workshop. George Mason University, Fairfax (1999)
24. Lebiere, C., Wallach, D., West, R. L.: A memory-based account of the prisoner's dilemma and other 2 × 2 games. In: Proceedings of the Third International Conference on Cognitive Modeling, pp. 185–193. Groningen (2000)
25. Lebiere, C., Gonzalez, C., Martin, M.: Instance-based decision making model of repeated binary choice. In: Proceedings of the Eighth International Conference on Cognitive Modeling, pp. 67–72. Ann Harbor (2007). https://doi.org/10.1184/R1/6571190.v1
26. Lebiere, C., Pirolli, P., Thomson, R., Paik, J., Rutledge-Taylor, M., Staszewski, J., Anderson, J.R.: A functional model of sensemaking in a neurocognitive architecture. Comput. Intell. Neurosci. **2013**, 921695 (2013). https://doi.org/10.1155/2013/921695

27. Mokkonen, M., Lindstedt, C.: The evolutionary ecology of deception. Biol. Rev. **91**(4), 1020–1035 (2016). https://doi.org/10.1111/brv.12208
28. Pita, J., Jain, M., Ordónez, F., Portway, C., Tambe, M., Western, C., Kraus, S.: ARMOR security for Los Angeles International Airport. In: Proceeding of the Twenty-Third AAAI Conference on Artificial Intelligence, pp. 1884–1885. Chicago (2008)
29. Pita, J., Jain, M., Ordóñez, F., Tambe, M., Kraus, S., Magori, R.: Effective solutions for real-world Stackelberg games: when agents must deal with human uncertainties. In: Proceedings of 8th International Conference on Autonomous Agents and Multiagent Systems (AAMAS) (2009)
30. Rowe, N.C., Rrushi, J.: Introduction to Cyberdeception. Springer, Cham (2016). https://doi.org/10.1007/978-3-319-41187-3
31. Sanner, S., Anderson, J.R., Lebiere, C., Lovett, M.C.: Achieving efficient and cognitively plausible learning in backgammon. In: Proceedings of the Seventeenth International Conference on Machine Learning. Morgan Kaufmann, San Francisco (2000). https://doi.org/10.1184/R1/6613298.v1
32. Schlenker, A., Thakoor, O., Xu, H., Fang, F., Tambe, M., Tran-Thanh, L., Vayanos, P., Vorobeychik, Y.: Deceiving cyber adversaries: a game theoretic approach. In: Proceedings of the 17th AAMAS (IFAAMAS), pp. 892–900. Stockholm (2018)
33. Shieh, E., An, B., Yang, R., Tambe, M., Baldwin, C., Meyer, G.: PROTECT: a deployed game theoretic system to protect the ports of the United States. In: Proceedings of the 11th International Conference on Autonomous Agents and Multiagent Systems (AAMAS), pp. 13–20. Valencia (2012)
34. Simon, H.A.: Rational choice and the structure of the environment. Psychol. Rev. **63**(2), 129–138 (1956). https://doi.org/10.1037/h0042769
35. Sinha, A., Fang, F., An, B., Kiekintveld, C., Tambe, M.: Stackelberg security games: looking beyond a decade of success. In: Proceedings of the 27th International Joint Conference on Artificial Intelligence, pp. 5494–5501. Stockholm (2018). https://doi.org/10.24963/ijcai.2018/775
36. Tambe, M.: Security and Game Theory: Algorithms, Deployed Systems, Lessons Learned. Cambridge University Press, Cambridge (2011). https://doi.org/10.1017/CBO9780511973031
37. Tesauro, G.: Connectionist learning of expert backgammon evaluations. In: Proceedings of the Fifth International Conference on Machine Learning, pp. 200–206. University of Michigan, Ann Arbor (1988). https://doi.org/10.1016/B978-0-934613-64-4.50026-8
38. Thomson, R., Lebiere, C., Bennati, S.: Human, model, and machine: a complementary approach to big data. In: Association for Computing Machinery Proceedings of the IARPA Workshop on Human Centered Big Data Research, pp. 27–31. Raleigh (2014). https://doi.org/10.1145/2609876.2609883
39. Thomson, R., Cranford, E.A., Lebiere, C.: Achieving active cybersecurity through agent-based cognitive models for detection and defense. In: Proceedings of the 1st International Conference on Autonomous Intelligent Cyber-defence Agents (AICA 2021) (2021)
40. West, R.L., Lebiere, C.: Simple games as dynamic, coupled systems: randomness and other emergent properties. J Cogn Syst Res. **1**(4), 221–239 (2001). https://doi.org/10.1016/S1389-0417(00)00014-0
41. Xu, H., Rabinovich, Z., Dughmi, S., Tambe, M.: Exploring information asymmetry in two-stage security games. In: *Proceedings of the National Conference on Artificial Intelligence*, pp. 1057–1063. Austin (2015)

Deceptive Signaling: Understanding Human Behavior Against Signaling Algorithms

Palvi Aggarwal, Edward A. Cranford, Milind Tambe, Christian Lebiere, and Cleotilde Gonzalez

1 Introduction

Security resources are often limited and, therefore, require careful planning to effectively use these resources. Just like in the physical world, in the cyber world defenders cannot protect all the resources in the network and they must deploy their limited defense resources effectively. Stackelberg Security Games (SSG) have been widely used to address this critical allocation of limited defense resources in airports, wildlife protection, etc. [10, 13, 14]. The Stackelberg Security Game (SSG) models the interaction between a leader (i.e., defender) and a follower (i.e., an attacker) and helps security agencies optimally allocate limited resources using the Strong Stackelberg Equilibrium (SSE) to mitigate security threats [15]. However, the optimal allocation of defense resources in naturalistic settings can be costly. Xu et al.[15] proposed an extension of the SSG framework to incorporate *signaling*, a security mechanism that can be cheaper and more flexible to deploy. With signaling, defenders strategically reveal some information about the defense strategy to the attacker to influence their decision-making without actually reallocating their defenses [1, 2]. Xu et al.[15] present the Strong Stackelberg Equilibrium with

P. Aggarwal (✉)
Carnegie Mellon University, Pittsburgh, PA, USA

Department of Computer Science, University of Texas at El Paso, El Paso, TX, USA
e-mail: paggarwal@utep.edu

E. A. Cranford · C. Lebiere · C. Gonzalez
Carnegie Mellon University, Pittsburgh, PA, USA
e-mail: cranford@cmu.edu; cl@cmu.edu; coty@cmu.edu

M. Tambe
University of Southern California, Los Angeles, CA, USA
e-mail: othakoor@usc.edu

T. Bao et al. (eds.), *Cyber Deception*, Advances in Information Security 89,
https://doi.org/10.1007/978-3-031-16613-6_5

Persuasion (peSSE) to optimally determine the type of signal to send (e.g., truthful or deceptive) depending on the actual status of the defense in the asset being protected. Theoretically, the use of signals improves the defender's utility against a perfectly rational attacker compared to strategies that do not use signaling. For a given target, the peSSE finds the optimal combination of deception (sending a deceptive message that the target is covered when it is not) and truth-telling (sending a truthful message that the target is covered) so that the attacker continues to believe the bluff. The goal of the peSSE is to reduce attacks on uncovered targets. Attackers earn a reward for successful attacks, suffer a loss for failed attacks, and earn zero for withdrawing. When a target is covered, the peSSE will always send a true signal. When uncovered, the peSSE will send a deceptive signal with a probability that brings the attacker's expected value of attacking, given a signal, to zero. This makes it equal to the utility of withdrawing the attack, and, based on standard game-theoretic assumptions of perfect rationality, the attacker will break ties in favor of the defender and withdraw.

In past research, we have proposed that peSSE is also suitable for cyber defense [8], where optimizing the probability of sending a deceptive signal can mitigate attacks on uncovered targets with little overhead. However, peSSE is based on the assumption of the attacker's perfect rationality, while humans exhibit, at best, bounded rationality [12]. To address this weakness of peSSE, researchers have begun to develop signaling algorithms for security against boundedly rational attackers [3]. However, these algorithms do not offer a substantial improvement over the peSSE in terms of reducing attacks and minimizing the loss of the defender. The main reason is that these algorithms assume rational behavior as specified by game theory optima such as Nash equilibria and human behavior systematically deviates from those theoretical descriptions. Human subjects exhibit learning –they do not generally compute the equilibria based on perfect knowledge of the interaction but rather have to painstakingly accumulate the information through experience and then make satisficing decisions by limited cognitive means. These deviations typically manifest themselves through systematic cognitive biases that reflect the interaction between limited cognitive mechanisms and the statistics of the task. Additionally, human behavior is dominated by individual differences in knowledge and capacity that manifest themselves as substantial variations in behavior.

To further address the weakness of peSSE, Gonzalez, Aggarwal, Cranford, and Lebiere [8] proposed a research framework for dynamic, personalized deception for cyber deception. This framework implements SSG algorithms for the distribution of limited defense resources with signaling theory (e.g., peSSE) to gain insights into human behavior from human-in-the-loop experiments, and cognitive modeling using instance-based learning theory (IBLT) to create personalized defense algorithms.

Using the framework proposed by Gonzalez et al. [8], in this chapter, we evaluate various signaling algorithms that consider rational human behavior and compare them with non-signaling algorithms. In addition, we compare the rational signaling algorithm against algorithms that consider bounded rationality in human behavior.

This chapter will summarize the results of various human experiments conducted to evaluate signaling algorithms.

2 Insider Attack Game

Insider Attack Game (IAG) is an online game that was developed to replicate a real-world scenario in the laboratory [4]. Following the two-stage SSG scenario, the IAG involves two stages. In the first stage, the defenders protect the nodes with limited resources, and in the second stage, the defenders send deceptive signals to the confused attacker. The IAG involves six nodes, and the defender could only protect two nodes. In the IAG, human participants play the role of employees who are insider attackers and plan to gain points by attacking various nodes. However, two security analysts (i.e., "defenders") monitor the computers. Attackers can earn points if they avoid the defenders but lose points if they are caught. Strong Stackelberg Equilibrium (SSE) optimizes the allocation of defenders and provides the monitoring probability (m-prob) of each computer based on its reward and penalty values [14]. An attacker observes the information on the computers and makes a move by selecting a computer to attack. In the second stage, after a computer is selected, defenders use the signaling techniques to strategically reveal potentially deceptive information to the attacker about whether the computer is being monitored [15]. The attacker follows by deciding whether to continue the attack or withdraw. A screenshot of the task interface is shown in Fig. 1. Attackers perform four rounds of 25 trials each, following an initial practice round of five trials. For each round, attackers are presented with six new computer targets, each with a different payoff (reward or penalty) structure. On a given trial, the two defenders monitor one computer each. Attackers can view information that describes the reward and penalty values of each target, as well as the monitoring probability (i.e., the average proportion of trials that the target is monitored). This information is provided to participants assuming that attackers are well prepared for the attack and gathered this information in the reconnaissance phase.

In each trial, attackers first select one of the targets to attack. After selection, they are presented with a signal (truthful or deceptive) on whether the computer is being monitored (Fig. 1b). If the message indicates that the computer is monitored, then the signal is present; otherwise, it is absent. The attacker must then decide whether to continue or withdraw the attack. An attack is considered successful and the attacker gain associated rewards if the computer was not monitored, otherwise, the attacker loses points. If attackers choose to withdraw the attack, they will receive zero points. Table 1 shows the rewards, penalties, and monitoring probabilities (m-prob) for each computer in each round. The monitoring probabilities for each target are derived by computing the SSE, which allocates defenses across a round in such a manner that the expected value of attacking each computer is positive and all equal. Each attacker experiences the same coverage and signaling schedule throughout the game. That is, the SSE allocates defenses across the 25 trials for each round, and so predetermines which targets are monitored during each trial.

Fig. 1 Insider Attack Game (**a**) Interface presenting 6 nodes and information on each node and (**b**) presenting signal message

Table 1 Payoff structure for each target in each round

Round	Target 1	Target 2	Target 3	Target 4	Target 5	Target 6
Round 1	[2, −1, 0.22]	[8, −5, 0.51]	[9, −9, 0.42]	[9, −10, 0.40]	[2, −6, 0.08]	[5, −5, 0.36]
Round 2	[5, −3, 0.41]	[8, −5, 0.48]	[7, −6, 0.41]	[8, −9, 0.37]	[5, −7, 0.27]	[2, −4, 0.05]
Round 3	[3, −3, 0.30]	[9, −4, 0.60]	[6, −6, 0.40]	[5, −8, 0.29]	[3, −6, 0.20]	[2, −2, 0.20]
Round 4	[4, −3, 0.37]	[6, −3, 0.51]	[7, −7, 0.40]	[5, −10, 0.24]	[5, −9, 0.26]	[3, −4, 0.23]

Note 1 The first number in brackets is the reward, the second number is the penalty, and the third is the probability that the computer is being monitored on any given trial, [payment, penalty, m-prob]

3 Signaling Algorithms

Research on Stackelberg Security Games (SSGs) led to the development of algorithms that have greatly improved physical security systems (e.g., protecting ports, scheduling air marshals, and mitigating poachers) through the optimal allocation of limited defense resources [9, 11, 13, 14]. Xu et al. [15] extended these models by incorporating elements of signaling, in which a defender (sender) strategically

Table 2 Signaling algorithms

Deception	Algorithm	Signal on	Attacker type	Adaptive
No-signal	peSSE	None	Rational	Non-adaptive
No-signal	Epsilon rationality	None	Rational	Non-adaptive
1-sided	peSSE	Uncovered nodes	Rational	Non-adaptive
1-sided	peSSE-FI	Uncovered nodes	Rational	Non-adaptive
1-sided	Epsilon rationality	Uncovered nodes	Rational	Non-adaptive
1-sided	Decision tree	Uncovered nodes	Boundedly rational	Adaptive
1-sided	Cognitive signaling	Uncovered	Boundedly rational	Personalized
2-sided	peSSE	Both	Rational	Non-adaptive
2-sided	Decision tree	Both	Boundedly rational	Adaptive
2-sided	Neural network	Both	Boundedly rational	Adaptive
2-sided	Epsilon rationality	Both	Rational	Non-adaptive

reveals information about their strategy to the attacker (receiver) to influence the attacker's decision-making [1, 2]. Their solution, the Strong Stackelberg Equilibrium with Persuasion (peSSE), improves defender utility against a perfectly rational attacker compared to strategies that do not use a signaling. For a given target, the peSSE finds the optimal combination of deceptive and truthful signals, so the attacker continues to believe the signal.

Table 2 summarizes the algorithms that we have used in experimental settings, particularly in the insider attack game (see next section). IAG was deployed under several experimental conditions to assess the effectiveness of deceptive signals in attack decision-making. The table highlights the high-level feature differences between defense algorithms. The level of signaling was varied at three levels: no-signal, signal uncovered nodes (1-sided signaling), and signals on both covered and uncovered nodes (2-sided signaling). Our research approach also considers the attacker type, i.e., signaling algorithms for both rational attacks and boundedly rational attackers. Finally, the algorithms vary between different levels of adaptability: non-adaptive, i.e., the signaling algorithm does not consider the past actions of the attacker, adaptive, i.e., learns the distribution of the attacker's actions and personalized, i.e., adapt the signal based on the individual attacker. Below is a brief summary of these algorithms.

No-signaling Algorithm is a baseline condition in which no signal is used. A signal is never presented to the attacker, regardless of whether a defender is present or absent (i.e., no deception was used). The no-signaling algorithm uses Stackelberg Security Games and calculates *Strong Stackelberg Equilibrium (SSE)* [15] to allocate defenders in the network.

1-sided Deception uses the Strong Stackelberg Equilibrium with Persuasion (peSSE) algorithm [15]. This algorithm improves defense against perfectly rational attackers compared to strategies that do not use a signaling. For a given target, the peSSE finds the optimal combination of bluffing (sending a deceptive message that the target is monitored when it is not) and truth-telling (sending a truthful message that the target is covered) so that a rational attacker would not attack in the presence of a signal. The peSSE algorithm exploits the information asymmetry between the defender and the attacker. Defenders have more accurate information about the network, whereas attackers could only observe the mixed strategy. Xu et al.[15] exploited this asymmetry by strategically injecting information to attackers through signaling. In this technique, the defender strategically reveals information about their strategy to the attacker to influence the attacker's decision-making. The peSSE signaling scheme presents signals with probabilities calculated according to the peSSE algorithm, as described above. The **peSSE-FI** (Full-Information) signaling scheme extends the assumption of perfect rationality by ensuring that attackers have full knowledge of the probabilities of deception available to them, in addition to monitoring probabilities. In another version of peSSE, **Epsilon Rationality** defenders consider an epsilon rational model for resource allocation. All the algorithms mentioned above are *non-adaptive*, as they do not consider the actions of the attacker to generate signals. A **Decision Tree** algorithm predicts the attacker's actions to generate signals. In the 1-sided version of this algorithm, the decision tree is considered adaptive, since the algorithm relies on attack prediction to generate defense.

2-sided Deception was first introduced by Cooney et al. [3]. They extended the peSSE by considering deceptive signals both on the covered and on the uncovered nodes. Cooney et al. [3] developed a **2-way peSSE** algorithm which lowers the overall frequency of showing a signal and introduces uncertainty for the rational attacker when no signal is shown. Cooney et al. [3] also focused on increasing the compliance of boundedly rational attackers by manipulating the frequency of signals. Two additional ML models, **decision tree (DT)** and a **neural network (NN)**, were used for identifying the *Goldilocks zone* and generating signals against a boundedly rational attacker.

A **Cognitive Signaling** algorithm is a different type of algorithm from the other algorithms in this list. This was developed by Cranford et al. [5, 6] using the attacker's "cognitive clone" in the insider attack game. A cognitive clone is a cognitive model that aims at emulating the decisions a human makes in a task. This model generates human attack predictions and these predictions are used to modify the signaling strategy dynamically and individually (i.e., based on the particular actions of an individual attacker) [6, 7]. The details of personalized and adaptive training are discussed in chapter 4 of this book.

4 Methods

4.1 Participants

One thousand seventy-nine participants participated in 11 experiments. The demographics of these participants is shown in Table 3. All participants were recruited

Table 3 Demographics

Condition	Sample size	Age	Gender (Female%)
NS	97	35.48	40%
NS-ER	99	33.94	28%
1-Sided-psse	100	34.55	42%
1-Sided-psse_FI	96	35.44	46%
1-Sided-DT_FI	98	36.02	45%
1-Sided-ER_FI	96	35.81	39%
1-Sided-CogSig	99	35.80	44%
2-Sided-pesse_FI	99	33.63	48%
2-Sided-LR	100	36.69	49%
2-Sided-NN	99	35.82	36%
2-SidedDT	96	34.71	33%

through Amazon Mechanical Turk (mTurk) and had a 90% or higher approval rate with at least 100 Human Intelligence Tasks (HIT) approved, resided in the USA, and had not participated in other conditions. Participants who completed the experiment and submitted the completion code received \$1 as a base payment. Participants could earn a bonus of up to \$ 4.50. We remove all participants for incomplete or duplicate participation. Demographic data within each condition are shown in Table 3.

4.2 Procedure

The experiment was conducted in Amazon Mechanical Turk. The experiment was advertised as "A fun game of decision-making to help keep our systems safe!!" Participants clicked the link of one of the experimental conditions and were presented for the first time with a consent form and asked a few demographic questions. After providing the informed consent, participants received instructions on how to play the game. Participants were told that they would take on the role of an employee in a company and that their goal was to steal proprietary information by attacking computers. They could receive points for attacking computers that were not monitored by one of the two defenders, as denoted by the number of yellow stars displayed on the targets (see Fig. 1a), but could lose points for attacking computers that were monitored by a defender, as denoted by the number of red stars displayed. Participants were informed that they would earn \$1 for completing the game and the questionnaire and would earn an additional \$0.02 per point accumulated throughout the game up to a maximum of \$4.50. After reading the instructions, the participants answered a few questions to test their knowledge of how to play the game and received feedback on the accuracy of their answers. After receiving the feedback, they could proceed to the game.

Participants started playing a practice round of five trials to become familiar with the interface and then played four rounds of the game for 25 trials per round.

The targets changed for each round as defined in Table 1. The location of the targets within the display was randomly assigned between participants but did not change within a round. Participants started a round by pressing a "continue" button indicating that they were ready to begin. For each trial, participants began by selecting one of the six targets with the click of a mouse. After clicking the target, one of two messages was displayed depending on the coverage and signaling schedule defined for the experimental condition (e.g., see Fig. 1b). One message reads, "This computer is being monitored! Do you want to access this computer?" if the computer was monitored or presented with a deceptive signal. The other message reads, "Do you want to access this computer?" if the computer was not monitored (in the No-Signal condition, this message was displayed every time regardless of coverage, and participants were never warned that the computer was being monitored). Participants responded by either clicking a "yes" or a "no" button. If participants responded "yes" and continued the attack, then they received the number of points denoted by the yellow stars if the target was not monitored but lost the number of points denoted by the red stars if the target was monitored. The total points earned in a round are displayed in the top right of the interface. If they responded "no" and withdrew their attack, then they receive zero points. In the full-information condition (psse-FI), for the second message above, participants were also told, "X% of time this computer appears as 'monitored' the analyst is NOT actually present." Where X was replaced with the percent of the time the signal is deceptive for that target.

After completing 25 trials, participants received feedback on their score for the round and their cumulative score across rounds. At the end of the fourth round, participants were provided their final score and then pressed an "ok" button to continue a post-game survey of 10 questions. These data were not analyzed and are not discussed further. After completing the survey, the participants were thanked for their participation and given a completion code. Participants had to return to the Mechanical Turk experiment website and enter their completion code to claim their reward. Participants were paid the $1 base rate plus their earned bonuses within 24 h of completing the experiment.

5 Results

In the following sections, we analyze the performance of various defense algorithms using two dependent variables: (1) the proportion of attack actions and (2) the utility of the defenders. Defenders are assigned -1 points if the attack was successful and 0 points if the attack was unsuccessful.

5.1 Is Signaling Effective?

We analyzed the proportion of attack actions in signaling algorithms and no-signaling algorithms. We combine all experimental conditions in signaling conditions (i.e., 1-sided deception, 2-sided deception, and cognitive signaling) and compare it with the no-signal condition (combining the decisions of both no-signal conditions). Figure 2(left) shows the proportion of attack actions under signal and no-signal conditions. We observe that when attackers were not provided any signals, they attacked almost all the time. However, attackers reduced attack actions when they provided signals, regardless of the type of signal or algorithm. The difference between signal and no-signal condition was significant (0.95>0.77; $F(1, 956) = 205.8$, $p < 0.0003$).

Next, we analyze the utility of the defender in no-signaling and signaling algorithms. Figure 2(right) shows that defenders benefit from using signaling algorithms compared to no-signaling algorithms. Defender's utility was significantly higher in signaling conditions (combined 1-sided, cognitive, and 2-sided signaling) compared to no-signaling conditions (−0.55>−0.65; $F(1, 956) = 13.01, p < 0.0003$). Overall, the use of signaling algorithms reduced the proportion of attacks and increased the defender's utility. Therefore, signaling is an effective strategy to deter cyberattacks. In the following sections, we evaluate different types of signaling algorithms for their effectiveness against attackers.

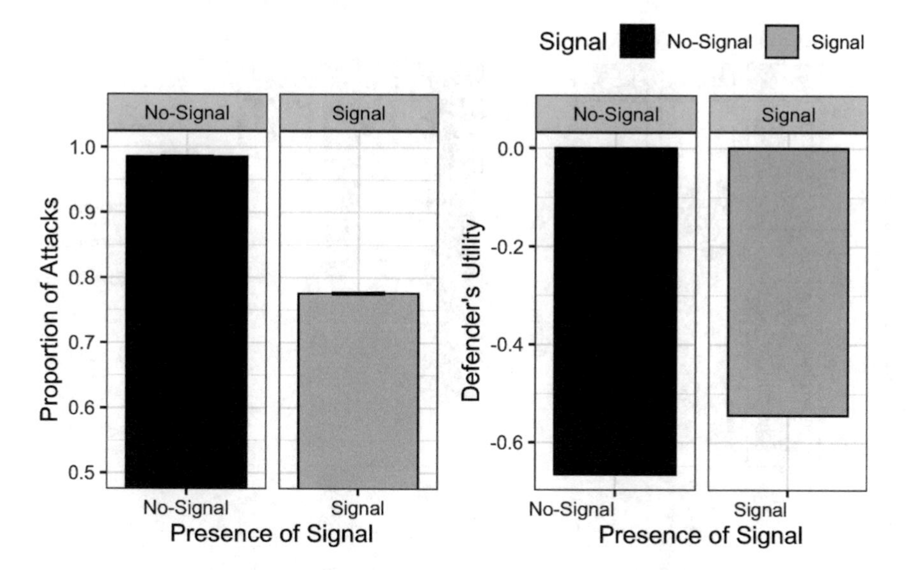

Fig. 2 Defender's Utility in different signaling and no-signaling algorithms

5.2 *Effect of Rational 1-sided and 2-sided Signaling Against No Signaling*

We analyze the effect of 1-sided and 2-sided signaling against the no-signal condition in Fig. 3. The proportion of attacks is significantly different in three types of algorithms ($F(2, 763) = 57.57, p < 0.0001$). The proportion of attack actions in Fig. 3(left) shows that both 1-sided and 2-sided deception are effective against the no-signal condition. Post hoc analyses using the Tukey HSD criterion for significance indicated that the average proportion of attacks in 1-sided signal condition is significantly lower (M = 0.82, SD = 0.21) than in the no-signal algorithm (M = 0.98, SD = 0.05). Similarly, we found that the average proportion of attacks in 2-sided signal condition is significantly lower (M = 0.76, SD = 0.18) than in the no-signal algorithm (M = 0.98, SD = 0.05). We observe that when attackers did not provide any signals, they attacked almost all the time. However, attackers reduced attack actions when they were provided signals irrespective of the type of signal or algorithm. We also found that the proportion of attacks was significantly lower in 2-sided signal condition (M = 0.76, SD = 0.18) compared to the 1-sided signal (M = 0.82, SD = 0.21).

Within the 1-sided deception condition, the proportion of attack actions is the lowest in the 1-sided-psse algorithm. However, the post hoc analysis suggests that there are no significant differences within 1-Sided signal algorithms. Within the 2-sided signal condition, the proportion of attacks is lowest in the 2-sided-NN

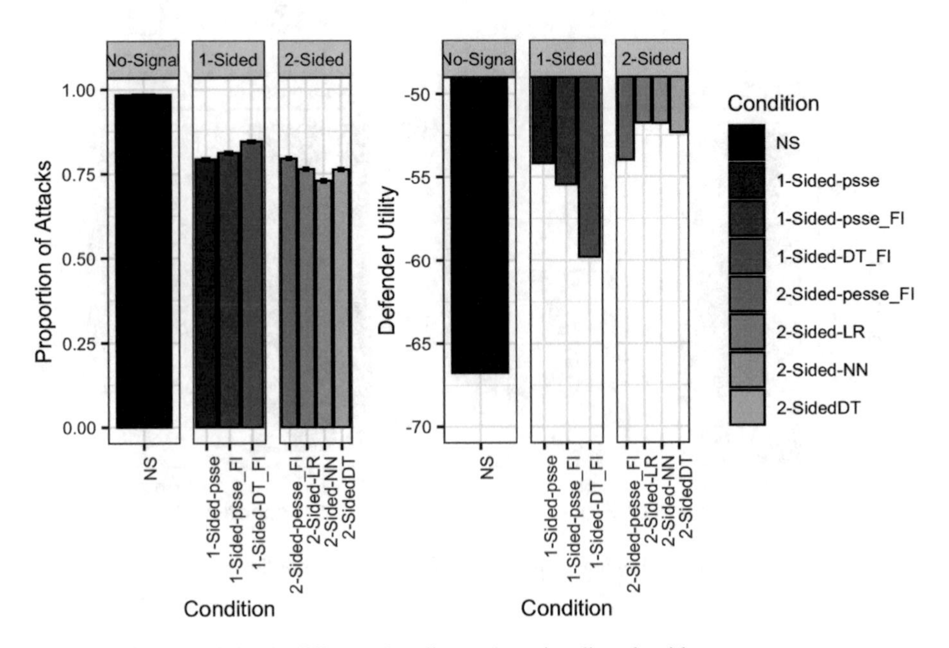

Fig. 3 Defender's Utility in different signaling and no-signaling algorithms

algorithm. Similar to the 1-sided signal condition, the post hoc analysis indicates that there are no significant differences with 2-sided signaling algorithms. The 2-sided-NN algorithm (M = 0.73, SD = 0.18) produced less number of attacks compared to the 1-sided-psse-FI (M = 0.80, SD = 0.18) and 1-sided-DT-FI (M = 0.85, SD = 0.15).

Defender's Utility We analyzed the defender's utility in various no-signaling and signaling algorithms. Figure 3(right) shows that defenders benefit from using signaling algorithms compared to those without signaling algorithms. Defenders' losses are significantly different in three types of algorithms ($F(2, 763) = 32.94, p < 0.0001$). We observe that overall the defender's utility for is lower for no-signal condition (M= −66.75, SD = 8.06) compared to 1-sided signaling (M= −58.05, SD = 14.62) and 2-sided signaling conditions (M= −53.96, SD = 14.84).

We also compared the utility of the defenders within each panel of Fig. 3(right). First, we compared the algorithms within 1-sided signaling algorithms. We observed that the DU was significantly lower in the 1-sided-DT-FI algorithm (M= −59.80, SD = 8.44) compared to 1-sided-psse (M= −54.19, SD = 14.49). Thus, the 1-sided-DT algorithm was the worst among all 1-sided ML algorithms when compared on the basis of defender's utility. We also compared the algorithms within a 2-sided deception panel and observed no statistical differences in DU between various algorithms.

5.3 Adaptive Signaling Using Cognitive Models

In the above sections, we analyzed the effect of signaling using algorithms that assume rational human behavior. We developed another algorithm for signaling that adapts the proportion of signals based on the attacker's actions. In this section, we compare adaptive 1-sided signaling with the best algorithm in 1-sided signal condition (i.e., 1-sided-psse) and 2-sided signal condition (i.e., 2-sided-NN). We observe in Fig. 4(left) that cognitive signaling slightly reduced the proportion of attacks compared to the 1-sided psse algorithm. However, we observed no statistical difference between the two algorithms. The attacks in 1-sided cognitive signaling are similar to the 2-sided-NN algorithm. Thus, adapting signals using cognitive models is helpful, however, we need to address the limitations of the cognitive model to further improve the performance of this algorithm. Future directions will be discussed later in this chapter. We also analyzed the defender's utility in three algorithms in Fig. 4(right). Although, we observe that with a 1-sided cognitive signaling algorithm, the defender's utility slightly increased, however, the difference was not significant. The defender's utility is the highest in the 2-sided-NN algorithm, however, this was not significantly different from 1-sided-pesse and 1-sided-CogSig algorithms.

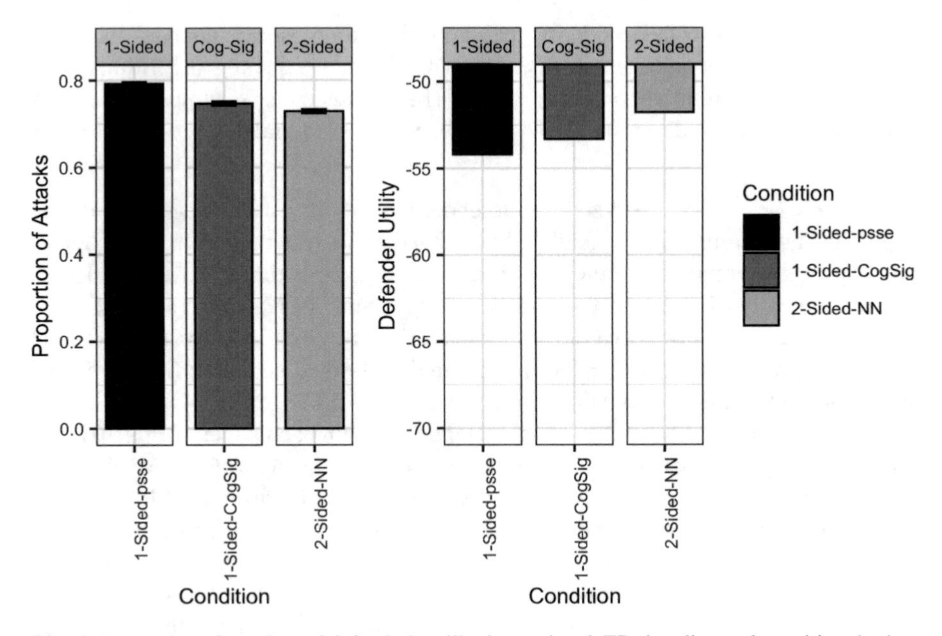

Fig. 4 Proportion of attacks and defender's utility in no-signal, ER signaling and cognitive signing

5.4 Discussion

With limited resources, it is difficult to protect all assets in a network. In this chapter, we showcase the effectiveness of deception in the form of signaling to reduce attacks and increase the utility of the defender. We compared signaling algorithms against no-signaling algorithms and observed that the proportion of attacks was reduced by approximately 20%. Thus, signaling helps create deterrence and reduce attack action. We evaluated various signaling algorithms, i.e., signaling only on uncovered nodes (1-sided signaling) and signaling on both covered and uncovered nodes (2-sided signaling). We observed that both 1-sided signaling and 1-sided signaling algorithms were effective over the no-signaling algorithms. Furthermore, 2-sided signaling reduced attacks even further compared to 1-sided signaling algorithms. This indicates that by increasing the levels of uncertainty, deceptive signals effectively deter attackers from attacking the network. We assume that fixed amounts of deception may not work for all the attackers. Therefore, we also conducted experiments using an adaptive 1-sided signaling algorithm that was developed using a cognitive model [6]. The 1-sided cognitive signaling slightly reduced the proportion of attacks and improved the defender's utility, but the difference was not significantly different from other 1-sided signaling algorithms. The cognitive signaling algorithm could be improved by adapting the signals to attacker actions or by creating uncertainty on the protected nodes as well. In the future, we plan to improve the signaling scheme by considering the cognitive models of humans in designing the signaling scheme.

Acknowledgments This research was sponsored by the Army Research Office and accomplished under grant number W911NF-17-1-0370 (MURI Cyberdeception). Some of the work discussed in this chapter was sponsored by the Combat Capabilities Development Command Army Research Laboratory and was accomplished under Cooperative Agreement Number W911NF-13-2-0045 (ARL Cyber Security CRA).

References

1. Battigalli, P.: Rationalization in signaling games: Theory and applications. Int. Game Theory Rev. **8**(01), 67–93 (2006)
2. Cho, I.K., Kreps, D.M.: Signaling games and stable equilibria. Q. J. Econ. **102**(2), 179–221 (1987)
3. Cooney, S., Wang, K., Bondi, E., Nguyen, T., Vayano, P., Winetrobe, H., Cranford, E.A., Gonzalez, C., Lebiere, C., Tambe, Milind: Learning to signal in the goldilocks zone: Improving adversary compliance in security games. In: Joint European Conference on Machine Learning and Knowledge Discovery in Databases. Springer (2019)
4. Cranford, E.A., Lebiere, C., Gonzalez, C., Cooney, S., Vayanos, P., Tambe, M.: Learning about cyber deception through simulations: Predictions of human decision making with deceptive signals in Stackelberg security games. In: CogSci (2018)
5. Cranford, E.A., Gonzalez, C., Aggarwal, P., Cooney, S., Tambe, M., Lebiere, C.: Towards personalized deceptive signaling for cyber defense using cognitive models. In: 17th Annual Meeting of the International Conference on Cognitive Modelling, Montreal, CA (2019)
6. Cranford, E.A., Lebiere, C., Aggarwal, P., Gonzalez, C., Tambe, M.: Adaptive cyber deception: Cognitively-informed signaling for cyber defense. In: Proceedings of the 53rd Hawaii International Conference on System Sciences (submitted). IEEE (2020)
7. Cranford, E.A., Gonzalez, C., Aggarwal, P., Tambe, M., Cooney, S., Lebiere, C.: Towards a cognitive theory of cyber deception. Cognitive Science **45**(7), e13013 (2021)
8. Gonzalez, C., Aggarwal, P., Lebiere, C., Cranford, E.: Design of dynamic and personalized deception: A research framework and new insights (2020)
9. Pita, J., Jain, M., Ordónez, F., Portway, C., Tambe, M., Western, C., Paruchuri, P., Kraus, S.: Using game theory for Los Angeles airport security. AI Magazine **30**(1), 43–43 (2009)
10. Pita, J., Jain, M., Tambe, M., Ordóñez, F., Kraus, S.: Robust solutions to Stackelberg games: Addressing bounded rationality and limited observations in human cognition. Artificial Intelligence **174**(15), 1142–1171 (2010)
11. Shieh, E., An, B., Yang, R., Tambe, M., Baldwin, C., DiRenzo, J., Maule, B., Meyer, G.: Protect: A deployed game theoretic system to protect the ports of the United States. In: Proceedings of the 11th international conference on autonomous agents and multiagent systems, vol. 1, pp. 13–20. Citeseer (2012)
12. Simon, H.A.: Rational choice and the structure of the environment. Psychological Review **63**(2), 129–138 (1956)
13. Sinha, A., Fang, F., An, B., Kiekintveld, C., Tambe, M.: Stackelberg security games: Looking beyond a decade of success. In: IJCAI, pp. 5494–5501 (2018)
14. Tambe, M.: Security and Game Theory: Algorithms, Deployed Systems, Lessons Learned. Cambridge University Press (2011)
15. Xu, H., Rabinovich, Z., Dughmi, S., Tambe, M.: Exploring information asymmetry in two-stage security games. In: Twenty-Ninth AAAI Conference on Artificial Intelligence (2015)

Optimizing Honey Traffic Using Game Theory and Adversarial Learning

Mohammad Sujan Miah, Mu Zhu, Alonso Granados, Nazia Sharmin, Iffat Anjum, Anthony Ortiz, Christopher Kiekintveld, William Enck, and Munindar P. Singh

1 Introduction

Advanced Persistent Threats (APTs) are a significant concern for enterprises. In APT attacks, advanced adversaries take slow and deliberate steps over months and even years to compromise critical resources (e.g., workstations and servers) in a network. A key step in the kill chain of APTs is reconnaissance. Historically, reconnaissance is largely active, for example, using network port scanning to identify which hosts are running which services. In response, many enterprises closely monitor their networks for scanning attacks. However, passive reconnaissance methods such as packet-sniffing and statistical traffic analysis are widely adopted in identifying network weaknesses. Although advanced encryption techniques limit the information available to traffic analysis, encrypted network traffic can still have observable characteristics like packet sizes and inter-arrival times that reveal useful information to attackers and is thus a potential threat for network security [8, 21]. Sophisticated adversaries possess knowledge about communication

M. S. Miah · N. Sharmin · C. Kiekintveld (✉)
Department of Computer Science, University of Texas at El Paso, El Paso, TX, USA
e-mail: msmiah@miners.utep.edu; nsharmin@miners.utep.edu; cdkiekintveld@utep.edu

M. Zhu · I. Anjum · W. Enck · M. P. Singh
Department of Computer Science, North Carolina State University, Raleigh, NC, USA
e-mail: mzhu5@ncsu.edu; ianjum@ncsu.edu; whenck@ncsu.edu; mpsingh@ncsu.edu

A. Granados
University of Arizona, Tucson, AZ, USA
e-mail: alonsog@email.arizona.edu

A. Ortiz
Microsoft AI for Good Research Lab, Redmond, WA, USA
e-mail: anthony.ortiz@microsoft.com

T. Bao et al. (eds.), *Cyber Deception*, Advances in Information Security 89,
https://doi.org/10.1007/978-3-031-16613-6_6

types and maintain databases of well-known traffic patterns and protocols such as UDP, TCP, VoIP, and ESP, among others. From the raw traffic, an adversary can determine likely features of the source and destination without needing to decrypt sensitive information. The adversary can also distinguish statistical characteristics used for communication by different protocols. There are currently no perfect methods to prevent traffic analysis completely.

Simultaneously, Software Defined Networking (SDN) technology is emerging as a powerful primitive for enterprise network security. SDN offers a global perspective on network communications between hosts. It can be used as an enhanced tool to identify network scanning, provide flexible access control to mitigate attackers bypassing defenses such as firewalls, and even prevent spoofing. However, SDN technology leads to increased functionality within network elements (e.g., switches), which makes them potential targets for attack. A compromised SDN switch is particularly dangerous because it can perform reconnaissance passively. As a result, defenders may have little or no indication that an APT is in progress.

In this chapter, we introduce our approach dubbed *Snaz* (for "snag and zap") to address the threat of passive network reconnaissance. *Snaz* uses honey traffic: fake flows deceptively crafted to make a passive attacker think specific resources (e.g., workstations and servers) exist and have specific unpatched, vulnerable software. *Snaz* assumes that the adversary knows about the possibility of honey traffic and uses game theory to characterize how best to send honey traffic. For doing so, we demonstrate how a defender can either successfully deflect an adversary or deplete the adversary's resources using an "optimal" amount of honey traffic. Here, optimality means causing the least possible disruption to normal user operations.

We model this defender–attacker interaction as a two-player non-zero-sum Stackelberg game. In this game, the defender sends honey traffic to confound the adversary's knowledge. However, if the defender sends too much honey traffic, the network may become overloaded. In contrast, the adversary wishes to act on information obtained using passive reconnaissance (e.g., a banner string indicating that a server is running a vulnerable version of the Apache web server). However, if the adversary acts on information in the honey traffic, it will unknowingly attack an intrusion detection node and be discovered. Thus, the game presents an opportunity to design an optimal strategy for defense.

Next, we present an empirical evaluation of the performance of our game model solutions under different conditions, as well as the scalability of the algorithm and some useful properties of the optimal solutions. Then, we show how we emulate *Snaz* in Mininet [9] and show the network overhead that results from honey traffic.

The remaining sections in this chapter discuss obfuscation technique that can effectively protect computer network infrastructure from attacks using traffic analysis. We go through the vulnerabilities of network traffic and the techniques that may be used to help attackers identify these flaws. Then, we introduce a novel algorithm for obfuscating network traffics that obtains state-of-the-art performance, i.e., is competitive with previous methods. The algorithm uses adversarial machine learning techniques to find realistic small perturbations that can improve security and privacy against traffic analysis. Finally, we provide a comparative analysis between our proposed approach and previous work.

2 Motivation and Related Work

Enterprise network administrators are increasingly concerned with Advanced Persistent Threats (APTs). In APTs, adversaries first obtain a small foothold within the network and then stealthily expand their penetration over the course of months and sometimes years. The past decade has provided numerous examples of such targeted attacks, e.g., Carbanak [16], OperationAurora [27]. Such attacks require significant planning. Initially, adversaries identify attack vectors including (1) vulnerable servers or hosts, (2) poorly configured security protocols, (3) unprotected credentials, and (4) vulnerable network configurations. To do so, they leverage network protocol banner grabbing, active port scanning, and passive monitoring [4, 24]. Examples of different types of desired information and corresponding attacks are shown in Table 1.

Software Defined Networking (SDN) has the potential to address operational and security challenges that large enterprise networks face [26]. SDN provides flexibility to programmatically and dynamically re-configuring traffic forwarding within a network [28] and provides opportunities for granular policy enforcement [23]. However, these more functional network switches form a large target for attackers as they can provide a foothold to perform data plane attacks using advanced reconnaissance and data manipulation and redirection [2]. Using one or more compromised switches, an adversary can learn critical information to mount attacks, including network topology and software and hardware vulnerabilities [5, 20].

Deception is an important tactic against adversary reconnaissance. A variety of approaches have applied game-theoretic analysis to cyber deception [38, 51]. Many of these previous works have focused on how to effectively use honeypots (fake systems) as part of network defense [7, 22, 39, 47]. This has included work

Table 1 Example of types of information used for attacks

Target type	Analysis space	Examples
Fingerprinting OS	TTL, Packet Size, DF Flag, SackOk, NOP Flag, Time Stamp	Windows 2003 and XP
Server software, version, service type	Default banners	Apache HTTP 2.2, Windows Server 2003
Network topology, forwarding logic	Flow-rule update frequency, controller-switch communication	Lack of TLS adoption, modified flow rules
Employee Credentials, personal information	Server–client traffic header and data	HTTP traffic, HTTPS traffic with weak TLS/SSL

on signaling games where the goal is to make real and fake systems hard to distinguish [29]. The work on security games (including games modeling both physical and cybersecurity) focuses on deception to manipulate the beliefs of an attacker [1, 19, 45, 49]. Another approach [42] proposes a game model of deception in the reconnaissance phase of an attack, though they do not consider honey flows. Stackelberg game models have been used to find optimal strategies for cyber-physical systems [13].

Network traffic obfuscation is another way of deception that can effectively deal with APT attacks. There are also numerous reasons why network administrators need to use traffic obfuscation. For example, sometimes Internet resources may become inaccessible due to an unavoidable circumstance, but an administrator may want to meet performance benchmarks by shaping the network traffic. Several previous articles have proposed network obfuscation systems. Encryption and padding in traffic features at a variety of levels, such as ciphertext formats, stateful protocol semantics, and statistical properties, are effective ways of preventing statistical traffic analysis [12, 40]. Guan et al. [17] show that sending dummy traffic with real traffic (called packet padding) can manipulate an adversary's observation to a particular traffic pattern and efficiently camouflage network traffic. However, this approach is usually inefficient and sometimes incurs immense network overhead. Another approach is to pad real packets to make them uniform size instead of creating a dummy packet, but it can also delay packet transmission. Wright et al. [48] propose a convex optimization algorithm to modify real-time VoIP and WEB traffic, which is optimal in terms of padding cost and reduces the accuracy of different classifiers. Later, Ciftcioglu et al. [8] propose a water-filling optimizing algorithm for optimal chaff-aided traffic obfuscation where packet morphing is performed by adding either chaff bytes or chaff packets and showing that the algorithm can maximize obfuscation given a chaff budget.

Machine learning techniques are quite common in classifying the various types of IP traffic [33]. Bar-Yanai et al. [3] present a classifier that is robust to the statistical classification of real-time encrypted traffic data. Mapping network traffic from different applications to the preselected class of service (COS) is still a challenging task. One approach uses predetermined statistical application signatures of connections, sessions, and application-layer protocols to determine the COS class for particular datagrams [10, 41].

Zander et al. [50] use unsupervised machine learning to classify unknown and encrypted network protocols where flows are classified based on their network characteristics. Though classification methods are effective for statistical traffic analysis, many machine learning algorithms are vulnerable to adversarial attacks. An attacker can generate adversarial samples by adding small perturbation to the original inputs intent to mislead machine learning models [14, 15]. They also train their own model with adversarial samples and transfer the samples to a victim model in order to produce incorrect output by the victim classifier [48]. Currently, no method is effective against adversarial examples [18, 34, 36]. Papernot et al. [36] introduce adversarial sample crafting techniques that exploit adversarial sample transferability across the machine learning.

Several mathematical and ML methods for crafting adversarial example exploit the gradient of the loss function or the target of classification [6, 14, 35, 37, 43]. Verma et al. [46] propose loss functions and the "Carlini–Wagner L_2" (also called CW) algorithm to craft network traffic using a post-processing operation to the generated distributions. However, the proposed approach sometimes creates invalid perturbations and distributions for each attack that do not match real-world settings. But, in our method, we impose more generalized constraints in generating adversarial network traffic samples; we generate a valid perturbation and distribution for every test sample that results in a more robust attack than in previous work.

3 *Snaz* Overview

Snaz is a deception system designed to mislead or delay passive reconnaissance by an adversary. Snaz provides this deception using *honey traffic* that is precisely controlled by the defender. The following sections will provide a high-level overview of the system and threat model.

3.1 Snaz Architecture

Snaz uses honey traffic to mislead adversaries using passive network reconnaissance. This deception consists of network flows with fake information, which we call *honey flows*. Traditionally, a network flow is defined as a 5-tuple: source IP, source port, destination IP, destination port, and protocol (e.g., TCP). For simplicity, we assume that honey flows include network flows in both directions to simulate real network communication.

Honey flows can fake information in network flow identifiers. For example, a honey flow can attempt to make the adversary believe a non-existent host has a specific IP address or a host is running a server on a specific port. Due to the flexible packet forwarding capabilities of SDN, the defender can route honey flows through any path it chooses, e.g., to tempt an adversary that has compromised a switch on a non-standard path. Honey flows can fake information in the packet payload itself. For example, network servers often respond with a banner string indicating the version of the software and sometimes even the OS version of the host. Attackers often use this banner information to identify unpatched vulnerabilities on the network. Honey flows can simulate servers with known vulnerabilities, making it appear as if there are easy targets. If at any point the adversary acts on this information (i.e., connects to a fake IP address), Snaz redirects the traffic to an intrusion detection node. Since the intrusion detection node does not normally receive network connections, the existence of any traffic directed toward it indicates the presence of an adversary on the network.

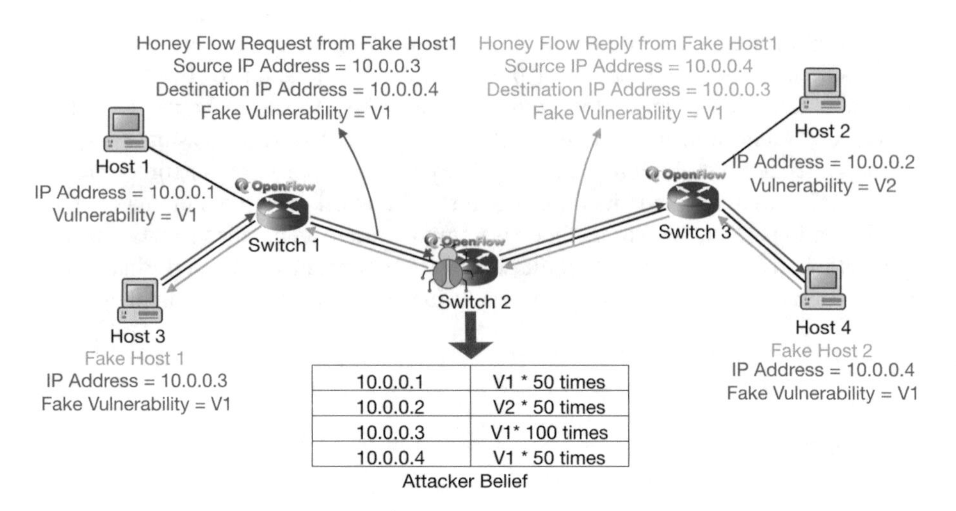

Fig. 1 Snaz uses honey traffic to mislead adversaries

Figure 1 shows a simplified example of honey flows causing an adversary to update its belief. The figure shows two real hosts: $Host$ 1 is with vulnerability type $V1$, and $Host$ 2 is with vulnerability type $V2$. The adversary has compromised $Switch$ 2 and observes all packets passing through it. Without honey traffic, the adversary can easily identify the vulnerabilities on the hosts (e.g., via banner strings) and attack them. In the figure, Snaz simulates the existence of two fake hosts ($Host$ 3 and $Host$ 4) using honey traffic. If the adversary is unaware of Snaz, it will probabilistically attack either $Host$ 3 or $Host$ 4 and be quickly detected. However, if the adversary is aware of Snaz (the scenario we consider in this chapter), it must keep track of all real and fake vulnerability information. How the defender and attacker act is the crux of our game-theoretic model in Sect. 3.3.

3.2 Threat Model and Assumptions

The adversary's goal is to compromise networked resources, e.g., workstations and servers, without detection. The adversary does not know what hosts are on the network or which hosts have vulnerabilities. It must discover vulnerable hosts using network reconnaissance. The adversary assumes that the defender has deployed state-of-the-art intrusion detection systems that can identify active network reconnaissance such as network port scanning. However, we assume the adversary has gained a foothold on one or more network switches (an upper bound of which is defined by the model). Using this vantage point, the adversary is able to inspect all packets that flow through the compromised switches. In doing so, it can learn (1) network topology and which port servers are listening to by observing network flow identifiers and (2) about the installed software versions by observing server and client banner strings.

We assume the adversary can map between banner strings and known vulnerabilities and their corresponding exploits. We conservatively assume this mapping can occur on the switch or can be done without the knowledge of the defender. We assume the adversary has the capability to initiate new network flows from the switch while forging the source IP address, as response traffic will flow back through the compromised switch and terminate as if it was delivered to the real host. Finally, we assume that the adversary is rational and is aware of the existence of Snaz and that honey traffic may be sent to fake hosts. However, the adversary does not know the specific configuration of Snaz, such as the distribution of honey traffic.

We assume the defender's network contains real hosts with exploitable vulnerabilities. The defender is aware of some, but not all of these vulnerabilities. For example, the defender's inventory system may indicate the existence of an unpatched and vulnerable server, but due to production requirements, the server is not yet patched. We further assume the defender can identify valuations of each network asset and approximate the valuation of the assets to the attacker (e.g., domain controllers that authenticate users are valuable targets). We assume that the SDN controller and the applications running on the controller are part of the trusted computing base (TCB). We further assume that the communication between the SDN controller and uncompromised switches is protected and not observable to the adversary (e.g., via SSL or an out-of-band control network). As a result, the adversary cannot alter the controller configuration or forwarding logic of uncompromised switches. Finally, we assume that the network utilization is not near maximum capacity during normal operation. However, exceeding honey traffic may cause congestion and cause network degradation.

3.3 Game Model

An important question we must answer to deploy *Snaz* effectively is how to optimize the honey traffic to be created by Snaz, including how much traffic to create of different types. This decision must balance many factors, including the severity of different types of vulnerabilities, their prevalence on the network, and the costs of generating different types of honey flows (e.g., the added network congestion). In addition, a sophisticated APT attacker may be aware of the possible use of this deception technique, so the decisions should be robust against optimal responses to honey traffic by such attackers. Finally, we note that many aspects of the environment can change frequently; for example, new zero-day vulnerabilities may be discovered that require an immediate response, or the characteristics of the real network traffic may change. Therefore, we require a method for making fast autonomous decisions that can be adjusted quickly.

We propose a game-theoretic model to optimize the honey flow strategy for *Snaz*. Our model captures several of the important factors that determine how flows should be deployed against a sophisticated adversary, but it remains simple enough that we can solve it for realistic problems in seconds (see Sect. 3.4 for details) allowing us to rapidly adapt to changing conditions. Specifically, we model the interaction as a two-player non-zero-sum Stackelberg game between the defender (leader) and an attacker (follower) where the defender (*Snaz*) plays a mixed strategy and the attacker plays pure strategy. This builds on a large body of previous work that uses Stackelberg models for security [44], including cyber deception using honeypots [39].

We now formally define the strategies and utilities of the players using the notation listed in Table 2. We assume that the defender is using *Snaz* as a mitigation for a specific set of i vulnerabilities that we label V_i. Every flow on the network indicates the presence of at most one of these types of vulnerabilities in a specific host. The real network traffic is characterized by the number of real flows R_i that indicate vulnerability type V_i. The pure strategies for *Snaz* are vectors that represent the number of honey flows that are created that indicate each type of vulnerability V_i; we write Φ_{ij} to represent the marginal pure action of creating j flows of type V_i. These fake flows do not need to interact with real hosts; they can advertise the existence of fake network assets (i.e., honeypots). The defender can play a mixed strategy that randomizes the number of flows of each type that are created, which we denote by Φ. To keep the game finite, we define the maximum number of flows that can be created of each type as R_i. The attacker's pure strategy a_i represents choosing to attack a flow of type V_i or not to attack. We assume the attacker cannot reliably distinguish real flow and honey flow, so an attack on a specific type corresponds to drawing a random flow from the set of all real and fake flows of this type.

The utilities for the players depend on which vulnerability type the attacker chooses, as well as on how many real and honey flows of that type are on the network. An attack on a real flow will result in a higher value for the attacker than on a honey flow of the same type and vice versa for the defender. Specifically, if the attacker chooses type V_i, it gains a utility $\upsilon_i^{a,r}$, which is greater than or equal to the

Table 2 Game notation

d	Defender
a	Attacker
$V_i \in V$	Set of i types of vulnerabilities in the network
R_i	Number of real flows indicating V_i
H_i	Upper bound on the number of honey flows indicating V_i
Φ_{ij}	Action of selecting $j \in [0, H_i]$ honey flows for V_i
Φ	Defender's mixed strategy as the marginal probabilities over $\{\Phi_{i0}, \ldots, \Phi_{iH_i}\}$
C_i	Cost of creating each honey flow that indicates V_i
$\upsilon_i^{a,r}$ $\upsilon_i^{a,h}$	The value the attacker gains from attacking a real or fake flow of type V_i
$\upsilon_i^{d,r}$ $\upsilon_i^{d,h}$	The value the defender loses from an attack against a real or fake flow of type V_i
a_i	Denotes the action of attacking a flow of type V_i where a_0 is the no-attack action, yielding 0 payoff

value for attacking a honey flow of the same type $\upsilon_i^{a,h}$ (which may be negative or 0). We assume that this component of the utility function is zero sum, so the defender's values are $\upsilon_i^{d,r} = -\upsilon_i^{a,r}$ and $\upsilon_i^{d,h} = -\upsilon_i^{a,h}$. The defender's utility function includes a cost term C_i that models the marginal cost of adding each additional flow of type V_i (for example, the additional network congestion which can vary depending on the type of flow). If the defender plays strategy Φ and the attacker attacks the V_i, the defender's expected utility is defined as follows:

$$U^d(\Phi, i) = P_i^r \upsilon_i^{d,r} + (1 - P_i^r)\upsilon_i^{d,h} - C^h. \tag{1}$$

Here, P_i^r denotes the probability of attacking a vulnerability of type V_i which can be calculated as follows:

$$RSMN(mc, P) = \sum_{j \in \{0,\ldots,H_i\}} \Phi_{ij}(R_i/(j + R_i)).$$

The overall cost C^h for playing Φ is given by Eq. 1:

$$C^h = \sum_{i \in V} \sum_{j \in \{0,\ldots,H_i\}} (\Phi_{ij} \times j \times C_i).$$

Analogously, the expected utility for the attacker is given by

$$U^a(\Phi, i) = P_i^r \upsilon_i^{a,r} + (1 - P_i^r)\upsilon_i^{a,h}. \tag{2}$$

3.3.1 *Snaz* Game Example

Consider a network with two types of vulnerabilities. Let the values be $\upsilon^{a,r} = (10, 20)$ and $\upsilon^{a,h} = (-5, -10)$ and the cost of creating a honey flow indicating each type of vulnerability be $C = (1, 0.5)$. The total number of real vulnerabilities for each type is $R = (5, 5)$, and the upper bound on honey flows is $H = (2, 3)$. Thus, at most two honey flows of 1st vulnerability type and three honey flows of the second vulnerability type can be created. Now, consider if the defender plays the following strategy Φ:

$$\Phi = \left\{ \Phi_1 = \begin{array}{|c|c|} \hline \Phi_{10} & 0 \\ \hline \Phi_{11} & 0.5 \\ \hline \Phi_{12} & 0.5 \\ \hline - & - \\ \hline \end{array}\ , \Phi_2 = \begin{array}{|c|c|} \hline \Phi_{20} & 0 \\ \hline \Phi_{21} & 0 \\ \hline \Phi_{22} & 0 \\ \hline \Phi_{23} & 1 \\ \hline \end{array} \right\}$$

In strategy Φ, the defender creates one honey flow 50% of the time and two honey flows 50% of the time with a Type 1 vulnerability. The defender also creates three honey flows 100% of the time with a Type 2 vulnerability. The attacker's best response is to attack vulnerability Type 2 with expected utility $U^a(\Phi, 2) = 8.75$, and the defender utility is $U^d(\Phi, 2) = -11.75$.

3.3.2 Optimal Defender's Linear Program

Our objective is to compute a Stackelberg equilibrium that maximizes the defender's expected utility, assuming that the attacker will also play its best response. To determine the equilibrium of the game, we formulate a linear program (LP) where the attacker's pure strategy a is a binary variable. We create a variable for each defender's pure strategy $\Phi_{i,j}$, the action of creating j honey flows for V_i. The following LP computes the defender's optimal mixed strategy for each type of vulnerability under the constraint that the attacker plays a pure strategy best response:

$$\max_{i \in V} \; U^d(\Phi, i)\, a_i \tag{3}$$

$$s.t. \quad a_i \in \{0, 1\}, \quad \Phi_{ij} \in [0, 1]$$

$$U^a(\Phi, i)\, a_i \geq U^a(\Phi, i')\, a_i \quad \forall\, i, i' \in V \tag{4}$$

$$\sum_{j \in \{0, \ldots, H_i\}} \Phi_{ij} = 1 \quad \forall \Phi_i \in \Phi \tag{5}$$

$$\sum_{i \in V} a_i = 1. \tag{6}$$

In the above formulation, the unknown variables are the defender's strategy $\{\Phi_{i0}, \ldots, \Phi_{iH_i}\}$ for each $\Phi_i \in \Phi$ and the attacker's action a_i. Equation 3 is the objective function of the LP that maximizes the defender's expected utility. The inequality in Eq. 4 ensures that the attacker plays a best response. Finally, Eq. 5 forces the defender's strategy to be a valid probability distribution.

3.4 Simulations and Model Analysis

We now present some results of simulations based on our game-theoretic model, as well as with an initial implementation of honey flow generation in an emulated network environment. We show that our game theory model can produce solutions that improve over simple baselines and can be calculated fast enough to provide

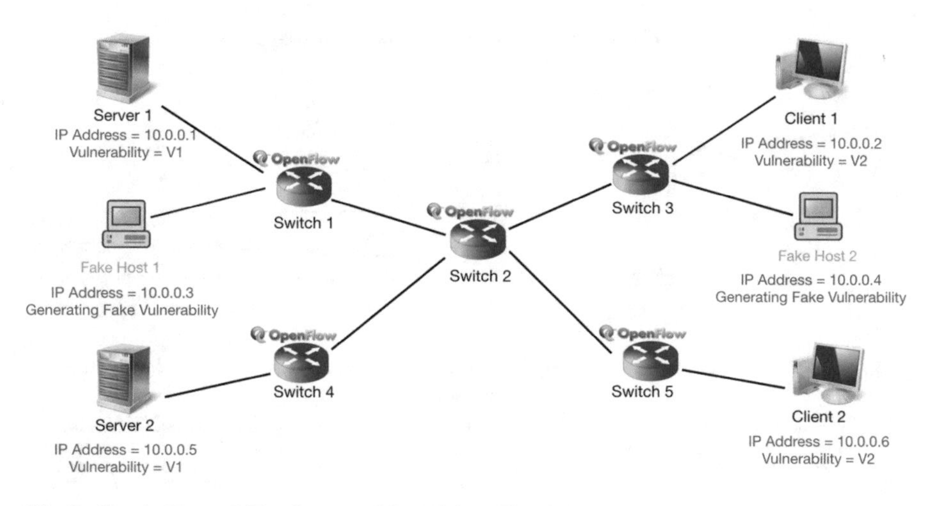

Fig. 2 Simple Network Topology used for Mininet Simulation

solutions for realistic networks. We also examine how the optimal solutions change based on the parameters of the model to better understand the structure of the solutions and the sensitivity to key parameters of the decision problem.

3.4.1 Preliminary Testbed Evaluation

We have constructed a preliminary honey flow system in the emulated environment of *Mininet* [30]. We want to show the possibility of generating plausible *honey traffic* in an emulated network. And we want to evaluate the effects of the deception in a more realistic context. We work on a small topology, as shown in Fig. 2. As shown in the figure, we consider four real and two fake hosts.

Some additional parameters are as follows:

- *Client 1* connects with *Server 1*, while *Client 2* connects with *Server 2*. In the simulation, each of these two clients sends 500 packets to the servers and receives corresponding replies.
- The *fake clients* are connected with the system and can send packets with fake vulnerabilities to each other.
- All the links in our simulation are 1 MBit/s bandwidth with a 10ms delay and 2% probability loss.
- The values assigned to the servers are 2 and real clients have a value of 1 for both attacker and defender.

We visualize the simulation results with two types of vulnerabilities in Fig. 3. In this test, we increase the amount of honey flows from the fake clients from 0 to 500 packets. The honey flows from fake client 1 are with vulnerabilities of type 1, while fake client 2 sends honey flows with vulnerabilities of type 2. Besides, we

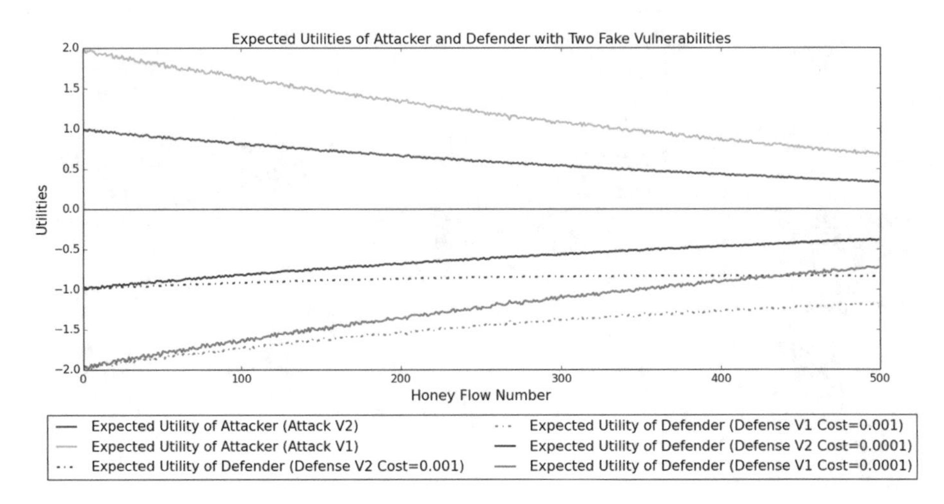

Fig. 3 Defender and attacker utility

experiment with different costs of honey flow generation. The dashed lines in Fig. 3 show the expected utilities when we generate one honey flow with a cost of 0.001 and solid lines represent that defender has to endure a cost of 0.0001 to generate each honey flow.

Although the simulation does not use the game-theoretic optimization, it conveys the idea that increasing the number of honey flows remarkably reduces the effectiveness of adversarial efforts. Meanwhile, we see that the cost of honey flow generation significantly affects defender utility.

3.4.2 *Snaz* Game Theory Solution Quality

Our next set of experiments focuses on evaluating the solution quality of the proposed Stackelberg game model for optimizing *Snaz* honey flows compared to some plausible baselines: (1) not generating honey flows at all and (2) using a uniform random policy for generating honey flows. We average the results over 100 randomly generated games, each with five types of vulnerabilities. We set the number of real flows for each type of vulnerability to 500. We set the upper bound on the number of honey flows for each type as (uniformly) randomly generated from [500,1000]. The values are described in the caption, and we vary the costs of creating flows as shown in Fig. 4. In these studies, We used the same important values for all five categories of vulnerabilities, which are described in the caption. In the graphs, the horizontal axes show various honeyflow generation costs and the vertical axes represent the defender's payoffs.

The results in Fig. 4 show that the game-theoretic solution significantly outperforms the two baselines in most settings, demonstrating the value of optimizing the honey flow generation based on the specific scenario. We note that the cost of

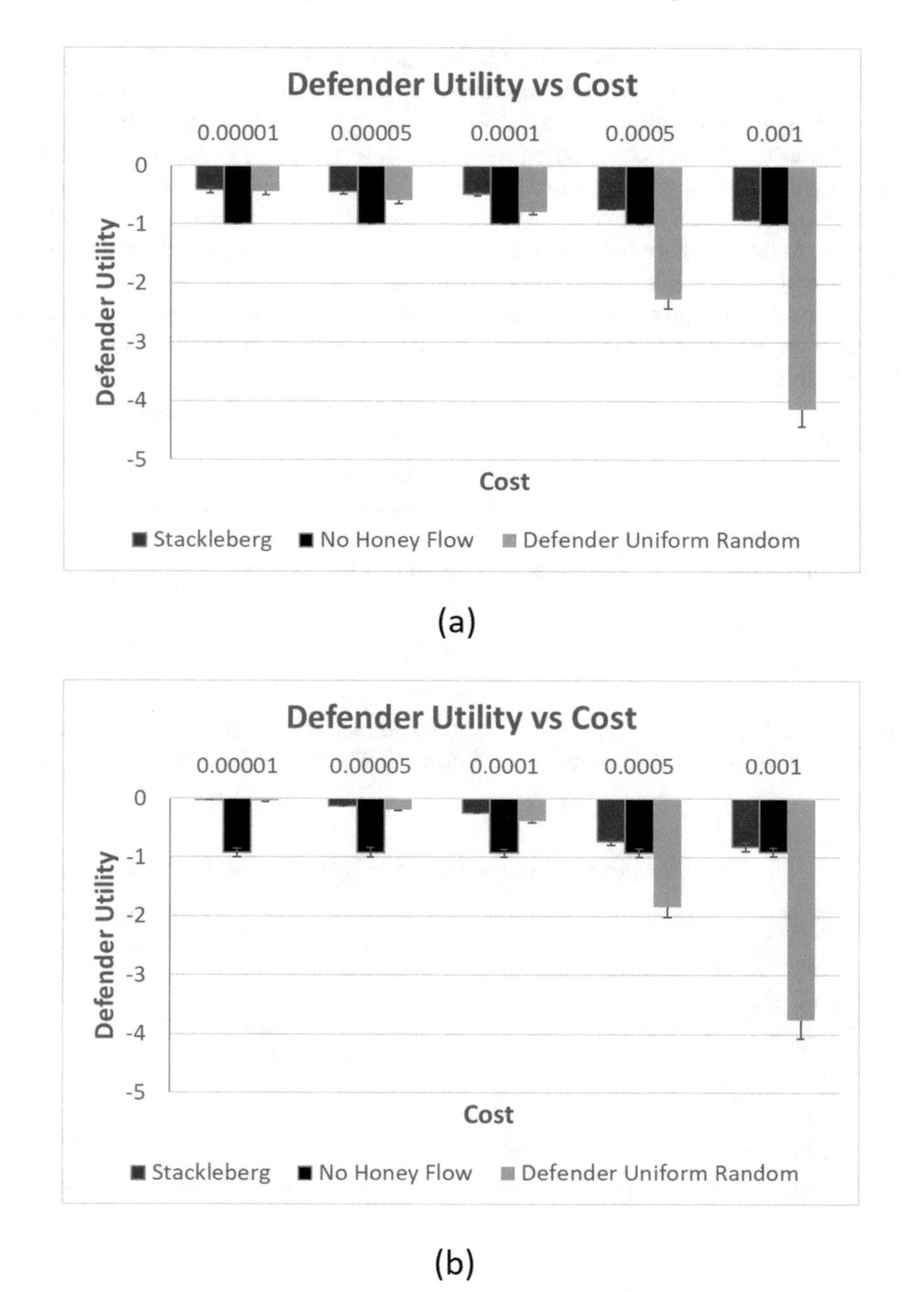

Fig. 4 Comparison of defender utility when the defender uses different values: (**a**) the value of attacking a fake vulnerability is zero and a real is one and (**b**) the value of attacking a fake vulnerability is the same as real value and the values are randomly generated from the interval [0.5, 1.0]

generating a honey flow has a significant impact on the overall result. When the cost is high, the game-theoretic solution is similar to not generating flows at all (since they are not highly cost-effective). Random honey flow generation can be detrimental for the defender. When the cost is low, the performance of the game-theoretic solution is similar to that of the uniform random policy; since flows are so cheap, it is effective to create a large number of them without much regard to strategy. At intermediate costs, which is the most likely scenario in real applications, the value of the strategic optimization is highest.

In our second experiment, we consider vulnerabilities with different values and examine the variation in the optimal solution as we vary the number of real flows. We use five vulnerabilities with the values of the real systems (0.8, 0.5, 0.9, 0.6, and 1.0) and attacking any fake system gives 0. Then, we set the upper bound of honey flows for each type of vulnerability that the defender can create to 1500. The honeyflow generation cost is 0.0005 for all types. The results in Fig. 5 show that the defender's strategy is to create more honey flows for the high-valued vulnerabilities. As the number of real flows increases, the cost of adding flows to create a high ratio is substantial and the overall number created drops for all types.

3.4.3 Solution Analysis

We now analyze how the ratio of honey flows to real flows changes in the optimal solution as we change the number of real flows. In Fig. 6, the network setup consists

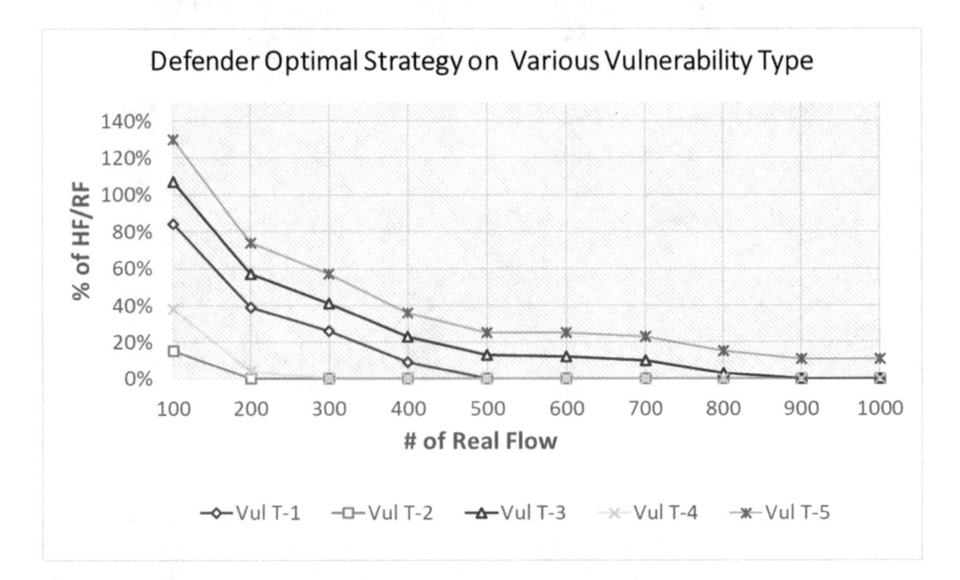

Fig. 5 Defender's optimal strategy as the number of real flows varies. The vertical axis of the graph shows the various ratios of honey and real flow for different types of vulnerabilities where the number of the real flow is variable

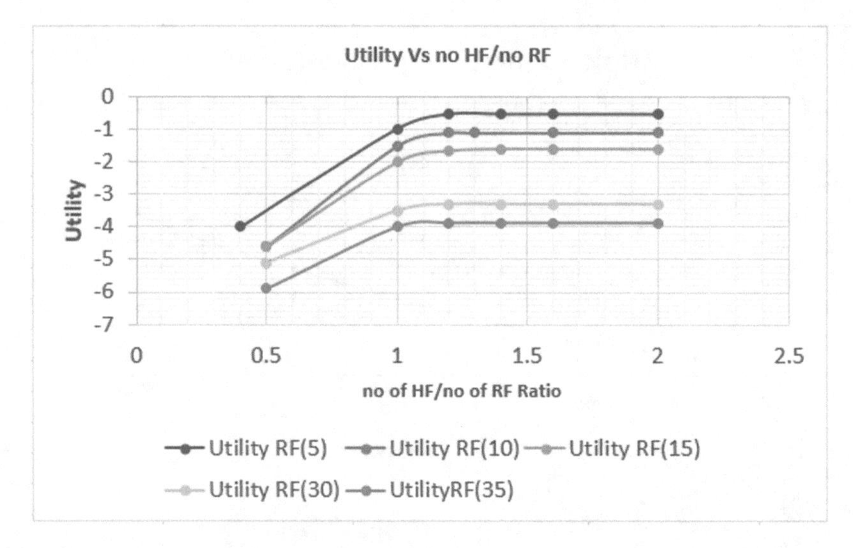

Fig. 6 The defender utility with varying honey flow ratios

of four vulnerabilities with values of (10, 20, 30, 40) and fake flows with values of (9, 18, 27, 32). The cost of generating each honey flow is 0.1. We show the defender's expected utility as we increase the ratio of honey flows to real flows. Each line represents a different number of real flows in the original game. We see that the defender utility increases as we add honey flows, but only up to a point; when the marginal value is less than the cost, the optimal solution is to stop adding additional flows. We see this in the shape of the curves.

3.4.4 Scalability Evaluation

In a practical application of *Snaz*, we would need to be able to calculate the optimal strategy quickly, since the network may change frequently, leading to changing game parameters. For example, the number of real flows will change over time, as will the number of hosts in the network. In addition, the values of traffic and vulnerabilities, and the specific vulnerabilities we are most interested in can change (e.g., due to the discovery of new vulnerabilities). We evaluate the scalability of the basic LP solution for this game as we increase the size of the game in two key dimensions: (1) by increasing the number of vulnerability types and (2) by increasing the number of flows.

We randomly generate games holding the other parameters constant to evaluate the solution time. The results are shown in Fig. 7. Though the solution time increases significantly as we increase the complexity of the game, we were able to solve realistic size games with a large number of flows and vulnerability types of interest within just a couple of seconds using this solution algorithm. This result signifies

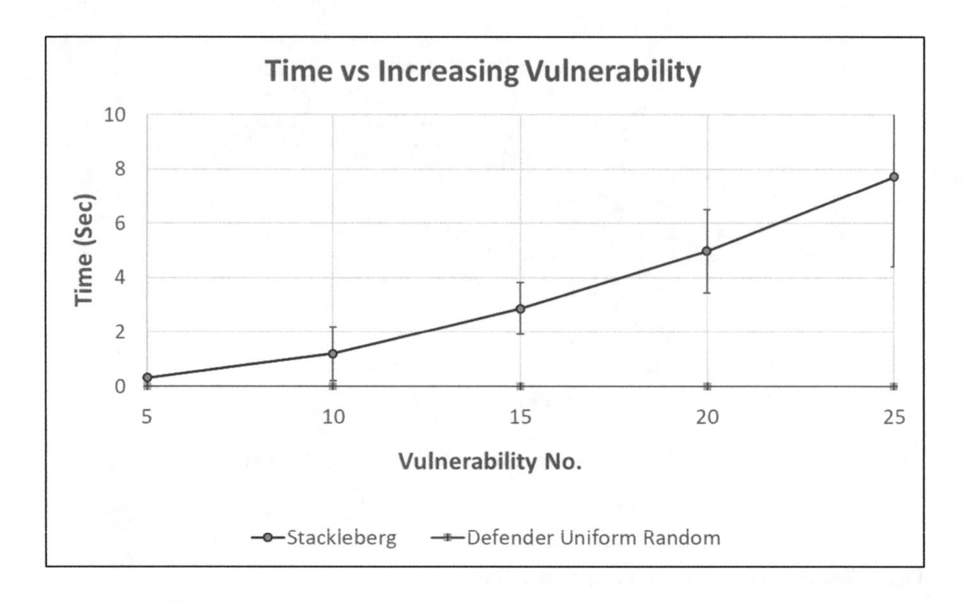

(a)

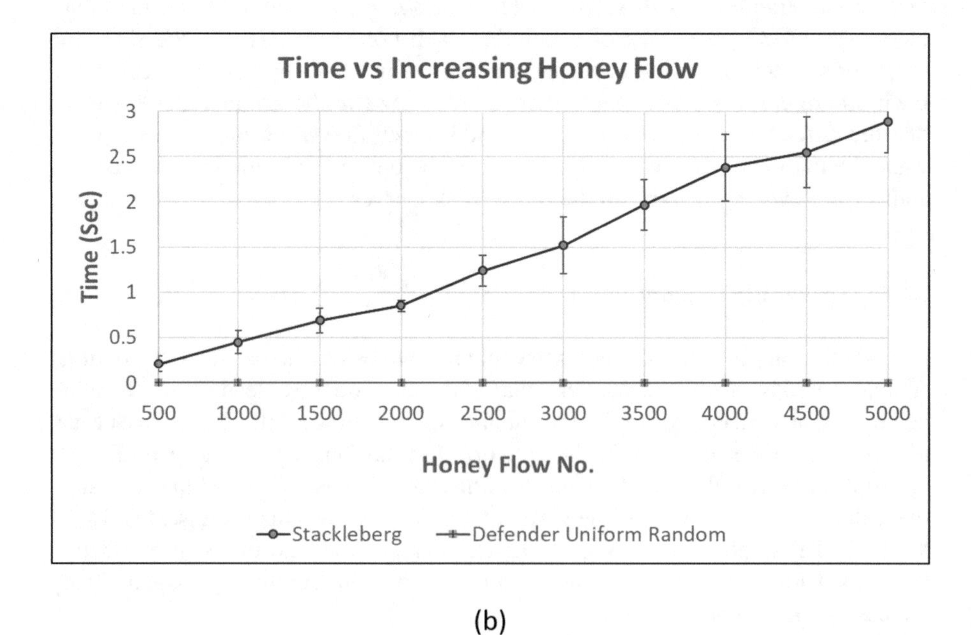

(b)

Fig. 7 Comparison of computational time when (**a**) varying the number of vulnerabilities and (**b**) varying the number of honey flows

that we can apply this to optimize honey flows in realistically sized networks with a fast response rate; with further optimization, we expect that the scalability could be improved significantly beyond this basic algorithm.

4 Decoy Traffic Generation Approach

One major limitation of the previous study is that we talked about using decoy traffic to confuse the adversary's information gathering through passive network reconnaissance, but we did not go into much detail about how to create realistic deceptive flows. The adversarial machine learning is a possible way of generating realistic honeyflows. To mimic the behavior of real traffic, we can use a GAN (Generative Adversarial Networks) generator that uses the real traffic and generate the fake data as a honeyflow. Generally, the GAN model is trained to generate its own vectors from latent space, which then decode into realistic network flows. But increasing the frequency of generating different types of honeyflows will incur increased network congestion and various types of generation-related cost. Therefore, instead of developing fake traffic from scratch, obfuscating network traffic can prevent statistical traffic analysis where an adversary can classify different applications and protocols from the observable statistical properties. The use of obfuscation also helps to reduce the risk of passive reconnaissance. In general, network traffic has a huge number of features but most of the traffic features are encrypted, which hides much information of the target system. However, it is still possible to analyze encrypted traffic based on the source, destination, routing, timing, quantity, and other characteristics. Obfuscating unencrypted features can significantly delay reconnaissance.

In the remaining of this chapter, we discuss the benefit of network traffic obfuscation and represent some adversarial learning techniques to obfuscate network traffic.

5 Network Traffic Obfuscation

Machine learning methodologies can help attackers to identify various applications and protocols. The performance of the classifier depends on the accuracy of collected information. Network traffic obfuscation is a technique where network traffic is manipulated (e.g., by adding dummy bytes with the packets to increase packet size) to limit the attacker's gathering of information by causing errors in the classification models. This obfuscation approach is effective at reducing the risk of passive reconnaissance where an attacker gathers traffic and uses statistical analysis to categorize different patterns (e.g., protocols, applications, and user information). A major challenge with this approach is to determine the optimal algorithm for masking the features of the traffic effectively, but within the constraints of feasible modifications and limited resources or network overhead.

Encryption and mimicry are two basic obfuscation methods, but they cannot remove fingerprints from metadata (e.g., packet size and inter-arrival timing). The adversaries can classify encrypted traffic based on statistical features, including packet and payload byte counts [11, 31]. Using mimicry, it is possible to shape a protocol to look like another, even though statistical fingerprints of the metadata would still be preserved [11, 41]. In this work, we consider data obfuscation methods that can be applied to network traffic. Our goal is to find more robust solutions for network administrators as defenders in performing statistical obfuscation while minimizing unnecessary overhead using AML.

We use Adversarial Machine Learning (AML) technique to solve this problem, where a defender seeks to protect the network from an adversary by finding realistic small perturbations that are added to the network traffic to reduce the accuracy of machine learning traffic classifiers. We begin by introducing the Restricted Traffic Distribution Attack (RTDA), an algorithm for realistic adversarial traffic generation that can be applied in real-world networks. Then, we calculate the average perturbation cost for a real system and provide a comparative analysis between our proposed approach and previous work. Our solution technique is a novel approach for generating adversarial examples that obtains state-of-the-art performance while considering more realistic constraints on perturbations.

5.1 Experimental Setup

This section describes our classification model and dataset, building on the previous work in [46]. In Sect. 5.3, we discuss our proposed approach in detail. Also, Table 3 shows the notation used below.

Table 3 Important notation

Notations	Description
τ	Original traffic
τ_θ	Modified traffic
δ	Perturbation amount
f_θ	Classification model
ρ	Application class set
x	Feature vector
x^{adv}	Adversarial feature vector
L_p	Distance metric

Table 4 Class composition and number used in this work from the dataset

Classification	Flow type	Number
Bulk	FTP	11,539
Database	Postgres, SQLNet, Oracle, Ingres	2648
Mail	IMAP, POP2/3, SMTP	28,567
Services	X11, DNS, Ident, LDAP, NTP	2099
P2P	KaZaA, BitTorrent, Gnutella	2094
WWW	WWW	328,091

5.1.1 Dataset

We perform experiments on the Internet Traffic Network dataset [32]. This dataset was generated by monitoring a research facility host with 1000 users connected via Gigabit Ethernet link. The objects to classify are traffic flows that represent the flow of one or more packets between the host and the client during a complete TCP connection. Each flow was manually classified. Table 4 shows the class information, flow types per class, and flow count. Following Verma et al. [46], we only include classes with at least 2000 samples in our training set.

5.1.2 Realistic Features

Each sample is composed of 249 features that were observed during generation time. In a real-time traffic transmission, the defender has the capability only to increase the size of the packets. Therefore, we do not use inter-arrival time as a feature. Our work shows that only using packet size is sufficient to attack a network. We select the 0, 25, 50, 75, and 100 percentiles of the IP packet sizes from both client to server and server to client. We normalized these features to the range (0, 1).

5.1.3 Classification Model

We replicate the training approach and neural network model used in the previous work [46]. The training model is a three-layer neural network with 300, 200, and 100 hidden units and applies a rectified linear function in every layer. We process the data by randomly dividing it into three datasets—5000 validation samples, 5000 test samples, and the remaining samples as training. Due to high class imbalance, we randomly sample the training set, so every class has an equal number of examples. We train the network using mini-batches of size 1000 for 300 epochs. The results are found in Table 5.

Table 5 Neural network accuracy per class

Class	Accuracy
Bulk	95%
Database	97%
Mail	95%
P2P	96%
Service	85%
WWW	91%

5.2 *Adversarial Settings*

We now formalize the models for the defender and the attacker. We also discuss some well-known approaches for generating adversarial examples.

5.2.1 Defender Model

We model the problem by considering an adversarial setting where a defender (d) tries to protect a network from an adversary (α). The goal of α is to observe d's network and classify its traffic flows (τ is an individual flow) by using statistical analysis, while d disguises τ by changing their features. A modified flow τ_θ can potentially result in α misclassifying τ_θ as relating to a different application or protocol class (σ) rather than the true one (ρ). We consider that d knows the attacker model f and observations O for training. That is, d is capable to create a substitute model f_θ for τ_θ. The transferability property of AML supports that any adversarial example that can fool a machine learning algorithm can also fool other machine learning algorithms irrespective of the implementation [36]. Therefore, d uses AML techniques to find an optimal way for generating τ_θ by considering that the traffic recipient has mechanisms for inverting the changes. However, in adding perturbations, d must adhere to the following constraints:

- The basic constraints on a protocol must be preserved. For example, packet size and timing cannot be the negative, and packet sizes must lie between the specified minimum and maximum.
- The network is constrained by performance benchmarks meaning that the network supports a maximum threshold of latency.
- The AML model should use small input perturbations for creating τ_θ since a large alteration of τ can break the protocol constraints and incur unnecessary network overhead.

5.2.2 Adversary Model

We assume that α observes a particular flow between a source and destination where the flow is always bidirectional. It has the required tools to analyze meta statistical signatures (e.g., packet size) and trains its classifier f_θ based on these features, using 0, 25, 50, 75, and 100 percentiles of the IP packets in both directions. The objective of α is to correctly classify the application set $\{\rho_1, \rho_2, \ldots, \rho_n\}$ observed in d's traffic τ, where n is the possible number of classes. Therefore, α determines a probability distribution over n classes by using $f_\theta(x)$, where x is a feature vector of $\{x_1, x_2, \ldots, x_n\}$ obtained from O.

5.2.3 Obfuscation Approaches

Let $C(x)$ be the classification of x by a model and $C^*(x)$ be the true class. Adversarial learning finds a perturbation δ such that when added to an input x, $C^*(x) \neq C(x+\delta)$. The value of δ should be small enough when added to x for producing $x^{adv} = x + \delta$, which implies that the difference between x^{adv} and x should be almost imperceptible. While many approaches can be used for generating adversarial examples, Szeged et al. [43] use the L-BFGS optimization procedure for generating an adversarial example x^{adv} when input x is given and formulates the problem as

$$\min ||x - x^{adv}||_2 + \lambda J(f_\theta(x^{adv}), t^{true}).$$

The first term sets the penalty for large perturbations to x and the second one penalizes when the classification deviates from the target class t^{true}. The loss function between t^{true} and output of the classifier $f_\theta(x^{adv})$ is denoted by J. $\lambda > 0$ is the model parameter.

The Carlini–Wagner L_2 attack is a robust iterative algorithm that creates adversarial examples with minimum perturbation [6]. This attack for a target class t is formalized as

$$\min ||\frac{1}{2}(tanh(w) + 1) - x||_2 + \lambda f_\theta(\frac{1}{2}(tanh(w) + 1), \text{ such that} C^*(x) \neq t,$$

where f_θ is defined by

$$f_\theta(x^{adv}) = \max(\max\{Z(x^{adv})_i : i \neq t\} - Z(x^{adv})_t, -k)$$

and $\delta = \frac{1}{2}(tanh(w) + 1) - x$ is the perturbation of the adversarial sample. Here, λ is chosen empirically through binary search and k controls the confidence of misclassification occurrence.

For generating untargeted adversarial perturbations, Goodfellow et al. [14] proposed a fast single-step method. This method determines an adversarial perturbation under L_∞ norm where the perturbation is bounded by the parameter ϵ that results in the highest increase in the linearized loss function. It can be obtained by performing one step in the gradient sign's direction with step-width ϵ.

$$x^{adv} = x + \epsilon \ sign(\Delta_x J(f_\theta(x^{adv}), t^{true})).$$

Here, L_∞ computes the maximum change to any of the coordinates:

$$||x - x^{adv}||_\infty = max(|x_1 - x_1^{adv}|, |x_2 - x_2^{adv}|, \dots, |x_n - x_n^{adv}|)$$

Szeged et al. [6] used L_∞ distance metrics to generate the CWL_∞ attack, where the optimization function is defined by the following:

$$\lambda \ minf_\theta(x + \delta) + ||\delta||_\infty$$

and $\delta = \frac{1}{2}(tanh(w) + 1) - x$.

This method has a lower success rate, but it is simple and computationally efficient [25].

5.3 *Restricted Traffic Distribution Attack*

We define an attack that can be translated readily to a real-life setting. To ensure the perturbation yields a valid distribution, we have constrained our attack in two ways: the attack is not allowed to reduce the packet size, and the generated distribution should preserve the monotonic non-decreasing property. We solve this problem by enforcing these constraints directly in the adversarial optimization framework.

Notice that it is possible to reduce the packet size in a distribution by inserting small dummy packets into the traffic, but this approach introduces a larger overhead into the network than only appending dummy bytes.

5.3.1 Perturbation Constraints

Given a distribution x, a general adversarial algorithm finds a perturbation δ for which it minimizes a distance metric L_p and changes the correct classification: $C^*(x) \neq C(x+\delta)$. This perturbation has no restrictions with respect to the direction that it modifies the original distribution. Instead, we clip every value below zero in the perturbation during learning:

$$\text{minimize } L_p(x, x + (\delta)^+), \text{ such that} C(x + (\delta)^+) = t,$$

where $(f)^+$ stands for $\max(f, 0)$ and t is not the correct label.

5.3.2 Distribution Constraints

Given a batch of adversarial distributions A, we define an operation that identifies every adversarial sample with decreasing consecutive features.

Let (A_i, A_{i+1}) be consecutive features in the batch A. We compute the following operation:

$$\text{diff} := (A_i - A_{i+1})^+.$$

We update our distribution based on this value: $A_{i+1} := A_{i+1} + diff$.

For a valid sample, the operation will result in 0, but for an invalid one, it will compute the difference between features, so after the update we automatically get non-decreasing features. This operation is sequentially applied to every pair of consecutive features in the same distribution during the optimization of the attack.

5.3.3 Framework

These restrictions in an attack should generate a valid adversarial distribution if convergence is possible. In this work, we choose the Carlini–Wagner attack for L_2 and L_∞ norm as our frameworks.

Implementation Details

We re-implement the Carlini–Wagner attack for L_2 and L_∞ norms. For the initial c, we select 10^{-3} and 10^{-1}, respectively, and search for five steps with 1000 as the maximum number of iterations. In our algorithm, we clip the perturbation before adding to the batch and then apply the series of operations to correct the distribution. We also replicate Verm et al.'s [46] method by applying a post-processing operation to the generated distributions from the CW L_2 attack.

5.4 Results

We test our two RTDA frameworks against previous adversarial approaches (Fig. 8).

We compare the attacks by evaluating how realistic the generated distributions are and the success rate for fooling the neural network. To evaluate how realistic an attack is, we compare the ratios of valid perturbations and valid distributions for each attack. The results are shown in Table 6 for realistic attacks and Table 7 for success rate per class.

Prior work did not consider the limitations of a perturbation in real-world settings. Our algorithm is more realistic than previous attacks. Both of our frameworks generate a valid perturbation and distribution for every test sample. Post-processing

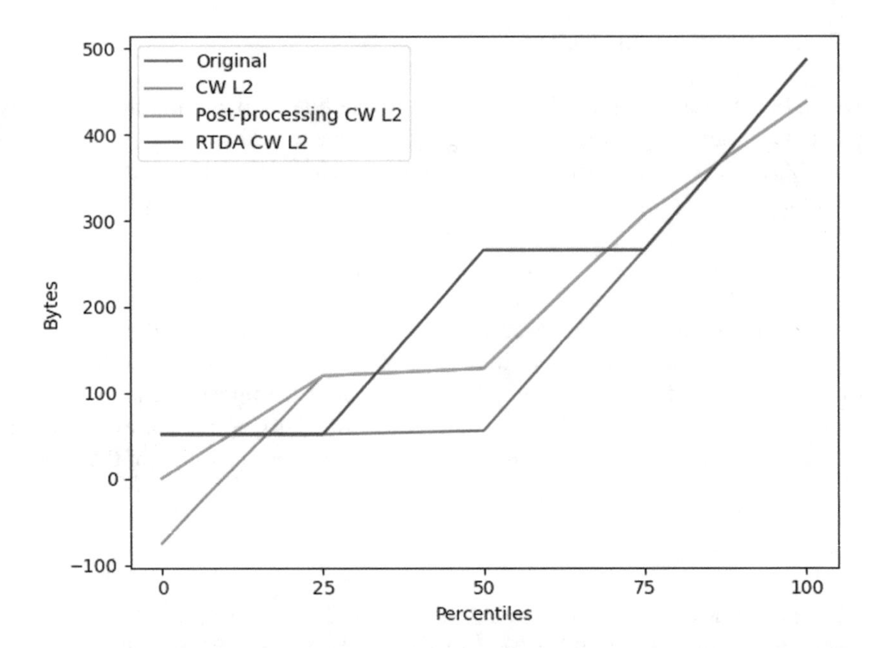

Fig. 8 A comparison of every L_2 adversarial example generated from the same distribution. Notice the negative packet size generated by CW L_2 and the reduction of the 0th and 100th percentile by post-processing CW L_2. In contrast, RTDA generates an adversarial by only increasing the 50th percentile

Table 6 Percentage of valid adversarial samples

	Valid perturbation	Valid distribution
RTDA CW L_2	100%	100%
Post-processing CW L_2	0%	100%
CW L_2	0%	20%
RTDA CW L_∞	100%	100%
CW L_∞	0%	22%

Table 7 Success rate per class (Fraction of instances for which an adversarial was found)

	Database	Bulk	Mail	P2P	Services	WWW
RTDA CW L_2	100%	100%	95%	93%	100%	95%
Post-processing CW L_2	75%	33%	29%	50%	53%	84%
CW L_2	100%	100%	100%	100%	100%	100%
RTDA CW L_∞	100%	100%	74%	72%	67%	100%
CW L_∞	100%	100%	100%	100%	100%	100%

has the disadvantage that resultant distributions may no longer be adversarial examples. Our approach directly finds attacks in the valid space enabling us to optimize for the best attacks. Therefore, RTDA substantially outperforms the success rate of the previous post-processing CW L_2 attack in every class. Even

Table 8 Average norm and perturbation

	L_p	δ mean (Bytes)
RTDA CW L_2	0.015	14.5
Post-processing CW L_2	0.012	15.4
CW L_2	0.011	12.9
RTDA CW L_∞	0.033	19.41
CW L_∞	0.026	15.74

with the additional constraints, our attacks L_2 and L_∞ are just 2% and 14% apart, respectively, from their unrestricted versions.

Both attacks have a larger norm in comparison to previous approaches. Surprisingly, RTDA L_2 has a smaller perturbation than the post-processed approach. On average, our attack can be applied to a system by increasing each packet by 14.5 bytes. Table 8 compares the corresponding norm and average perturbation for each approach.

6 Conclusion

This chapter discussed two techniques to employ deceptively crafted traffic flows to improve network security. In the first work, we introduced *Snaz*, a technique that uses honey traffic to confound the knowledge gained by the adversary through passive network reconnaissance. We defined a Stackelberg game model for optimizing one of the key elements of *Snaz*, the quantity and type of honey flows to create. This model balances cost and value trade-offs in the presence of a sophisticated attacker but can still be solved fast enough to be used in a dynamic network environment.

In the second part of the chapter, we proposed a novel network traffic obfuscating approach that is robust against network traffic attackers, where we leverage adversarial attacks as a mechanism to obfuscate network traffic. Our adversarial machine learning algorithm outperforms previous approaches, achieving state-of-the-art results and reducing the perturbation's network overhead.

Acknowledgments The Army Research Office supported this work under award W911NF-17-1-0370. The views and conclusions contained in this document are those of the authors and should not be interpreted as representing the official policies, either expressed or implied, of the Army Research Laboratory or the U.S. Government. The U.S. Government is authorized to reproduce and distribute reprints for government purposes without standing any copyright notation.

References

1. An, B., Tambe, M., Ordonez, F., Shieh, E., Kiekintveld, C.: Refinement of strong Stackelberg equilibria in security games. In: Twenty-Fifth AAAI Conference on Artificial Intelligence (2011)

2. Antikainen, M., Aura, T., Särelä, M.: Spook in your network: Attacking an SDN with a compromised OpenFlow switch. In: Bernsmed, K., Fischer-Hübner, S. (eds.) Secure IT Systems, pp. 229–244. Springer International Publishing, Cham (2014)
3. Bar-Yanai, R., Langberg, M., Peleg, D., Roditty, L.: Realtime classification for encrypted traffic. In: International Symposium on Experimental Algorithms, pp. 373–385. Springer (2010)
4. Bartlett, G., Heidemann, J., Papadopoulos, C.: Understanding passive and active service discovery. In: Proceedings of the 7th ACM SIGCOMM Conference on Internet Measurement (IMC), pp. 57–70. ACM (2007)
5. Benton, K., Camp, L.J., Small, C.: OpenFlow vulnerability assessment. In: Proceedings of the Second ACM SIGCOMM Workshop on Hot Topics in Software Defined Networking, HotSDN '13, pp. 151–152. ACM, New York, NY, USA (2013). https://doi.org/10.1145/2491185.2491222
6. Carlini, N., Wagner, D.: Towards evaluating the robustness of neural networks. In: 2017 IEEE Symposium on Security and Privacy (SP), pp. 39–57. IEEE (2017)
7. Carroll, T.E., Grosu, D.: A game theoretic investigation of deception in network security. Secur. Commun. Networks **4**(10), 1162–1172 (2011)
8. Ciftcioglu, E., Hardy, R., Chan, K., Scott, L., Oliveira, D., Verma, G.: Chaff allocation and performance for network traffic obfuscation. In: 2018 IEEE 38th International Conference on Distributed Computing Systems (ICDCS), pp. 1565–1568. IEEE (2018)
9. De Oliveira, R.L.S., Schweitzer, C.M., Shinoda, A.A., Prete, L.R.: Using Mininet for emulation and prototyping software-defined networks. In: 2014 IEEE Colombian Conference on Communications and Computing (COLCOM), pp. 1–6. IEEE (2014)
10. Duffield, N.G., Roughan, M., Sen, S., Spatscheck, O.: Statistical, signature-based approach to IP traffic classification (Feb 9 2010), US Patent 7,660,248
11. Dusi, M., Crotti, M., Gringoli, F., Salgarelli, L.: Tunnel hunter: Detecting application-layer tunnels with statistical fingerprinting. Computer Networks **53**(1), 81–97 (2009)
12. Dyer, K.P., Coull, S.E., Shrimpton, T.: Marionette: A programmable network traffic obfuscation system. In: 24th USENIX Security Symposium (USENIX Security 15), pp. 367–382. USENIX Association, Washington, D.C. (Aug 2015). https://www.usenix.org/conference/usenixsecurity15/technical-sessions/presentation/dyer
13. Feng, X., Zheng, Z., Mohapatra, P., Cansever, D.: A Stackelberg game and Markov modeling of moving target defense. In: International Conference on Decision and Game Theory for Security, pp. 315–335. Springer (2017)
14. Goodfellow, I.J., Shlens, J., Szegedy, C.: Explaining and harnessing adversarial examples. Preprint (2014). arXiv:1412.6572
15. Goodfellow, I., Papernot, N., McDaniel, P., Feinman, R., Faghri, F., Matyasko, A., Hambardzumyan, K., Juang, Y.L., Kurakin, A., Sheatsley, R., et al.: cleverhans v0. 1: an adversarial machine learning library. Preprint (2016). arXiv:1610.00768
16. Group-IB: Fox-IT: Anunak: Apt against financial institutions (2014)
17. Guan, Y., Fu, X., Xuan, D., Shenoy, P.U., Bettati, R., Zhao, W.: NetCamo: camouflaging network traffic for QoS-guaranteed mission critical applications. IEEE Trans. Syst. Man Cybern. A Syst. Hum. **31**(4), 253–265 (2001)
18. He, W., Wei, J., Chen, X., Carlini, N., Song, D.: Adversarial example defense: Ensembles of weak defenses are not strong. In: 11th {USENIX} Workshop on Offensive Technologies ({WOOT} 17) (2017)
19. Horák, K., Zhu, Q., Bošanskỳ, B.: Manipulating adversary's belief: A dynamic game approach to deception by design for proactive network security. In: International Conference on Decision and Game Theory for Security, pp. 273–294. Springer (2017)
20. Jero, S., Bu, X., Nita-Rotaru, C., Okhravi, H., Skowyra, R., Fahmy, S.: BEADS: Automated attack discovery in OpenFlow-based SDN systems. In: Proceedings of the International Symposium on Research in Attacks, Intrusions, and Defenses (RAID), vol. 10453, pp. 311–333. LNCS (2017)

21. Karthika, C., Sreedhar, M.: Statistical traffic pattern discovery system for wireless mobile networks. Comput. Sci. Telecomm. **45**(1), 63–70 (2015)
22. Kiekintveld, C., Lisý, V., Píbil, R.: Game-theoretic foundations for the strategic use of honeypots in network security. In: Cyber Warfare, pp. 81–101. Springer (2015)
23. Kim, H., Feamster, N.: Improving network management with software defined networking. IEEE Commun. Mag. **51**(2), 114–119 (2013). https://doi.org/10.1109/MCOM.2013.6461195
24. Kondo, T.S., Mselle, L.J.: Penetration testing with banner grabbers and packet sniffers. J. Emerg. Trends Comput. Inf. Sci. **5**(4), 321–327 (2014)
25. Kurakin, A., Goodfellow, I., Bengio, S.: Adversarial examples in the physical world. Preprint (2016). arXiv:1607.02533
26. Levin, D., Canini, M., Schmid, S., Schaffert, F., Feldmann, A.: Panopticon: Reaping the benefits of incremental SDN deployment in enterprise networks. In: 2014 USENIX Annual Technical Conference (USENIX ATC 14), pp. 333–345. USENIX Association, Philadelphia, PA (2014)
27. Matthews, T.: Operation Aurora – 2010's major breach by Chinese hackers (2019). https://www.exabeam.com/information-security/operation-aurora/
28. McKeown, N., Anderson, T., Balakrishnan, H., Parulkar, G., Peterson, L., Rexford, J., Shenker, S., Turner, J.: OpenFlow: Enabling innovation in campus networks. SIGCOMM Comput. Commun. Rev. **38**(2), 69–74 (2008). https://doi.org/10.1145/1355734.1355746
29. Miah, M.S., Gutierrez, M., Veliz, O., Thakoor, O., Kiekintveld, C.: Concealing cyber-decoys using two-sided feature deception games. In: Proceedings of the 53rd Hawaii International Conference on System Sciences (2020)
30. Mininet: Mininet an instant virtual network on your laptop (or other pc) (2018). https://mininet.org/
31. Mohajeri Moghaddam, H., Li, B., Derakhshani, M., Goldberg, I.: SkypeMorph: Protocol obfuscation for tor bridges. In: Proceedings of the 2012 ACM Conference on Computer and Communications Security, pp. 97–108 (2012)
32. Moore, A.W., Zuev, D.: Internet traffic classification using Bayesian analysis techniques. In: Proceedings of the 2005 ACM SIGMETRICS International Conference on Measurement and Modeling of Computer Systems, pp. 50–60 (2005)
33. Nguyen, T.T., Armitage, G.: A survey of techniques for internet traffic classification using machine learning. IEEE Commun. Surv. Tutor. **10**(4), 56–76 (2008)
34. Ortiz, A., Fuentes, O., Rosario, D., Kiekintveld, C.: On the defense against adversarial examples beyond the visible spectrum. In: MILCOM 2018-2018 IEEE Military Communications Conference (MILCOM), pp. 1–5. IEEE (2018)
35. Ortiz, A., Granados, A., Fuentes, O., Kiekintveld, C., Rosario, D., Bell, Z.: Integrated learning and feature selection for deep neural networks in multispectral images. In: Proceedings of the IEEE Conference on Computer Vision and Pattern Recognition Workshops, pp. 1196–1205 (2018)
36. Papernot, N., McDaniel, P., Goodfellow, I.: Transferability in machine learning: from phenomena to black-box attacks using adversarial samples. Preprint (2016). arXiv:1605.07277
37. Papernot, N., McDaniel, P., Jha, S., Fredrikson, M., Celik, Z.B., Swami, A.: The limitations of deep learning in adversarial settings. In: 2016 IEEE European Symposium on Security and Privacy (EuroS&P), pp. 372–387. IEEE (2016)
38. Pawlick, J., Colbert, E., Zhu, Q.: A game-theoretic taxonomy and survey of defensive deception for cybersecurity and privacy. ACM Comput. Surv. (CSUR) **52**(4), 82 (2019)
39. Píbil, R., Lisý, V., Kiekintveld, C., Bošanský, B., Pěchouček, M.: Game theoretic model of strategic honeypot selection in computer networks. In: International Conference on Decision and Game Theory for Security, pp. 201–220. Springer (2012)
40. Pinheiro, A.J., Bezerra, J.M., Campelo, D.R.: Packet padding for improving privacy in consumer IoT. In: 2018 IEEE Symposium on Computers and Communications (ISCC), pp. 00925–00929 (2018)

41. Roughan, M., Sen, S., Spatscheck, O., Duffield, N.: Class-of-service mapping for QoS: a statistical signature-based approach to IP traffic classification. In: Proceedings of the 4th ACM SIGCOMM Conference on Internet Measurement, pp. 135–148 (2004)
42. Schlenker, A., Thakoor, O., Xu, H., Fang, F., Tambe, M., Tran-Thanh, L., Vayanos, P., Vorobeychik, Y.: Deceiving cyber adversaries: A game theoretic approach. In: Proceedings of the 17th International Conference on Autonomous Agents and MultiAgent Systems, pp. 892–900. International Foundation for Autonomous Agents and Multiagent Systems (2018)
43. Szegedy, C., Zaremba, W., Sutskever, I., Bruna, J., Erhan, D., Goodfellow, I., Fergus, R.: Intriguing properties of neural networks. Preprint (2013). arXiv:1312.6199
44. Tambe, M.: Security and Game Theory: Algorithms, Deployed Systems, Lessons Learned. Cambridge University Press (2011)
45. Thakoor, O., Tambe, M., Vayanos, P., Xu, H., Kiekintveld, C.: General-sum cyber deception games under partial attacker valuation information. In: AAMAS, pp. 2215–2217 (2019)
46. Verma, G., Ciftcioglu, E., Sheatsley, R., Chan, K., Scott, L.: Network traffic obfuscation: An adversarial machine learning approach. In: MILCOM 2018-2018 IEEE Military Communications Conference (MILCOM), pp. 1–6. IEEE (2018)
47. Wagener, G., Dulaunoy, A., Engel, T., et al.: Self adaptive high interaction honeypots driven by game theory. In: Symposium on Self-Stabilizing Systems, pp. 741–755. Springer (2009)
48. Wright, C.V., Coull, S.E., Monrose, F.: Traffic morphing: An efficient defense against statistical traffic analysis. In: NDSS, vol. 9. Citeseer (2009)
49. Yin, Y., An, B., Vorobeychik, Y., Zhuang, J.: Optimal deceptive strategies in security games: A preliminary study. In: Proc. of AAAI (2013)
50. Zander, S., Nguyen, T., Armitage, G.: Self-learning IP traffic classification based on statistical flow characteristics. In: International Workshop on Passive and Active Network Measurement, pp. 325–328. Springer (2005)
51. Zhu, M., Anwar, A.H., Wan, Z., Cho, J.H., Kamhoua, C., Singh, M.P.: A survey of defensive deception: Approaches using game theory and machine learning. IEEE Commun. Surv. Tutor. (COMST) **23**(3), 1–35 (2021). https://doi.org/10.1109/COMST.2021.3102874

Mee: Adaptive Honeyfile System for Insider Attacker Detection

Mu Zhu and Munindar P. Singh

1 Introduction

An advanced persistent threat (APT) is a challenging form of cyberattack in which an attacker carries out long-term plan to become an inside attackers. Chen et al. [9] define six stages of an APT attack: (1) reconnaissance, (2) delivery, (3) initial intrusion, (4) command and control (C2), (5) lateral movement, and (6) data exfiltration. In the first three stages, an attacker gathers user information, such as accounts and passwords via phishing, and creates backdoors in a compromised device. Then, the attacker can remotely control and access the victim's device without being detected by traditional cybersecurity technologies, such as Intrusion Detection System (IDS) and firewall. Recent research focuses on blocking an APT [17, 22, 23]. In contrast, our research focuses on the detection of insider attacks.

Following Salem et al. [21], insider threats are divided into two types: *traitors*, who misuse their legitimate credentials, and *masqueraders*, who impersonate a legitimate user (and, generally, know less than a traitor about where the victim's valuable information resides). The masquerader-type insider attack is difficult to detect as the attacker leverages legal authentications to intrude devices and harvest sensitive data. As a result, traditional defensive approaches are insufficient to detect and respond to insider attacks. This work uses defensive deception against masquerader-type insider attacks to fill this gap.

Defensive deception [15] is an effective method to mitigate and detect attacks, such as reconnaissance [2, 5] and insider attacks [7, 8]. Where traditional cybersecurity focuses on attacker actions, defensive deception focuses on anticipating such

M. Zhu · M. P. Singh (✉)
Department of Computer Science, North Carolina State University, Raleigh, NC, USA
e-mail: mzhu5@ncsu.edu; mpsingh@ncsu.edu

T. Bao et al. (eds.), *Cyber Deception*, Advances in Information Security 89,
https://doi.org/10.1007/978-3-031-16613-6_7

actions [3]. Deception has two goals: *confusion*, i.e., wasting the adversary's effort and hiding sensitive information, and *detection*, i.e., identifying malicious actions.

Honeyfiles, or decoy documents, are a lightweight defensive deception technology [8, 24, 29] comprising two major tasks: (1) generate content in a honeyfile [1, 26, 27] and (2) decide number and placement of honeyfiles [8, 11, 24]. Bowen et al. [8] propose the *Decoy Document Distributor* (D^3) *system*, which generates and places decoy documents in a file system. Salem and Stolfo [20] analyze how legitimate users and adversaries are affected by the number and location of honeyfiles. Raising the number of honeyfiles on a device increases the probability of the adversary being confused and detected but at the cost of resource consumption and increasing the false-positive rate of detecting an attacker by confusing legitimate users.

How can a defender decide how many honeyfiles to place and where to place them to increase the effectiveness of detecting and disrupting attackers while reducing the impact on regular users? We introduce a honeyfile approach named Mee geared toward a large-scale enterprise network. Mee demonstrates **decentralized deployment**, that is, by any user. It also demonstrates **centralized control**, that is, the defender analyzes suspicious behavior across the network to determine the number and placement of honeyfiles for each device. This approach helps a defender accomplish the following:

- Detect an attack in progress.
- Detect a compromised device because an attacker who enters a device's filespace needs to explore it to locate valuable data and is likelier to touch a honeyfile than legitimate users who are familiar with their file systems.

We consider the attacker a masquerader-type adversary who can penetrate devices but does not identify the locations of valuable files on that device. Thus, the attacker needs to explore the victim's device and search for valuable files. Assume that the attacker is aware of the existence of the honeyfile system but cannot distinguish honeyfiles from regular files.

Our contributions include the following:

- We propose Mee as a novel honeyfile system for detecting insider attackers in enterprise networks that dynamically adjusts the number of honeyfiles placed on each device.
- We describe a Bayesian game model to analyze the optimal strategies for the attacker and the defender.
- We simulate and compare Mee with the traditional honeyfile approach and show that Mee is more effective at detecting insider attackers and has a more negligible impact on legitimate users.

Organization

The rest of this chapter is organized as follows. Section 3 describes our problem in terms of details about insider attacker and the threat model. Section 4 introduces the design of our honeyfile system. Section 5 describes our scenario and model design.

Section 7 compares Mee with the traditional honeyfile system. Section 8 describes the conclusions and our future directions.

2 Related Work

Yuill et al. [28] define defensive deception as the defender plan actions to mislead the attacker to take or not take specific action for improving the system security. Almeshekah and Spafford [4] improve the concept of cyber deception by considering *confusion* and refining the definition as "planned actions taken to mislead or confuse attackers and thereby cause them to take (or not take) specific actions that aid computer-security defenses." Pawlick et al. [19] and Zhu et al. [30] collect and introduce current deception technologies. The authors investigate the game theory and machine-learning-based deception research and analyze their (dis)advantages. Zhu et al. [31] design a deceptive network flow via the generative adversarial network (GAN) to disturb the attacker's reconnaissance.

The honeyfile system is an intrusion detection mechanism based on deception. Specifically, honeyfiles are decoy or deceptive files intended to lure the attackers' access. Yuill et al. [29] design a file server with a honeyfile system and show how honeyfiles confuse and detect threats. Gómez-Hernández et al. [11] leverage honeyfile-based security system, named R-lock, to against ransomware. Ben Whitham [27] design a high-interaction system that can analyze the selected document and generate a corresponding honeyfile to mimic the selected document.

Game theory applies well in defensive deception to assist the players (e.g., attackers or defenders) in searching for the optimal solution. Pawlick et al. [18] use signaling games to model interactions between the defender and APT attackers. Wan et al. [25] and Cho et al. [10] use a hypergame model to show how differences in perceptions of the players affect decision-making. Anwar and Zhu et al. [6] design a hypergame model to help the defender select high-interaction honeypots or low-interaction honeypots against network reconnaissance. Although much deception research applies game theory to assist defenders in planning strategies, most of them only consider a defender and an attacker as the players. This work leverages the Bayesian game to model the interaction between the attacker, the defender, and the legitimate users. We aim to search for the optimal solution that confuses the attacker more and disturbs the users less.

The Bayesian game is a game model in that all players do not know their opponents' types, actions, or payoffs, and each player knows its type, a set of actions, and corresponding payoffs. Hence, each player has a subjective prior probability distribution of their opponent's type [12]. La et al. [14] proposed a two-player attacker-defender Bayesian game in an IoT network considering honeypot as the deception method. Mao et al. [16] used honeypots in a non-cooperative Bayesian game with imperfect, incomplete information where an attacker is a leader and a defender is a follower. Huang and Zhu [13] discussed several deception techniques, such as honeypots and fake personal profiles. In addition, the authors propose a static Bayesian game to detect the stealthy and deceptive characteristics of an attacker.

3 Problem Statement

We focus on masquerader-type insider attackers in an enterprise network environment. The attacker can gather authorization of device owners and search for the valuable files in target devices without triggering traditional defensive technologies, such as IDS. By considering deployed honeyfiles, the attacker may have three results after penetrating a device:

Success: Viewing or transferring valuable files from a device.

Invalid: Not finding a valuable file, i.e., suffering wasted effort but no additional loss.

Defeat: Triggering alarms by touching honeyfiles. The defender would update its belief about the network's security level while cleaning or replacing the compromised device, because of which the attacker would lose a compromised device and have wasted its effort.

An adversary does reconnaissance before attacking to investigate whether a device contains valuable files to save energy. Thus, an insider attacker has clear intent regarding what information is valuable and targets devices accordingly. For example, an attacker who sought a final exam in a university would target several specific professors in a particular department, not a random device. Therefore, we name the device owner's identity, such as professor, student, and CEO, as *organizational role*. One insider attacker may pay more attention to several specific organizational roles.

4 Design of Mee System

The use of honeyfile has two purposes: to confuse the attacker with false information and detect any unauthorized access to connected devices. The number of honeyfiles in one device can significantly change the effectiveness of a honeyfile system. However, current honeyfile research mainly focuses on the honeyfile system in a single device, which cannot provide adequate observation to assist the defender in estimating the security situation and changing the number of honeyfiles. In contrast, we consider all devices in an entire enterprise network. Specifically, we have a central controller, named *Mee controller*, that receives information (e.g., about accesses to honeyfiles) from and instructs clients, named *Mee clients*, which reside on individual devices. Following instructions from the Mee controller, Mee client increases or decreases the number of honeyfiles.

4.1 Mee Client Design

The Mee client is an application endpoint that resides on each device. A Mee client generates and deletes honeyfiles and detects suspicious behaviors on a honeyfile that include opening, modifying, deleting, and transferring. When anyone acts on a honeyfile, the corresponding Mee client sends the alarm to the Mee controller and receives the controller's instructions.

Honeyfiles need to be placed and named well to avoid confusing legitimate users and attract attention from the adversary. Salem and Stolfo [20] empirically studied decoy documents. They invited 52 students to download and install decoy documents and monitored the students' behaviors to figure out the suitable placement and names of decoy documents that minimize false positives. Although we mainly focus on the number of honeyfiles in each device, we incorporate some current honeyfile research, such as which honeyfile names the attacker interests more. Thus, we assign a *sensitivity*, $H_s \in [1, 3]$, to each honeyfile. A higher value represents a greater attraction for both the adversary and a legitimate user.

Legitimate users may touch honeyfile by accident. However, even if the users accidentally act on, such as open or read, honeyfiles, they have a lower chance to edit or transfer honeyfiles. To capture this feature, we assign *seriousness* of action on a honeyfile to reflect how much of a security threat the action can be. We consider two levels of seriousness:

Weak: Open or close a honeyfile.
Strong: Edit, transfer, or apply tools such as `zip` and `tar`.

When anyone (e.g., the user or attacker) touches honeyfiles, the Mee client collects corresponding sensitivity and seriousness values and sends them to the Mee controller.

4.2 Mee Controller Design

The Mee controller represents the defender across the network. It receives alarms from Mee clients and analyzes them to update its beliefs about the security level of each device. Then, it instructs each client to create or delete a specified number of honeyfiles on its device. Upon receiving an alarm, the controller determines whether the described access is by a legitimate user or an attacker. We define HN_j as the number of honeyfiles on device j. This number is adjusted based on the defender's beliefs about how secure a device is and the attacker's goals. To assist the Mee controller in making the optimal decision, we describe the following measures of the alarm analysis to assess network security situations.

Device Group and Group Risk Level
We assume that the insider attacker targets files placed on devices belonging to specific organizational roles, which we model as *groups*. For example, in a univer-

sity, we may have groups for professors, students, and accountants. Meanwhile, one device can be included in multiple groups. For example, one device may belong to groups "professor" ($G_{\text{professor}}$) and "engineering" ($G_{\text{engineering}}$). Below, G_i is the set of devices in the group i, and HG_i denotes the number of honeyfiles in the group i, overloading the notation for the number of honeyfiles in device d. We use R_i to represent the *risk level* of group i. Higher R_i represents a higher probability of being the target group of an attacker. If someone touches a honeyfile on a device in group i, the defender raises the corresponding R_i.

Group Risk Level Update

Recall the sensitivity of a honeyfile and the seriousness of an action from Sect. 4.1. Equation 1 computes the change of risk level of group g due to a single action a on a specific honeyfile, h.

We define the group risk update ($\Delta risk_g(h, a)$) as the product of the honeyfile sensitivity and the action seriousness divided by the number of honeyfiles in the group.

$$\Delta \text{risk}_g(h, a) = \frac{\text{sensitivity}_h * \text{seriousness}_a}{HG_g}. \tag{1}$$

Through the group risk level updating, the Mee controller evaluates the security situation for each group. To compare a group's security situation to the rest of the network, we introduce R_{-i} with a negative in the subscript to denote the average group risk level of all groups except group i. Equation 2 shows how we calculate R_{-i}, where NG denotes the number of groups in a network.

$$R_{-i} = \frac{\sum_{j \neq i} R_j}{NG - 1}. \tag{2}$$

Based on R_{-i} and R_i, we use the classification as *Dangerous*, *Medium*, and *Safe* to separate groups (as shown in Equation 3).

$$\textit{Classification} = \begin{cases} \text{Dangerous if } R_i > R_{-i} * 2 \\ \text{Medium if } R_{-i} < R_i < R_{-i} * 2 \\ \text{Safe if } R_i < R_{-i}. \end{cases} \tag{3}$$

Via the group risk level, the Mee controller can evaluate the security situation of all the groups. If the insider attacker tends to compromise the device based on the roles of the device owner, the Mee controller increases the number of honeyfiles in each device whose group obtains a high group risk level.

4.3 Communication Between Mee Client and Controller

As we discussed above, Mee clients and controller exchange messages by which the controller evaluates the network security situation and asks Mee clients to adjust the number of honeyfiles in the corresponding device. Therefore, we name *Honeyfile Alarm* as the messages that transfer from all Mee clients to the controller. Each honeyfile alarm contains a tuple $\langle CN, HN, AS\rangle$, where CN represents the client name, HN represents the honeyfile that is acted on, and AS represents the action upon the triggered honeyfile. In contrast, *Command* is the message that is from Mee controller to each Mee client, which contains a tuple as $\langle CN, I\rangle$. Here, CN represents the target host of the command, and I represents the instructions that the Mee client needs to follow with.

The Mee system is shown in Fig. 1. If anyone acts on a honeyfile, the Mee client sends honeyfile alarm to Mee controller. The alarm contains information that includes the location of the triggered honeyfile, corresponding honeyfile sensitivity, and action seriousness, which are used to update the belief of the Mee controller, such as each group's risk level. All alarms are stored in Mee controller as well. Based on the group risk and alarm history, the Mee controller sends the command to Mee clients to adjust the number of honeyfiles in the device or check the device if necessary.

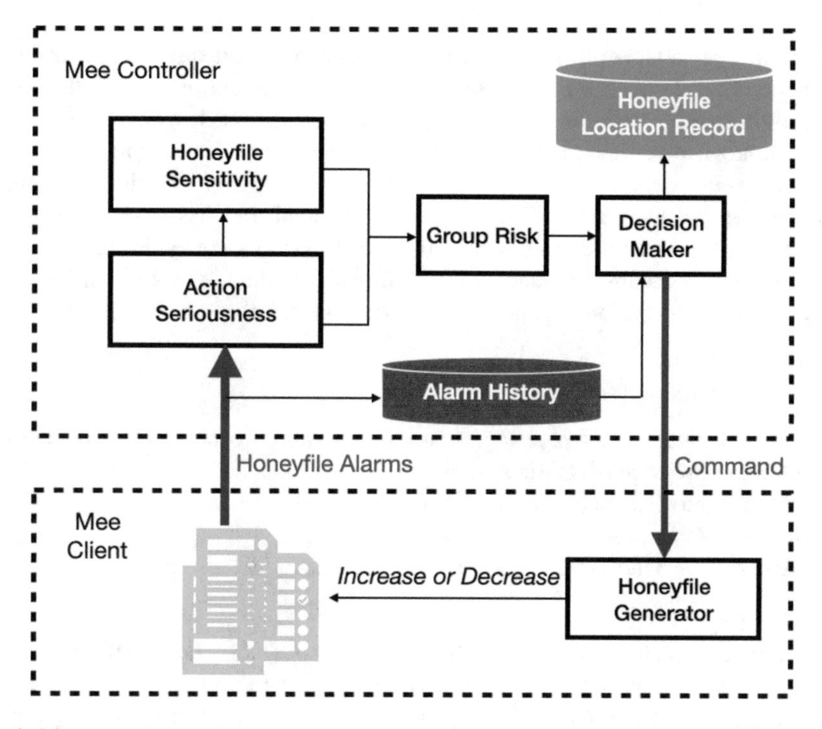

Fig. 1 Mee system structure

5 Scenario and Model

In this section, we describe the model of insider attackers, the defender, and users, and also the interaction between them.

5.1 Network and Node Model

An enterprise network environment includes connected end-user devices, such as computers and laptops. As introduced in Sect. 4.2, these devices are separated into groups based on their respective owners' roles. Each group's group risk level indicates its security situation.

All connected personal devices (e.g., laptop) have an installed Mee client, which can generate or delete honeyfile. Let NF_i denote the number of honeyfiles in device i. Following the concept of file sensitivity, each created honeyfile has a value to represent the attraction for attackers and regular users.

5.2 Attacker Model

We assume an attacker conducts reconnaissance before it attempts to penetrate a device. Specifically, the adversary gathers full perception of all the users' basic information, including their organizational roles (such as CEO, professor, administrator), to guide it to select target devices. We also assume an insider attacker can access and explore a victim device without triggering traditional defensive technologies, such as IDSs. The attacker is rational and has knowledge of the existence of the honeyfile system but cannot distinguish a honeyfile from a regular file. However, as the insider attacker is unfamiliar with the file location in the victim device, it needs to explore a penetrated device to search for valuable files.

Attacker Actions
We consider the following action set:

Penetrate Device: Leverage legitimate authorization to penetrate a device (but without knowledge of its file system).
Search: Explore device i to search for valuable files.
File Read: Open, read, or close a file in device i.
File Transfer or Modify: Transfer, edit, or delete a file.

Attacker Payoff
We separate the attacker payoff into *Effectiveness (EA):* the benefit which the attacker gets from an action; *Action Cost (AC):* cost to attacker to deploy action i; and *Impact of Failure (IF):* cost to attacker if the attacker touches a honeyfile.

5.3 *Defender Model*

Mee combines decentralized deployment of honeyfiles with centralized control to adjust the number of honeyfiles in each connected device. Specifically, the Mee controller monitors the devices and decides how many honeyfiles to place on each device. The defender's goal is to maintain group risk levels and detect a compromised device.

Defender Actions
We design four actions for the defender:

Check: Inform a device's Mee client to check for existing backdoors or update OS and application to avoid vulnerabilities. Upon doing so, set the number of honeyfiles on the device resets to the initial value and the group risk level value to the R_{-i} (defined in Sect. 4.2) value.
Increase the number of honeyfiles on a device.
Decrease the number of honeyfiles on a device.
No change: Maintain current strategy and save resources.

Note that in Mee, the defender chooses an action based on alarms from deployed honeyfiles as well as the network security situation. For example, if the defender receives an alarm from a device in the dangerous group, the defender has a higher probability to choose the check action.

Defender Payoff
The defender's payoff has four parts:

Effectiveness (ED) Reward when a malicious action is detected, e.g., by recovering the compromised device and misleading the attacker with honeyfile.
Defense Cost (DC) Cost of deploying an action.
Failure in Protecting Real File (FR) Punishment upon failing in protecting real files.
Impact to Legitimate User (IN) False positive if a user accidentally opens, closes, transfers, or modifies a honeyfile.

5.4 *Model of Legitimate User*

We model the behaviors of legitimate users to capture Mee's impact on them. Therefore, we design the user action set as:

Login: Access a device.
Search: Explore a device, e.g., open a folder and search for a file.
Read a file.
Transfer, modify, or delete a file.

No matter what organizational roles are, the users may access their devices and read or modify their target files at any time. Therefore, given our threat and system models, there are two differences between insider attackers and legitimate users.

First, the login behaviors from regular users across the network are relatively randomly distributed. As a result, if there is no insider attacker in the environment, all devices have the same chance to be logged in by their owner. In contrast, the insider attacker would compromise the device based on the roles of the device owner. Second, users can quickly locate a target file as they are familiar with the file system, but a masquerader, lacking such knowledge, would have a relatively random movement to search for valuable files. Therefore, an insider attack has a greater chance to act on a honeyfile than a legitimate user.

6 Honeyfile Game with Mee

We consider the honeyfile game as a two-player dynamic Bayesian game. The two players update their beliefs according to the game's evolution. One player, named by *player a*, is the defender who can deploy honeyfiles within connected devices to detect insider attackers. Another player, denoted by *player b*, has two potential types: insider attacker and regular user. Although the type of the defender is the common knowledge to the players, the defender does not know another player's type. However, it can generate its belief by observing honeyfile alarms from each device. Let $t \in T = [0, 1]$ represent the player b's type, where $t = 0$ represents a legitimate user and $t = 1$ represents an attacker.

In the defender's belief, we use $\sigma_a = p$ to denote the probability that player b is an attacker ($t = 1$) and $\sigma_u = 1 - p$ to denote the probability that player b is a legitimate user ($t = 0$). An alarm from a Mee client represents an observation from player b. The defender has no knowledge of which device is compromised. Therefore, the honeyfile alarms may represent the detection of an attacker or false alarms from legitimate users. To make an optimal decision, the defender needs to estimate player b's type based on its belief, including the corresponding group risk level and the history of previous alarms. With the definition of action seriousness and the two players' action sets, Fig. 2 explains the decision tree between the two players.

Utility Function

Let $w \in [1, 3]$ denote the worth of a regular file, where a higher number represents a more valuable file. For simplicity, we stipulate that an insider attacker obtains a gain of w if it reads a file and obtains $2 * w$ if it transfers or modifies a file. Let c_c denote the cost of compromising a device, c_r represent the cost of reading a file, and c_t represent the cost of transferring or modifying a file. Recall the definition of honeyfile sensitivity and action seriousness. We use h_p to denote the punishment of an attacker when it acts on a honeyfile. For simplicity, we stipulate that the punishment of reading a honeyfile is h_p, and the punishment of transferring or

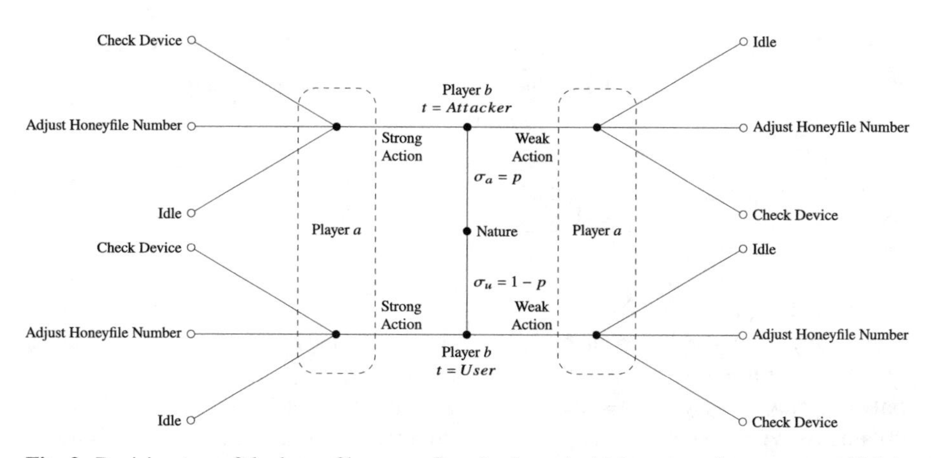

Fig. 2 Decision tree of the honeyfile game: *Step 1:* player b obtains a type from nature, which is the root of the decision tree, as its private information. *Step 2:* After player b triggering a honeyfile alarm, if player b is an attacker, the path of decision tree goes to the upside. On the other hand, the path goes down. *Step 3:* The defender (player a) chooses its action based on the perspective of player b's type

Table 1 Player b is an insider attacker. The tuples below include ⟨defender's payoff, attacker's payoff⟩

	Read a file	Transfer or modify a file
Check device	$\alpha * r_{cd} - c_{cd}, \beta w - (1-\beta)h_p - r_{cd} - c_r$	$\alpha * r_{cd} - c_{cd}, 2\beta w - (1-\beta)h_p - r_{cd} - c_t$
Increase honeyfile	$-c_a, \beta w - (1-\beta)h_f - c_r$	$-c_a, 2\beta w - 2(1-\beta)h_f - c_t$
Decrease honeyfile	$-c_a, \beta w - (1-\beta)h_f - c_r$	$-c_a, 2\beta w - 2(1-\beta)h_f - c_t$
No change	$0, \beta w - (1-\beta)h_f - c_r$	$0, 2\beta w - 2(1-\beta)h_f - c_t$

modifying a honeyfile is $2 * h_p$. Let $\beta \in [0, 1]$ represent the probability that an insider attacker estimates an actual file. The attacker calculates β based on its history, such as how many honeyfiles and actual files it touches.

We use c_a to represent the cost when the defender adjusts (e.g., increases and decreases) the number of honeyfiles. Let c_{cd} represent the cost of checking device action and r_{cd} represent the reward if the defender successfully recovers a compromised device. Meanwhile, $-r_{cd}$ represents the loss of an insider attacker if the defender recovers a compromised device. Let α represent the probability that the defender thinks the device is compromised.

Table 1 shows the expected payoff when an insider attacker acts on a file. Note that the payoff in Table 1 does not include the cost of compromising a device. As a rational player, an insider attacker expects that the reward from real files in a device is larger than the cost of compromising the device. In other words, if NR represents a set of real files that an attacker acts on (e.g., reads and transfers), it expects $c_c < \sum_{nr \in NR}(w_{nr} - c_{r(nr)}) + \sum_{nr \in NR}(2w_{nr} - c_{t(nr)})$. Table 2 shows the payoff of the defender when player b is a legitimate user.

Table 2 Player b is a legitimate user. The tuples below include ⟨defender's cost, user's cost ⟩

	Read a file	Transfer or modify a file
Check device	$\alpha * r_{cd} - c_{cd}, 0$	$\alpha * r_{cd} - c_{cd}, 0$
Increase honeyfile	$-c_a, 0$	$-c_a, 0$
Decrease honeyfile	$-c_a, 0$	$-c_a, 0$
No change	$0, 0$	$0, 0$

Dynamic Game

The honeyfile game is repeatedly played between player a and player b. Let k_t represent the timeline of one game, where $t = 0, 1, 2, \ldots$. Here, at any one time, only one player b (e.g., an insider attacker or a legitimate user) is active. Player b first obtains its type from nature as private information. Upon one player b finishing its objective, such as an attacker gathering enough valuable files or a user obtaining the target file, this player b exits the game, and another player b comes in to participate. The player a (e.g., the defender) keeps accumulating its beliefs throughout the game. Specifically, the defender updates its belief about the type of player b after receiving honeyfile alarms. It chooses actions based on all the gathered information, such as group risk levels and honeyfile locations, from the beginning of the game.

7 Implementation and Evaluation

We adopt these metrics to quantify performance:

Attacker payoff ($\langle EA, AC, IF \rangle$) We evaluate the attacker's payoff based on the definition in Sect. 5.2.

Defender payoff ($\langle ED, DC, FR, IN \rangle$) The defender's payoff is calculated using the definition in Sect. 5.3.

Accuracy measures: To compare the accuracy of Mee with the traditional honeyfile system, we calculate false-positive rate (FPR), true-positive rate (TPR), and the area under receiver operating characteristic (ROC) curve.

7.1 Simulation Settings

We compare Mee with the traditional honeyfile system via a simulation. In the testbed, we deploy 114 devices and separate them into 20 groups. The installed Mee client in each connected device can create or remove honeyfiles following the Mee controller's command. Meanwhile, the Mee controller can estimate the network security situation via receiving honeyfile alarms from Mee clients.

Adversary Setting

The action set of an insider attacker is {*Penetrate*, *Search*, *Read*, *Transfer*} (as introduced in Sect. 5.2). The insider attacker selects several groups as its *target groups* and has a higher probability of penetrating the devices within the target groups rather than randomly selecting a device to attack. Specifically, in all the following tests, an attacker has a ten percent probability of randomly choosing a target device to compromise (without considering the target group) and a ninety percent probability of selecting a device in the target group(s).

Assume that an insider attacker can penetrate any device across the network and explore the compromised machine to search for valuable files. However, not every compromised device contains a valuable file. In each simulation, the attacker starts from an initial budget, representing the expected cost for searching and gathering the valuable files on every device. Suppose the cost of searching for valuable files within a compromised device is more than the initial budget. The attacker can choose to abandon the current device and penetrate another device instead.

User Setting

To calculate the defender's cost and false-positive rate of the honeyfile alarm, we model the behaviors of legitimate users in our simulation. A user has an action set [*Login*, *Search*, *Read File*, *Transfer or Modify File*]. Besides, the user obtains a full map of their file systems, which can assist the legitimate user in accessing a target file faster than the attacker. However, every user has a ten percent probability of choosing an incorrect action or action target. Thus, the user may act on a honeyfile and generate a false alarm.

Although the last three actions of the user model are the same as the corresponding actions in the attacker model, we emphasize the differences between legitimate users and the adversary. At first, each legitimate user has a clear target file and has complete knowledge of the file system. Thus, the user can locate any file in the device without a random search. And then, users access the corresponding devices at any time, so the distribution of login behaviors across the network is random. In contrast, the insider attacker tends to perform the compromise action relying on its target group(s) but randomly explores the file system.

Defender Setting

We simulate both the traditional honeyfile system, in which the number of honeyfiles in each device is static, and Mee, in which dynamic adjusts the number of honeyfiles in each device. The defender has an action set [*Check Device*, *No Change*] for the traditional honeyfile system, and an action set [*Check Device*, *No Change*, *increase honeyfile decrease honeyfile*] for Mee (as introduced in Sect. 5.3).

Honeyfile Generation

Following Gómez-Hernández et al. [11], we monitor honeyfiles using `inotify -tools`, which is a library and a set of command-line programs for Linux that monitors and acts upon file system events. Specifically, our client uses `inotify-tools` to detect actions, such as open, close, and modify, on a honeyfile. Then, the Mee client sends such information to the Mee controller.

For simplicity, we consider only `txt` file in our research, i.e., files with the `txt` extension. For the traditional honeyfile system simulation, we deploy the same number of honeyfiles in each device. Then, we change the number of honeyfiles in each device and observe the defender's performance, the attacker's cost, and the false/true-positive rate. For Mee, we deploy 40 honeyfiles in each device as the initial statement. The defender can modify the number of honeyfiles based on its belief of network security situation.

7.2 *Comparing Mee with the Traditional Honeyfile System*

A traditional honeyfile system does not adjust the number of honeyfiles in each device and maintain group risk levels. To simulate a traditional honeyfile system, we deploy a fixed number of honeyfiles in each connected device. As discussed in Sect. 7.1, we separate the devices into 20 groups and posit that the insider attacker prefers to attack Groups 3 (G3) and 4 (G4). We evaluate the performance of the traditional honeyfile system and Mee based on various numbers of honeyfiles and the group risk level update.

Adjusting the Number of Honeyfiles with Mee

Mee changes the number of honeyfiles in each connected device to reduce unnecessary overhead and the impact on legitimate users. Figure 3a records the variance of the number of honeyfiles in each group. We assign ten insider attackers and 200 legitimate users in this test. Because the attacker's target groups are G3 and G4, Mee automatically adjusts and deploys more honeyfiles in these two groups than others. The average numbers of honeyfiles in G3 and G4 are 29 and 30, whereas the average number of honeyfiles in other groups is 18.

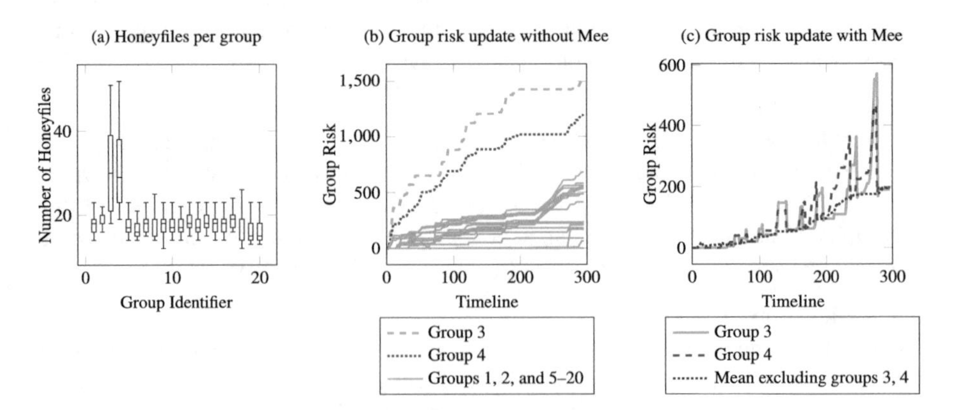

Fig. 3 Mee's performance: (**a**) numbers of honeyfiles in different groups; (**b**) group risk updates without Mee; (**c**) group risk updates with Mee

Group Risk Level of Traditional Honeyfile System and Mee
Figure 3b shows the result of group risk level update with the traditional honeyfile system. The X-axis represents the timeline in one test, and the Y-axis represents the group risk level. Each player joins in and performs as a legitimate user or an insider attacker in each time slot. The group risk levels of G3 and G4 (dotted red and dashed green lines) are much higher than other groups (solid gray lines), representing the tendency of the attacker's movement to be captured via the group risk level updating.

With the same setting, Mee controller receives and analyzes honeyfile alarms from Mee clients to maintain the group risk level. Figure 3c shows the group risk level updating with Mee. Anyone who triggers the honeyfile alarm makes group risk levels rise. Suppose Mee controller confirms that a group has an abnormal (e.g., much higher than other groups) group risk level. It can apply *check* action and reset the corresponding group risk level to R_{-i} (introduced in Sect. 5.3). In Fig. 3c, the red dash line and solid green line represent the group risk levels of G3 and G4 (e.g., the attacker's target groups). We use the blue dotted line for the rest of the group's risk levels to denote their average value.

Defender and Attacker Payoffs with Traditional Honeyfiles and with Mee
We then calculate the defender's payoff by considering ten insider attackers and 200 legitimate users. The players' payoff calculations follow the setting of Sect. 5. For the traditional honeyfile system, we increase the number of honeyfiles in each device from 0 to 100. As shown in Fig. 4a, the defender payoff (blue dashed line) keeps increasing before the number of honeyfiles in each device is less than forty and decreasing later. The results show that growing the number of honeyfiles can assist the defender in increasing its payoff because the defender can detect the insider attackers more effectively. But the redundancy of honeyfiles may disrupt legitimate users—and disrupt them more and more as the number of honeyfiles is increased, thereby generating more false-positive alarms that only confuse the defender to check a safe device. Note that Mee system can automatically adjust the number of honeyfiles in the system. The straight (orange) line represents the defender payoff when we deploy Mee with the same attacker and user settings.

With the same setting as above, Fig. 4b records the attacker's payoff given a certain number of honeyfiles in the environment. The X-axis represents the number of honeyfiles in each device, and the Y-axis represents the attacker's payoff. With a traditional honeyfile system, more honeyfiles can significantly reduce the attacker's payoff. However, compared with Fig. 4a, the overhead of honeyfiles also disturbs the defender's performance. Also, the straight orange line represents the attacker payoff when we apply Mee.

TPR and FPR with the Traditional Honeyfile System and with Mee
We calculate the false-positive rate (FPR), true-positive rate (TPR), and the area under the ROC curve for the traditional honeyfile system and Mee. The area under the ROC curve is defined as $ROCarea = TPR * (1 - FPR)$.

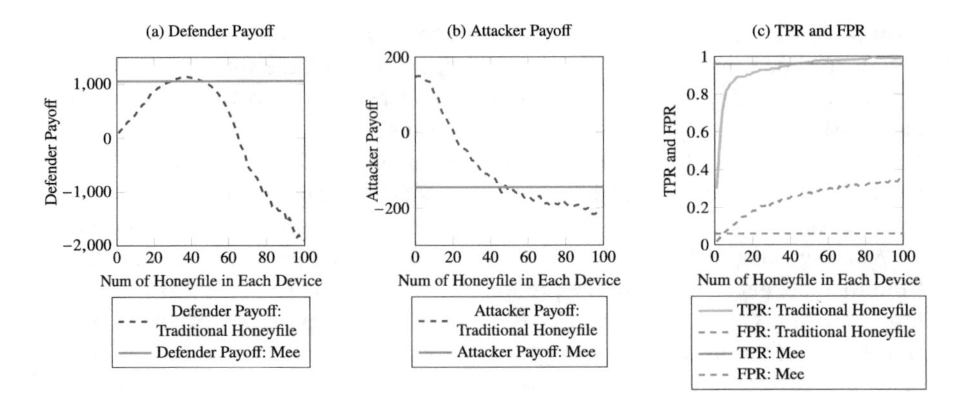

Fig. 4 Comparison between the traditional honeyfile system and Mee

We consider positive, negative, true positive, false positive as metrics for the defender:

Positive: Defender detects that an insider attacker is present.
Negative: Believes that the honeyfile alarm is triggered by a legitimate user.
True positive: Detection of the insider attacker rather than a legitimate user.
False positive: Misunderstanding of honeyfile alarm triggered by a user.

With the same test settings as above, Fig. 4c shows the TPR and FPR of the traditional honeyfile system and Mee. The solid green line and dashed brown line show the TPR and FPT of the traditional honeyfile system, while the orange solid and dashed lines represent Mee. With the traditional honeyfile system, increasing the number of honeyfiles in each device can significantly raise the TPR but also brings the high FPR. Mee maintains the TPR at a high level and reduces the FPR.

We then test the performance of the traditional honeyfile system and Mee with the different number of insider attackers. Following the definition of action seriousness in Sect. 4.1, with the different number of actions on honeyfile, we define several situations for insider attacker detection with a traditional honeyfile system. Figure 5 shows the ROC obtained with different definitions of detection. For example, the red marks represent the defender detection is triggered if anyone acts on the honeyfile in one device with at least two weak actions or one strong action. Within each detection definition, we increase the number of insider attackers in one test from 1 to 100. As shown in Fig. 5, when weak action=2 and strong action=1, the ROC value concentrates within high TPR and FPR, which indicates that the detection is extra sensitive. Even if a traditional approach detects insider attackers, it generates many false-positive alarms that confuse the defender and impact legitimate users. Meanwhile, increasing the number of attackers does not significantly influence the performance of the traditional honeyfile system. When weak action=3 and strong action=2 (shown as green marks in Fig. 5), although the FPR reduces, the TPR is lower than that when weak action=2 and strong action=1, which represents that a large number of insider attackers escape detection. Orange marks in Fig. 5 represent

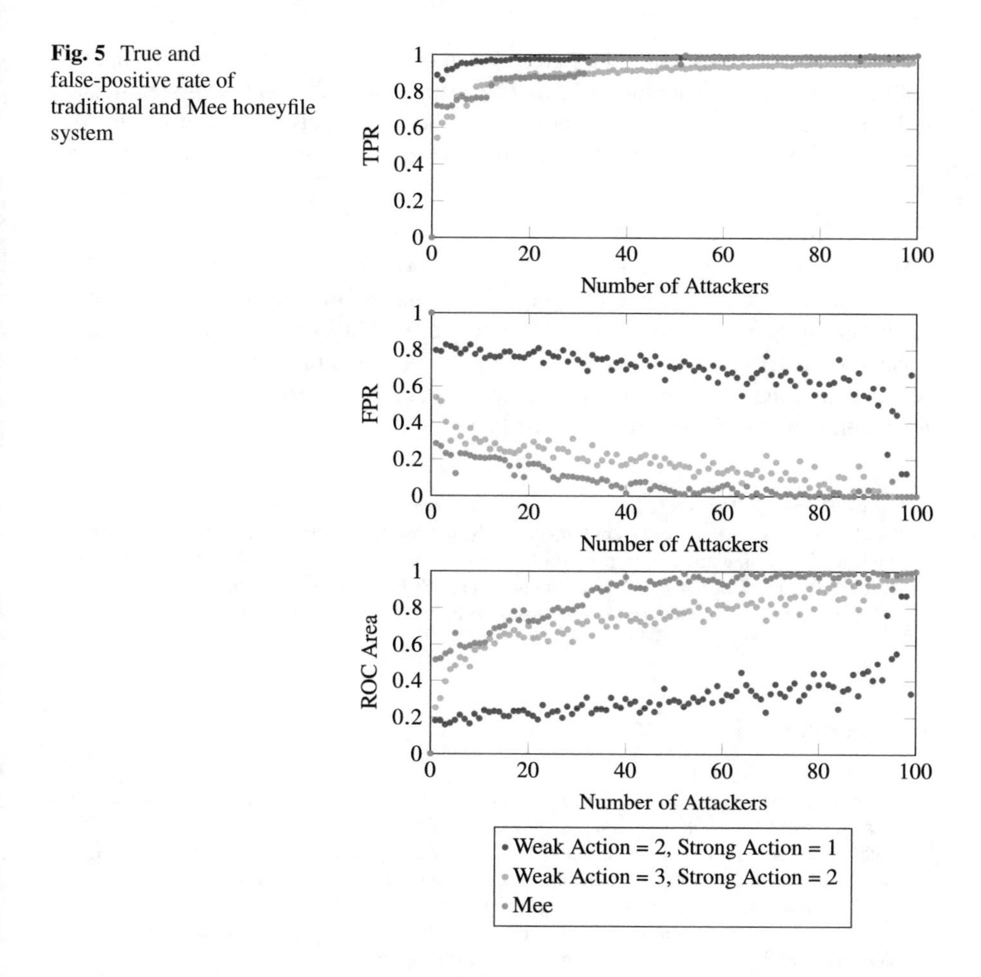

Fig. 5 True and false-positive rate of traditional and Mee honeyfile system

the performance of Mee, showing that Mee can reduce the FPR and maintain a high level of the TPR. From the figure showing ROC area, it becomes clear that Mee yields better performance than previous approaches.

8 Conclusion and Future Work

Defensive deception technologies are commonly used to delay and detect stealthy attacks. Honeyfile systems are a simple and lightweight deception technology that is widely applied. We design Mee as a novel honeyfile system to confuse and detect the insider attacker. Mee leverages *centralized* control and *decentralized* deployment to adjust the number of honeyfiles in each connected device. With Mee, the defender can reduce the overhead and the false-positive alarms from the legitimate users. We use game theory to model the interaction between the defender, the attacker, and legitimate users. To measure the performance of Mee, we simulate and compare it with the traditional honeyfile system. The results show that Mee can significantly reduce the average number of honeyfiles in the whole network. As a result, Mee avoids deploying unnecessary honeyfiles and decreasing the impact on legitimate users.

Future Work
This research assumes each device has an equal workload. Hence, all the honeyfiles in the network obtain the same probability that legitimate users act on them. However, in the real world, some users may have a higher likelihood of touching a honeyfile, such as those using the device more frequently than others. Therefore, we will consider the environment with a more realistic model and model the legitimate user with more details in our future work.

Acknowledgments The Army Research Office supported this work under award W911NF-17-1-0370. The views and conclusions contained in this document are those of the authors and should not be interpreted as representing the official policies, either expressed or implied, of the Army Research Laboratory or the U.S. Government. The U.S. Government is authorized to reproduce and distribute reprints for Government purposes without standing any copyright notation.

References

1. Abay, N.C., Akcora, C.G., Zhou, Y., Kantarcioglu, M., Thuraisingham, B.: Using deep learning to generate relational HoneyData. In: Autonomous Cyber Deception, pp. 3–19. Springer (2019)
2. Achleitner, S., La Porta, T., McDaniel, P., Sugrim, S., Krishnamurthy, S.V., Chadha, R.: Cyber deception: Virtual networks to defend insider reconnaissance. In: Proceedings of the 8th ACM CCS International Workshop on Managing Insider Security Threats, pp. 57–68 (2016)
3. Almeshekah, M.H., Spafford, E.H.: Planning and integrating deception into computer security defenses. In: Proc. New Security Paradigms Workshop, pp. 127–138 (2014)
4. Almeshekah, M.H., Spafford, E.H.: Cyber security deception. In: Cyber Deception, pp. 23–50. Springer (2016)
5. Anjum, I., Zhu, M., Polinsky, I., Enck, W.H., Reiter, M.K., Singh, M.P.: Role-based deception in enterprise networks. In: Proceedings of the 11th ACM Conference on Data and Application Security and Privacy (CODASPY), pp. 65–76. ACM, Online (Apr 2021). https://doi.org/10.1145/3422337.3447824
6. Anwar, A.H., Zhu, M., Wan, Z., Cho, J.H., Singh, M.P., Kamhoua, C.A.: Honeypot-based cyber deception against malicious reconnaissance via hypergame theory. In: IEEE Global Communications Conference: Communication & Information Systems Security, Rio de Janeiro, Brazil (Dec 2022)
7. Ben Salem, M., Stolfo, S.: Combining baiting and user search profiling techniques for masquerade detection. J. Wireless Mobile Netw. Ubiquit. Comput. Depend. Appl. **3**(1) (Mar 2012)
8. Bowen, B.M., Hershkop, S., Keromytis, A.D., Stolfo, S.J.: Baiting inside attackers using decoy documents. In: Proc. Int'l Conf. on Security and Privacy in Communication Systems, pp. 51–70. Springer (2009)
9. Chen, P., Desmet, L., Huygens, C.: A study on Advanced Persistent Threats. In: Proc. IFIP Int'l Conf. on Communications and Multimedia Security, pp. 63–72. Springer (2014)

10. Cho, J.H., Zhu, M., Singh, M.P.: Modeling and analysis of deception games based on hypergame theory. In: Autonomous Cyber Deception, pp. 49–74. Springer (2019)
11. Gómez-Hernández, J.A., Álvarez-González, L., García-Teodoro, P.: R-Locker: Thwarting ransomware action through a honeyfile-based approach. Comput. Secur. **73**, 389–398 (2018)
12. Harsanyi, J.C.: Games with incomplete information played by "Bayesian" players, I–III Part I. The basic model. Management Science **14**(3), 159–182 (1967)
13. Huang, L., Zhu, Q.: Dynamic Bayesian games for adversarial and defensive cyber deception. In: Autonomous Cyber Deception, pp. 75–97. Springer (2019)
14. La, Q.D., Quek, T.Q., Lee, J., Jin, S., Zhu, H.: Deceptive attack and defense game in honeypot-enabled networks for the Internet-of-Things. IEEE Internet Things J. **3**(6), 1025–1035 (2016)
15. Lu, Z., Wang, C., Zhao, S.: Cyber deception for computer and network security: Survey and challenges. Preprint (2020). arXiv:2007.14497
16. Mao, D., Zhang, S., Zhang, L., Feng, Y.: Game theory based dynamic defense mechanism for SDN. In: Proc. Int'l Conf. on Machine Learning for Cyber Security, pp. 290–303. Springer (2019)
17. Marchetti, M., Pierazzi, F., Guido, A., Colajanni, M.: Countering advanced persistent threats through security intelligence and big data analytics. In: 2016 8th International Conference on Cyber Conflict (CyCon), pp. 243–261. IEEE (2016)
18. Pawlick, J., Colbert, E., Zhu, Q.: Modeling and analysis of leaky deception using signaling games with evidence. IEEE Trans. Inf. Foren. Secur. **14**(7), 1871–1886 (2018)
19. Pawlick, J., Colbert, E., Zhu, Q.: A game-theoretic taxonomy and survey of defensive deception for cybersecurity and privacy. ACM Comput. Surv. (CSUR) **52**(4), 1–28 (2019)
20. Salem, M.B., Stolfo, S.J.: Decoy document deployment for effective masquerade attack detection. In: Proc. Int'l Conf. on Detection of Intrusions and Malware, and Vulnerability Assessment, pp. 35–54. Springer (2011)
21. Salem, M.B., Hershkop, S., Stolfo, S.J.: A survey of insider attack detection research. In: Insider Attack and Cyber Security, pp. 69–90. Springer (2008)
22. Shan-Shan, J., Ya-Bin, X.: The APT detection method based on attack tree for SDN. In: Proceedings of the 2nd International Conference on Cryptography, Security and Privacy, pp. 116–121 (2018)
23. Tankard, C.: Advanced persistent threats and how to monitor and deter them. Network Security **2011**(8), 16–19 (2011)
24. Voris, J., Jermyn, J., Boggs, N., Stolfo, S.: Fox in the trap: Thwarting masqueraders via automated decoy document deployment. In: Proceedings of the Eighth European Workshop on System Security, pp. 1–7 (2015)
25. Wan, Z., Cho, J.H., Zhu, M., Anwar, A.H., Kamhoua, C., Singh, M.P.: Foureye: Defensive deception against advanced persistent threats via hypergame theory. IEEE Trans.Network Serv. Manag. (2021)
26. Whitham, B.: Minimising paradoxes when employing honeyfiles to combat data theft in military networks. In: Proc. Military Communications and Information Systems Conf. (MilCIS), pp. 1–6. IEEE (2016)
27. Whitham, B.: Automating the generation of enticing text content for high-interaction honeyfiles. In: Proc. 50th Hawaii Int'l Conf. on System Sciences (2017)
28. Yuill, J.J.: Defensive computer-security deception operations: Processes, principles and techniques. Ph.D. thesis, North Carolina State University (2007)
29. Yuill, J., Zappe, M., Denning, D., Feer, F.: Honeyfiles: Deceptive files for intrusion detection. In: Proc. 5th Annual IEEE Information Assurance Workshop, pp. 116–122 (2004)
30. Zhu, M., Anwar, A.H., Wan, Z., Cho, J.H., Kamhoua, C.A., Singh, M.P.: A survey of defensive deception: Approaches using game theory and machine learning. IEEE Commun. Surv. Tutor. **23**(4), 2460–2493 (2021). https://doi.org/10.1109/COMST.2021.3102874
31. Zhu, M., Xi, R., Sharmin, N., Miah, M., Kiekintveld, C., Singh, M.P.: Honeyflow: Decoy network traffic generation via Generative Adversarial Network. In: IEEE Global Communications Conference: Communication & Information Systems Security, Rio de Janeiro, Brazil (Dec 2022)

HoneyPLC: A Next-Generation Honeypot for Industrial Control Systems

Efrén López Morales, Carlos E. Rubio-Medrano, Adam Doupé, Ruoyu Wang, Yan Shoshitaishvili, Tiffany Bao, and Gail-Joon Ahn

1 Introduction

Industrial Control Systems (ICSs) are widely used by many industries including public utilities such as the power grid, water, and telecommunications [48]. These utilities are integral to people's daily life, and any interruption to them may cause significant damage and losses. The increasingly interconnected nature of modern ICS makes them more vulnerable than ever to cyberattacks. For example, a cyberattack that targets a power grid would potentially lead to blackouts in a city or across an entire geographical region. Regrettably, this proposition is no longer a fiction. The number of attacks targeting ICS has been steadily increasing since the infamous Stuxnet malware first showed the world that ICS networks are not secure [14]. Also, in 2015, a cyberattack targeting the Ukrainian power grid successfully took down several of its distribution stations. The ensuing outages left approximately 225,000 people without access to electricity for several hours [7].

E. L. Morales · C. E. Rubio-Medrano (✉)
Texas A&M University—Corpus Christi, Corpus Christi, TX, USA
e-mail: elopezmorales@islander.tamucc.edu; carlos.rubiomedrano@tamucc.edu

A. Doupé · R. Wang · Y. Shoshitaishvili · T. Bao
Arizona State University, Tempe, AZ, USA
e-mail: doupe@asu.edu; fishw@asu.edu; yans@asu.edu; tbao@asu.edu

G.-J. Ahn
Arizona State University, Tempe, AZ, USA
Samsung Research, Seoul, Republic of Korea
e-mail: gahn@asu.edu

T. Bao et al. (eds.), *Cyber Deception*, Advances in Information Security 89,
https://doi.org/10.1007/978-3-031-16613-6_8

1.1 The Problem: Preventing Attacks Targeting ICS via PLCs

One of the key components of ICS networks is Programmable Logic Controllers, better known as PLCs [48]. PLCs are commonly found in supervisory control and data acquisition or SCADA systems. These systems are used to control separated assets that require centralized data acquisition which are a type of ICS [48]. Figure 1 illustrates these relationships. PLCs control mission-critical electrical hardware such as pumps or centrifuges, effectively serving as a bridge between the cyber and the physical worlds. Because of their critical role, PLCs have been recently targeted by cyberattacks, which attempt to disrupt their proper functioning in an effort to affect their corresponding ICS as a whole. As an example, PLCs were the primary target of the Stuxnet malware as they controlled critical physical processes in a nuclear facility. To better understand cyberattacks against ICS and PLCs, several honeypots have been proposed [5, 15, 16, 24, 39, 51]. However, current honeypot implementations for ICS fail to provide the necessary features to capture data for most recent and sophisticated attack techniques. For example, a common limitation exhibited by most of the existing approaches is their low-interaction nature: they usually rely on basic and shallow simulations of network protocols, which usually lack complex functionality that limits the attack vectors and makes them easy to discover by attackers. These shortcomings heavily restrict the value of the attack data that can be gathered by these ICS honeypots.

1.2 Challenges for Solving the Problem

Providing a solution to these issues comes with a set of unique challenges. First, it is difficult to achieve meaningful, step-by-step protocol simulation that can eventually

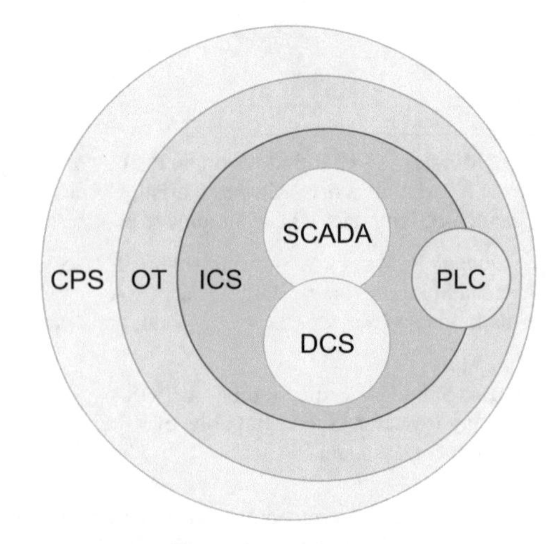

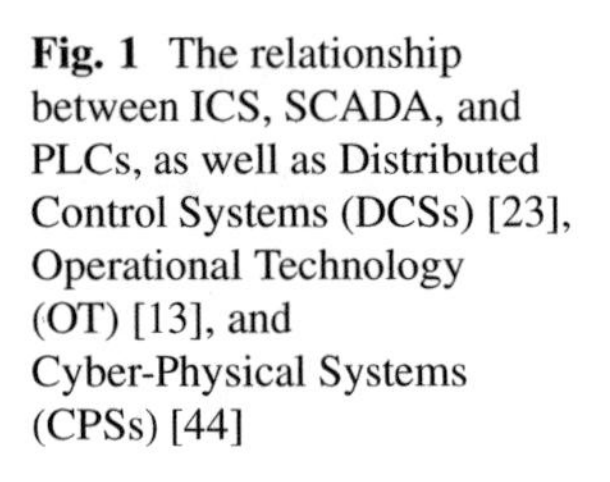

Fig. 1 The relationship between ICS, SCADA, and PLCs, as well as Distributed Control Systems (DCSs) [23], Operational Technology (OT) [13], and Cyber-Physical Systems (CPSs) [44]

result in high-level, deceiving interactions between honeypots and attackers. These *inadequate simulations* complicate concealing the true nature of honeypots up to the point accurate and valuable data, e.g., the actual malicious ladder logic code itself can be retrieved from attackers for further analysis. Second, several network protocols largely used in ICS, e.g., S7comm [51], are *proprietary*, in the sense that no detailed documentation on them is publicly available, which prevents an effective understanding of the protocol, including hidden configuration parameters as well as implicit, undocumented assumptions, which can ultimately reveal the true nature of a honeypot to an attacker. Moreover, existing PLCs used in practice vary in terms of configuration settings, supported protocols, and the way they are customized for different application domains. Creating a general framework that can effectively support such *heterogeneity* of PLCs devices, regardless of their brand and model, without requiring the edition of large and clumsy configuration files, represents a non-trivial challenge.

1.3 Proposed Approach: A Next-Generation Honeypot for ICS

To alleviate the aforementioned concerns targeting ICS worldwide and effectively tackle the research challenges just discussed, this chapter presents HoneyPLC: a high-interaction, extensible, and malware-collecting honeypot modeling PLCs, which is specifically crafted for ICS. HoneyPLC includes *advanced simulations* of the most common network protocols found in PLCs, namely, the TCP/IP Stack, S7comm, HTTP, and SNMP, addressing the challenges introduced by inadequate simulations and protocol closeness as discussed before. As an example, our TCP/IP Stack simulation benefits from the introduction of a novel technique called *fingerprint reversing*, which allows for accurately modeling TCP, ICMP, and UDP probes at runtime, providing an effective, customized response to each interaction as initiated by an attacker, largely increasing the level of engagement and subsequent deception. In addition, our simulation of the S7comm protocol, which is core to PLC communications, provides a level of simulation that is able to trick even proprietary tools such as the Siemens Step7 Manager [4]. Moreover, HoneyPLC also provides *enhanced extensibility features*, allowing for PLCs of different models and manufacturing brands to be effectively simulated, thus addressing the PLC heterogeneity challenge just discussed. We have successfully tested this feature using five *real* PLCs, allowing for HoneyPLC to currently support *out of the box* the Siemens S7-300, S7-1200, and S7-1500, the Allen-Bradley MicroLogix 1100, and the ABB PM554-TP-ETH PLCs. HoneyPLC also implements an advanced simulation of the internal memory blocks featured by modern PLCs, allowing for the *automated capture and storage of malicious ladder logic programs*, which can be later analyzed to reveal new attacking techniques.

The features just discussed are, to the best of our knowledge, exclusive to HoneyPLC and also significantly advance the *state of the art* for ICS honeypots. This positions HoneyPLC as a convenient and flexible tool that can serve as a reliable basis for the analysis and understanding of emerging threats and attacks, as well as the subsequent development of protection techniques for ICS.

1.4 Contributions to Scientific Literature

Overall, this chapter makes the following contributions:

1. It provides a summary of the limitations and shortcomings of existing ICS Honeypots and discusses how they address (or not) emerging malware threats, as well as new ICS technology, e.g., new PLC models and ICS network protocols.
2. It presents HoneyPLC, a high-interaction honeypot for PLCs, which not only solves many of the limitations of related approaches but also provides convenient support for further understanding and eventually defeating emerging threats for ICS.
3. It introduces the HoneyPLC PLC Profiler Tool, which allows for the effective simulation of many different PLCs regardless of their model and manufacturer.
4. Finally, experimental evidence is provided showing that HoneyPLC is not only effective at engaging and deceiving *state-of-the-art* tools for network reconnaissance but also outperforms existing honeypots in the literature, achieving a performance level comparable to *real* PLC devices.

1.5 Source Code Availability and Chapter Roadmap

In an effort to further open and produce reproducible science, HoneyPLC and all our experimental results are available online.[1] This chapter is an extended version of a paper that appeared at the Proceedings of the 2020 ACM SIGSAC Conference on Computer and Communications Security (ACM CCS'20) [25], and it is organized as follows: Sect. 2 introduces detailed information about PLCs, honeypots, ICS-specific malware, as well as similar approaches found in the literature. Section 3 elaborates on the lack of support of such existing approaches for handling emerging threats for ICS, resulting in a problem that is then addressed in Sect. 4. Later, Sect. 5 presents experimental evidence of the suitability of HoneyPLC for being deployed in practice by precisely describing testing environments, procedures, and results. Subsequently, Sect. 6 delves into a discussion about how our approach ranks up against current literature and outlines what future research could be undertaken as a result of this work. Finally, Sect. 7 concludes this chapter.

[1] https://github.com/sefcom/honeyplc.

2 Background and Related Work

Before diving into the details of HoneyPLC, we present some background on the tools and technologies that are addressed in further sections, namely, PLCs themselves, network reconnaissance tools, malware specifically tailored for disrupting ICS, and honeypots that have been developed for protecting ICS environments.

2.1 Programmable Logic Controllers

A Programmable Logic Controller (PLC) is a small industrial computer designed to perform logic functions based on input provided by electrical hardware such as pumps, relays, mechanical timers, switches, etc. PLCs have the capability of controlling complex industrial processes, making them ubiquitous in ICS and SCADA environments [47]. Some popular PLC manufacturers include Siemens [45], Allen-Bradley [2], and ABB [1]. Internally, PLCs have programmable memory blocks that store instructions to implement different functions, for example, input and output control, counting, logic gates, and arithmetic calculations.

2.2 Network Reconnaissance Tools

In practice, the process of *network reconnaissance* involves identifying the topology of a network, the protocols used, the different devices that may be connected through it, etc. Since such a process is essential for carrying out successful attacks to ICS and PLCs, we now present a set of tools for network reconnaissance that are widely used in practice, which were used to evaluate HoneyPLC as it is discussed in Sects. 3 and 5.

2.2.1 Nmap

Nmap or "Network Mapper" [26] is a popular open-source utility that is able to detect the operating system and services that a particular device is running by sending raw IP packets over the network. Once a given detection scan is completed, Nmap can either report a single OS match or a list of potential OS guesses, each guess with its own confidence percentage rate, in the range of 0 to 100, where 0 denotes the complete absence of confidence and 100 denotes a complete confidence on the projected guess result.

2.2.2 PLCScan

PLCScan [43] is a reconnaissance tool used to scan PLC devices in a given network. PLCScan reveals PLCs that implement the S7comm protocol over TCP port 102 or the Modbus protocol over TCP port 502. It is written as a command line Python script and lists PLC information including basic hardware, serial number, name of the PLC, and firmware version.

2.2.3 Shodan

Shodan is a search engine and crawler [27] specifically tailored for devices exposed across the Internet, e.g., webcams, routers, and ICS devices, among others. The Shodan Honeyscore (part of the Shodan API [27]) is a tool that checks whether a device is a honeypot or not. Given an IP address, the Shodan Honeyscore calculates the probability that the host is a honeypot, in a range between 0.0 and 1.0, where 0.0 means that the host is definitively a *real* system and 1.0 means the host is definitively a honeypot. According to Shodan's creator, the following criteria are used for calculating Honeyscores [28]: (1) too many open network ports, (2) a service not matching the environment, for example, an ICS device running on AWS EC2, (3) known default settings of known honeypots, (4) if a host was initially classified as a honeypot, then it is highly likely that it remains a honeypot today, even though its configuration may look real, (5) a Machine Learning classification algorithm (not disclosed), and, finally, (6) the same configuration being used across multiple honeypots.

2.3 Exemplary ICS Malware

Recently, a series of dedicated malware instances have attempted to disrupt the functioning of ICS environments, and some of them have been successful and have ultimately resulted in costly damages. With that in mind, we now present a summary of the malware that is most relevant to the problem addressed by our proposed HoneyPLC approach.

2.3.1 Stuxnet

The first ever-documented cyber-warfare weapon, Stuxnet, was a turning point in the history of cybersecurity [12], targeting PLC models 315 and 417 made by Siemens to modify their inner ladder logic code while concealing itself from ICS administrators [21]. The malware would first spread itself via USB sticks and the local network, looking for vulnerable Windows workstations. Later, it would proceed to infect the Step7 and WinCC Siemens proprietary software by hijacking a Dynamic Link Library (DLL) file used to communicate with the PLCs. Finally, the malicious ladder logic payload would be dropped only on the aforementioned models based on specific manufacturer numbers and memory blocks.

2.3.2 Pipedream Toolkit

Pipedream is the seventh documented malware that specifically targets ICS [11]. It is not a single-purpose malware but a modular framework that includes multiple exploits that target different ICS devices. These devices include Open Platform Communications Unified Architecture (OPC UA) servers, Schneider Electric PLCs, and OMRON PLCs. Pipedream is believed to have been developed by a nation state or a state-sponsored group and was classified as an advanced persistent threat or APT by the Department of Energy or DOE [8].

2.3.3 Dragonfly

Also known as Havex malware [37], Dragonfly was a large-scale cyberespionage campaign that targeted ICS software in the energy sector in the United States and Europe. In order to infect its targets, three different attack vectors were used. First, a spam campaign that used spear phishing targeted senior employees in energy companies. Second, Watering Hole attacks [37] that compromised legitimate energy sector websites were deployed to redirect the target to another compromised website that hosted the Lightsout exploit, which ultimately dropped the Oldrea or Karagany malwares [10] in the target's host. The third and final attack vector used was a dedicated *trojanized* software (legitimate software that is turned into malware), the attackers leveraged to successfully compromise various legitimate ICS software packages, ultimately inserted their own malicious code. Once a host was infected, the Havex malware leveraged legitimate functionality available through the OPC protocol to draw a map of the industrial devices present in the ICS network. This kind of data would be highly valuable when designing future attacks. Dragonfly was entirely focused on spying and gathering information on ICS networks.

2.3.4 Crashoverride

Otherwise known as Industroyer [46], CRASHOVERRIDE is a sophisticated malware designed to disrupt ICS networks used in electrical substations. It shows in-depth knowledge of ICS protocols used in the electrical industry that would only be possible with access to specialized industrial equipment. CRASHOVERRIDE dealt with physical damage by opening circuit breakers and keeping them open even if the grid operators tried to close them back to restore the system. It is believed to have been the cause of the power outage in Ukraine in December of 2016 [14].

2.4 *Honeypots for ICS*

Honeypots are computer systems that purposefully expose a set of vulnerabilities and services that can be probed, analyzed, and ultimately exploited by an attacker [33], allowing for all possible interaction data to be monitored, logged, and stored for future analysis. A summary of existing ICS honeypots is shown in Table 1.

2.4.1 Low-Interaction Honeypots

Low-interaction honeypots offer the least amount of functionality to an attacker [29, 33]. The services exposed by this kind of honeypot are usually implemented using simple scripts and finite state machines. Because of their limited interaction, attackers may not be able to complete their attack steps or may even realize that their target is a fake system. On the other hand, low-interaction honeypots cannot be fully compromised as they are not real systems, which greatly reduces maintenance costs and time invested in configuration and deployment. Gaspot [50] is a low-interaction honeypot written as a Python script that simulates a gas tank gauge. It can be modified to change temperature, tank name, and volume. The SCADA HoneyNet Project was the first honeypot implementation specifically built for ICS [39, 49]. This project was aimed at developing a software framework capable of simulating ICS devices like PLCs using Python scripts. Conpot [16] is also a low-interaction ICS honeypot implementation that simulates a Siemens S7-200 PLC and can be manually modified to simulate other PLCs by editing an XML file.

2.4.2 High-Interaction Honeypots

High-interaction honeypots lie on the other side of the spectrum, as they strive to offer the same level of interaction as a real system [29]. CryPLH is a high-interaction honeypot that simulates an S7-300 Siemens PLC [5] and includes HTTP, HTTPS, S7comm, and SNMP services running on a Linux host that has been modified to accept connections on specific ports. The S7comm protocol is simulated by showing an incorrect password response and the TCP/IP Stack is simulated via the Linux kernel. S7commTrace [51] provides a high-interaction simulation of the S7comm protocol and supports the Siemens S7-300 PLC. Antonioli et al. [3] proposed a high-interaction honeypot that leverages the MiniCPS framework to simulate the Ethernet/IP protocol and a generic PLC. HoneyPhy [24] provides a novel physics-aware model to simulate a generic analog thermostat and the DNP3 protocol.

Table 1 Comparison of existing PLC Honeypots in the literature and HoneyPLC

Keys: ○ = No coverage; ◑ = Limited coverage; ● = Optimal coverage

Approach/ Feature	Extensibility	TCP/IP stack simulation	*Out-of-the-Box* PLCs	ICS network services	Ladder Logic capture	Physics interaction	Logging
Gaspot [50]	○	○	◑	○	○	◑	●
SCADA HoneyNet [39]	○	●	◑	●	○	○	●
Conpot [16]	◑	○	◑	●	○	○	●
Digital Bond's Honeynet [49]	○	○	◑	●	○	○	●
DiPot [6]	◑	○	◑	●	○	○	●
SHaPe [20]	◑	○	◑	◑	○	○	●
CryPLH [5]	○	◑	◑	●	○	○	○
S7commTrace [51]	◑	○	◑	◑	○	○	●
Antonioli et al. [3]	◑	◑	◑	●	○	◑	●
HoneyPhy [24]	◑	○	◑	◑	○	◑	○
HoneyPLC	●	●	●	●	●	○	●
Sections addressing feature	4.2, 5.2	4.3, 5.3, 5.4	4.2, 5.2	4.3, 5.6	4.4, 5.7	6	4.5

3 Limitations of Existing Honeypots

Despite the benefits of honeypots previously discussed, existing honeypots, shown in Table 1, fail to provide the necessary features to capture data on sophisticated attacks, thus exhibiting the following limitations:

L-1 **Limited Extensibility.** A common limitation in the current literature is the narrow extensibility support for the many different PLC devices and network services that are used in ICS in practice and have already been targeted by recent attacks. As an example, Stuxnet and the Kemuri attack targeted different kinds of PLCs, whereas CRASHOVERRIDE targeted different network services, as was discussed in Sect. 2. Following Table 1, several approaches in the literature provide limited extensibility capabilities, which mostly include the manual edition of XML files to support additional PLCs. This process, besides being tedious and time-consuming, may be highly error-prone and may ultimately reveal the true nature of a honeypot to attackers if implemented incorrectly. This is aggravated by the fact most of the approaches in the literature support only one or two PLC models only. In contrast, HoneyPLC currently provides *out-of-the-box* support for 5 PLCs of three major brands, as detailed in Sect. 5.2.

L-2 **Limited Interaction.** Current approaches mostly provide limited functionality when it comes to TCP/IP Stack simulations, as well as native ICS network protocols, as described in Sect. 2. This is a serious limitation that stops current approaches from extracting value from adversarial interactions and malware. As an example, CRASHOVERRIDE leveraged advanced ICS protocol features that are not supported by low-interaction honeypots. This would ultimately result in the loss of highly valuable data. Even high-interaction honeypots fail to provide advanced enough protocol simulations. For example, CryPLH [5] implements the S7comm protocol using a Python script that only simulates an incorrect password screen. HoneyPLC solves this limitation by providing extended support for various networks protocols, as we will discuss in Sect. 4.3 and evaluate through experiments in Sects. 5.3–5.6.

L-3 **Limited Covert Operation.** The moment an attacker discovers the true nature of a honeypot, it is game over, as the attacker might stop interacting with it altogether and stop revealing her attack methods. Therefore, honeypots should aim to fool widely used network reconnaissance tools, e.g., Nmap, introduced in Sect. 2.2, to maintain their covert operation. In such regard, the SCADA HoneyNet Project [39] is the only approach in the literature that provides a convincing deception to attackers. Also, Linux Kernel simulations, implemented by several approaches in the literature, e.g., CryPLH, fail to deceive Nmap. Other work fails to attempt or even mention such a crucial feature. To overcome this, HoneyPLC provides advanced network simulations intended to deceive reconnaissance tools, as shown in Sect. 4.3.

L-4 **No Malware Collection.** The highly specialized nature of ICS devices calls for better analysis, dissection, and understanding techniques specifically tailored for emerging malware trends. In such regard, honeypots are a great tool to collect and analyze malware [34]. However, as shown in Table 1, there exist no honeypots for ICS in the literature that can provide such functionality. To solve this, HoneyPLC provides a novel feature to capture ladder logic, as described in Sects. 4.4 and 5.7.

4 HoneyPLC: A Convenient High-Interaction Honeypot For PLCs

Having described the limitations of existing approaches, we now present HoneyPLC, an extensible, high-interaction, and malware-collecting honeypot for ICS. HoneyPLC provides advanced protocol simulations, e.g., TCP/IP, S7comm, HTTP, and SNMP, achieving an interaction level comparable to *real* PLCs, ultimately introducing low-to-moderate levels of risk as well as low maintenance costs. We start by providing an illustrative use case scenario, which exemplifies how the different inner modules and components of HoneyPLC interact with an attacker at runtime when an attempt to compromise a PLC is made. Later, we elaborate on how HoneyPLC solves each of the limitations highlighted in Sect. 3.

4.1 Illustrative Use Case Scenario

For illustrative purposes, we present an example use case scenario featuring HoneyPLC, which is based on the architectural design graphically shown in Fig. 2. After this case scenario has been completed, HoneyPLC may have been able to collect crucial information about the attack inside its logging infrastructure: (1) the public IP address of the attacker, (2) the specific PLC memory blocks the attacker was targeting and, best of all, the critical piece, and (3) the ladder logic program he/she has injected. Later on, such a malware sample can be analyzed at the byte level to get a better understanding of the malicious instructions that the attacker wanted the PLC to execute. In Sect. 6, we elaborate on this idea as a part of our future work.

4.1.1 Initial Setup

As it will be further discussed in Sect. 4.2, HoneyPLC can be extended to simulate PLCs of different models, communication protocols, and/or manufacturer brands. With that in mind, the very first step when using HoneyPLC includes choosing the PLC Profile featuring the desired real-life PLC that will be exposed to attackers as a honeypot. This process is shown in Fig. 2 (Step 1). PLC Profiles can be chosen

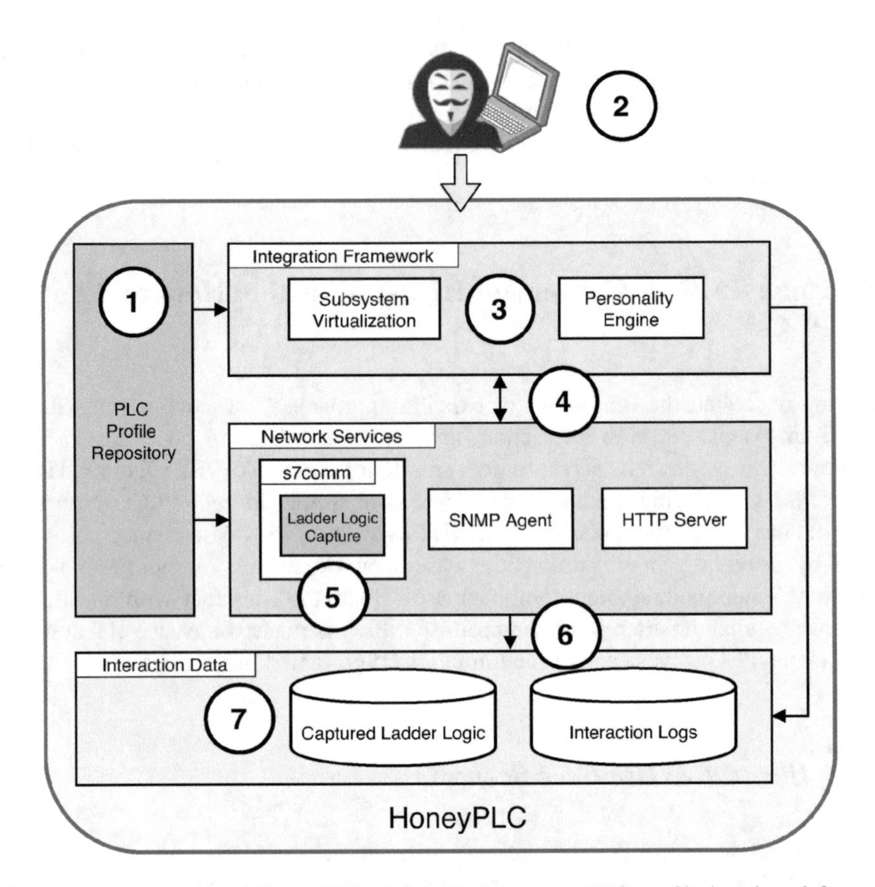

Fig. 2 The architecture of HoneyPLC. Before deployment, a PLC profile is selected from a repository (1). Later, at runtime, an attacker may initiate contact via a dedicated protocol, e.g., S7comm (2). Communications are then processed by the Personality Engine (3), later forwarded to the S7comm server (5), and are eventually logged by the interaction data framework (6). Finally, all code injected by the attacker is captured within the repository module (7)

from a dedicated repository included as a part of HoneyPLC. For the rest of this case scenario, let us assume the S7-1200 model is selected.

4.1.2 Fingerprinting

Once HoneyPLC is deployed, an attacker may try to fingerprint it using a reconnaissance tool such as Nmap or PLCScan (Fig. 2 (Step 2)). When initial contact is established, all the TCP/IP requests will be handled by the HoneyPLC's Personality Engine, which in turn is based on features provided by the Honeyd [9] tool, as it will be further discussed in Sect. 4.3 (Fig. 2 (Step 3)). Since the S7-1200 PLC model was selected in the beginning, the Personality Engine will use the appropriate fingerprint contained within the PLC Profile to reply to communications started by Nmap. At this point, Nmap may confirm to the attacker that she is dealing with a PLC and not a honeypot, as we show in Sect. 5.

4.1.3 Reconnaissance

In a subsequent step, an attacker might try to initiate an S7comm connection to check what PLC memory blocks are available. As mentioned in Sect. 2, such a process is crucial when attempting to modify the inner ladder logic code of a PLC. The connection is first handled by the HoneyPLC's Network Services module and later forwarded to a dedicated S7comm server (Fig. 2 (Step 4)). The S7comm server then replies with the requested information, and the Integration Framework forwards the replies to the attacker. In the meantime, the S7comm server is logging all the interactions, including the attacker's source IP address and memory block requests made to the PLC.

4.1.4 Code Injection

At this point, when the attacker identifies a PLC memory block suitable for injection, he/she uses an S7comm application like PLCinject [41] to load ladder logic code into the PLC, effectively overwriting any preexisting code and introducing a custom-made malicious payload (Fig. 2 (Step 5)). As a result, the HoneyPLC's S7comm server will write the code into the dedicated HoneyPLC repository, which is managed by the Interaction Data module (Fig. 2 (Steps 6 and 7)).

4.1.5 Confirmation and Farewell

Finally, the attacker has two options. First, he/she can continue interacting with HoneyPLC, e.g., trying to download the MIB via the SNMP protocol to get more information about the network configuration or any banner present. Second, she might stop interacting altogether, at which point HoneyPLC's work is over.

4.2 Supporting PLC Extensibility

As described in Sect. 3, existing approaches in the literature provide limited support for the large variety of PLC models currently in the market, which limits their suitability for being used in practice. To solve this issue, this section starts by describing how different PLC models are supported by HoneyPLC by means of so-called *PLC Profiles* and then moves on to describe how other models in the market can be supported by developing new PLC Profiles by means of the HoneyPLC PLC Profiler Tool.

4.2.1 PLC Profiles

The PLC Profile Repository, shown in Fig. 2 (Step 1), is a collection of PLC Profiles that hold all the required data to simulate a given PLC. It communicates with the Integration Framework and Network Services modules to customize the PLC that HoneyPLC is simulating at any given time and addresses the lack of extensibility discussed in Limitation L-1. In turn, a PLC Profile is a collection of three discrete datasets, which allow HoneyPLC to simulate a particular PLC device by means of highly customized simulations of network interactions, as it will be discussed in Sect. 4.3.

- *SNMP MIB.* A Management Information Base (MIB) is a standard used by SNMP agents. Because most PLC devices implement a simple SNMP agent, a custom MIB is needed for HoneyPLC to provide a realistic SNMP simulation.
- *Nmap Fingerprint.* A plain text file with the Nmap fingerprint to effectively simulate the TCP/IP Stack of a particular PLC device. As it will be detailed later in this section, this fingerprint allows HoneyPLC to effectively engage and deceive well-known reconnaissance tools such as Nmap.
- *Management Website.* Some PLC devices provide a light webserver with a splash screen and some configuration options. Because of this, a PLC Profile includes a copy of such website, including, but not limited to, image, HTML, and CSS files.

4.2.2 PLC Profiler Tool

The HoneyPLC Profiler Tool automates the creation of new HoneyPLC Profiles. It interfaces with three different applications: Nmap, (Sect. 2.2), snmpwalk [35], and wget [36]. To obtain the profile for a target PLC, the HoneyPLC Profiler requires the IP address of the PLC device as the only input. Then, the Profiler runs a series of queries to obtain the three discrete sets of data from the target PLC described before: an SNMP MIB, a website directory, and an Nmap fingerprint. First, snmpwalk is used for reading all the available Object IDs (OIDs) from the public community string, creating an identical MIB to the one used by the PLC. OIDs may include, among other important configuration settings, the unique identifier of the PLC, as well as its base IP address. Second, Nmap's OS detection is used to get the TCP/IP stack fingerprint of the target PLC, in a process that includes scanning all well-known TCP and UDP ports. This fingerprint will be later leveraged by HoneyPLC's Integration Framework to provide meaningful TCP/IP interactions as a response to requests initiated by an attacker. Third, wget is used to download a complete copy of the splash screen or administration website, if any. Finally, the HoneyPLC Profiler will create a custom directory that can be used by HoneyPLC, inside its dedicated PLC Profile Repository, shown in Fig. 2 (1), to simulate the target PLC.

4.3 *Supporting Operational Covertness*

As described in Sect. 3, being able to engage attackers without revealing a honeypot nature is crucial for obtaining valuable information on the vectors, techniques, and goals being used for compromising PLCs. To this end, this section describes how HoneyPLC supports *meaningful* network interactions leveraging the TCP, IP, S7comm, SNMP, and HTTP protocols, which are widely used by PLCs in practice.

4.3.1 TCP/IP Simulation

Within HoneyPLC's Integration Framework, depicted in Fig. 2, a sophisticated TCP/IP Stack simulation is implemented by leveraging Honeyd [33], a popular framework for honeypot simulation, as well as Nmap, discussed in Sect. 2.2. The process is depicted in Fig. 3. Initially, when a new PLC is to be modeled by HoneyPLC, Nmap is used to generate a detailed TCP/IP Stack fingerprint for it. Next, such a fingerprint is integrated with the Honeyd fingerprint database, by appending it to Honeyd's nmap-os-db text file. Later, at runtime, when a tool

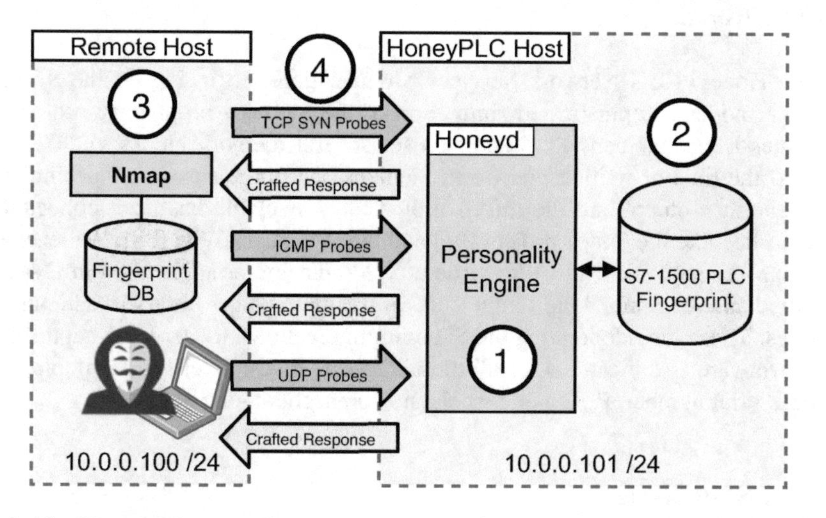

Fig. 3 The HoneyPLC personality engine: first, a PLC Profile is selected from the repository, including its Nmap fingerprint (1). When an attacker tries to fingerprint HoneyPLC using Nmap, such a tool will send a series of Probes to determine the OS or Device (2). HoneyPLC will then reply with appropriately crafted responses that simulate a real PLC, thus effectively deceiving Nmap and the attacker (3)

like Nmap tries to fingerprint a HoneyPLC host, HoneyPLC Personality Engine, leveraging Honeyd, will respond with the appropriate fingerprint information. To achieve this, the Engine reads a particular fingerprint from Nmap's database and *reverses* it, which means that when Honeyd simulates a particular device, it introduces its IP/TCP Stack peculiarities: TCP SYN packet flags, IMCP packet flags, and timestamps. The generation of accurate Nmap fingerprints imposed a variety of challenges. First, PLC devices of different manufacturers and models use different UDP and TCP ports that are not standard or may not be properly defined within the device manuals, e.g., port 2222 for the MicroLogix 1100 PLC. The lack of heterogeneity required us to perform a manual inspection, which was time-consuming and error-prone. Second, we analyzed the Nmap reports that contain the fingerprint results and modified the format to be compatible with the Honeyd fingerprint database. Third, an extensive analysis of the Nmap reports containing the fingerprint results was also required, such that important changes can be introduced for producing better results, i.e., changes in the overall format to make the newly produced fingerprint compatible with the Honeyd fingerprint database. Additionally, the creation of accurate Honeyd templates brought its own set of challenges. For HoneyPLC to provide enhanced interaction capabilities, which can engage attackers for extended periods of time (as we further describe in Sect. 4.3), we significantly improved the standard simulation scripts included within Honeyd. Specifically, we used the subsystem virtualization feature provided by Honeyd: this feature facilitates the integration of the different HoneyPLC components.

4.3.2 S7comm Server

Within HoneyPLC's Network Services Module, depicted in Fig. 2, the S7comm server provides a sophisticated simulation of the Siemens proprietary protocol. It simulates a real Siemens PLC and exposes several memory blocks via TCP port 102. At the time of writing this work, Siemens had not released the specifications of S7comm protocol and the information that is available has been collected by third parties like the Snap7 project [31] and the Wireshark Wiki [38]. We leveraged the Snap7 framework [31, 40] to write an S7comm server application in C++. We modified and recompiled the source code of the main Snap7 library to add our own features. These include logging the S7comm interactions, ladder logic capture, and PLC firmware specifications for all three Siemens PLC models, for example, CPU model, serial number, PLC name label, and copyright among others.

4.3.3 SNMP Server

Within HoneyPLC's Network Services Module, the SNMP Agent implements an advanced simulation of the SNMP protocol along with believable MIB data, effectively allowing HoneyPLC to reply to any external SNMP server query. SNMP is commonly used in practice to monitor network connected devices and listens

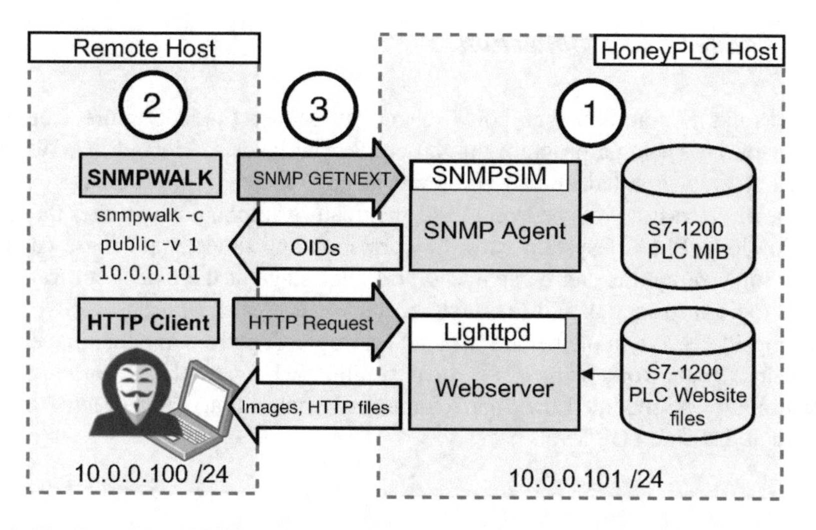

Fig. 4 The HoneyPLC SNMP and the Webserver agents. The MIB database and the website HTTP files, obtained from a PLC profile, are first loaded by each agent (1). Then, the attacker may use SNMPWalk as well as an HTTTP client to establish connection with HoneyPLC (2). Later, each agent will reply to each request using the information obtained from the PLC profile (3)

to requests over UDP port 161. Since *real* PLCs do implement SNMP agents, implementing this sub-component adds to the deception capabilities of HoneyPLC. Our simulation process, shown in Fig. 4 (top), can be described as follows: in practice, a typical SNMP setup includes a *Manager* as well as an *Agent* module. The SNMP Manager continually queries the Agent for up-to-date data. an SNMP Agent exposes a set of data known as Management Information base or MIB. In order to simulate the SNMP protocol, we use the light Python application *snmpsim*, which simulates an SNMP Agent based on real time or archived MIB data. When an SNMP request is received by HoneyPLC, the SNMP Agent replies with an OID as a real PLC would do.

4.3.4 HTTP Server

Finally, the HoneyPLC's HTTP server provides an advanced simulation of the HTTP server of the Real PLCs and serves websites found in real PLCs, as illustrated in Fig. 4 (bottom). As an example, most Siemens PLC devices include an optional HTTP service to manage some of its internal configuration features. This functionality was in turn implemented with lighttpd [19], a lightweight webserver to handle all HTTP quests. When an HTTP request hits HoneyPLC, its Integration Framework relays the request to the lighttpd server. Later, the webserver replies with the website data from a HoneyPLC profile.

4.4 Ladder Logic Collection

HoneyPLC's S7comm Server holds the novel Ladder Logic Capture feature. It writes any ladder logic program that an attacker uploads to HoneyPLC. When an adversary uploads a ladder logic program to any of the S7comm Server memory blocks, while trusting it to be a real PLC, this feature automatically writes them into the file HoneyPLC filesystem with the corresponding timestamp. These captured ladder logic programs can be analyzed at a later stage at the byte level to expose ladder logic instructions and then extract new attack patterns used by adversaries targeting PLCs. We implemented the Ladder Logic Capture component leveraging the Snap7 framework using C++, in a similar fashion as the S7comm Server. Additionally, we modified the Snap7 framework main library files to integrate this feature at the Linux OS level.

4.5 Implementing Record Keeping via Logging

The Interaction Data component holds all of the interaction data gathered by HoneyPLC. It maintains two kinds of data. First, it contains all logs produced by our S7comm servers, the SNMP agent, and the HTTP server. Second, it contains all the ladder logic programs that get injected via the S7comm server. This component communicates directly with the Network Services component. We configured Honeyd, lighttpd, snmpsim, and the S7comm Server to automatically log all interactions. The S7comm Server writes to the file system all interactions including IP address of originating host, timestamp, and memory block ID in the case of reading or writing. Next, snmpsim logs IP information what OIDs were accessed and timestamps. Finally, the lighttpd webserver includes all the major features of a modern webserver with detailed logging that includes IP address information, accesses website files, and timestamps. All of them log every interaction all the time.

5 Evaluation

As shown throughout Sect. 4, HoneyPLC is designed to effectively deceive attackers into believing that they are dealing with *real* PLCs. This section starts by enumerating a set of experimental questions, which are based on the limitations of existing approaches as presented in Sect. 3. Then we present a series of experiments designed to provide affirmative answers to each question backed up by experimental evidence. For this purpose, we used the following PLC models: Siemens S7-300, S7-1200, and S7-1500, as well as the Allen-Bradley MicroLogix 1100 and the ABB PM554-TP-ETH, which are shown in Fig. 5, as these models are common in

Fig. 5 PLCs procured for experimental purposes including, from left to right, Siemens S7-300, S7-1500, S7-1200, Allen-Bradley MicroLogix 1100, and ABB PM554-TP-ETH

practice. As an example, a query[2] on Shodan [27], shows more than a 1700 Internet-facing PLCs across several different countries. For each experiment, we describe its environmental setup, the methodologies used, and the results obtained. Table 2 shows a summary of the experiments we performed comparing HoneyPLC with other honeypots in the literature whose implementation was either available online or was obtained from their authors upon request. A description of the obtained results is provided next, and an extended discussion comparing HoneyPLC with related work is shown in Sect. 6.

5.1 Experimental Questions

As an initial step, we now enumerate the research questions we have attempted to collect evidence for by means of the experiments shown later in this section. For each question, we describe how it relates to the limitations described in Sect. 3 and what subsections presented later address it.

Q-1 **Can HoneyPLC support different *real* PLCs?**
Since current approaches provided limited support for various types of PLCs being widely used by ICS in practice, we were interested in exploring the capabilities of HoneyPLC to model different PLCs using the PLC Profiler Tool described in Sect. 4.2. This question is related to Limitation L-1, as discussed in Sect. 3. We strive to answer to this question in Sects. 5.2 and 5.2.5.

Q-2 **Can HoneyPLC conceal its honeypot nature from attackers?**

[2] https://www.shodan.io/search?query=siemens+port%3A102.

Table 2 Experimental comparison of PLC Honeypots

Keys: ○ = No coverage; ◑ = Limited coverage; ● = Optimal coverage

Experiment	Conpot [16]	SCADA HoneyNet [39]	Gaspot [50]	S7comm trace [51]	HoneyPLC
Nmap (Sect. 5.3)	◑	●	◑	◑	●
PLCScan (Sect. 5.3)	●	◑	N/A	◑	●
Honeyscore (Sect. 5.4)	●	○	○	○	●
Step 7 Manager (Sect. 5.5)	○	○	N/A	○	●
PLCinject (Sect. 5.7)	○	◑	○	○	●

Table 3 PLC devices supported by ICS Honeypots

Approach	Supported PLC devices
Gaspot [50]	Veeder Root Guardian AST
SCADA HoneyNet [39]	Siemens CP 343-1
Conpot [16]	Siemens S7-200, Allen Bradley LOGIX5561
Digital Bond's Honeynet [49]	Modicon Quantum PLC
DiPot [6]	Siemens S7-200
SHaPe [20]	IEC 61850-Compliant PLC
CryPLH [5]	Siemens S7-300
S7commTrace [51]	Siemens S7-300
Antonioli et al. [3]	Generic PLC
HoneyPhy [24]	Generic Analog Thermostat
HoneyPLC	Siemens S7-300, S7-1200, S7-1500, Allen-Bradley MicroLogix 1100, ABB PM554-TP-ETH

More specifically, can HoneyPLC fool widely used reconnaissance tools? Also, we were interested in obtaining evidence regarding the interactions HoneyPLC may have obtained when deployed in the *wild*, i.e., via an Internet connection. This question is related to Limitations L-2 and L-3. We elaborate on this question in Sects. 5.3, 5.4, and 5.6.

Q-3 **Can HoneyPLC effectively capture Ladder Logic code?**
Since capturing Ladder Logic code represents a highly desirable feature for analyzing threats to ICS, we were interested in exploring the capabilities of HoneyPLC, as described in Sect. 4, to properly carry out such task. This question is related to Limitation L-4 and is addressed in Sect. 5.7.

5.2 *Case Study: PLC Profiling*

As mentioned in Sect. 3, current state-of-the-art honeypots for PLCs have been modeled over a limited number of PLCs, as shown in Table 3, and support for any extensions is quite limited. Therefore, we were interested in exploring the capabilities of HoneyPLC to support PLCs of different models and manufacturers.

5.2.1 Profiling Siemens PLCs

First, we evaluate the ability of HoneyPLC to support PLCs manufactured by Siemens which are very common both in industry deployments and in academic research [42].

5.2.2 Environment Description

For our first case study, we procured three Siemens PLCs: the S7-300, the S7-1200 and the S7-1500 models, which are shown in Fig. 5. Each PLC was connected to a special power supply and data or Ethernet cables. Additionally, we used the Siemens Step7 Manager, tools to configure IP addressing. We also deployed the HoneyPLC Profiler Tool and Python 3 in a laptop host where we connected our PLCs.

5.2.3 Methodology

We connected each PLC model to our experimental laptop host and used our command line-based HoneyPLC Profiler Tool to create the PLC Profiles for the three PLCs. To launch the tool, we input the PLC IP address and the name of PLC Profile directory. While the HoneyPLC Profiler Tool starts querying data from the PLC progress messages are shown including error messages, if any. We encountered some difficulties while developing and testing the Profiler Tool. First, we had to expand the number of ports scanned to obtain a better Nmap fingerprint, so that Nmap reports it with a higher confidence. We also had to make adjustments to download the PLC websites to include images and correct HTML paths. Also, it was necessary to manually modify the PLC profile HTML files to correct broken links.

5.2.4 Results

Overall, we were successful in creating all three PLC profiles. These profiles were saved in our experimental laptop host file system and were later used in the other experiments depicted in this section. The HoneyPLC Profiler Tool took approximately 5 min to create each profile and we only had to make some small manual modifications to some HTML files, as mentioned before. For PLCs produced by Siemens, the retrieval of their corresponding profiles may be facilitated if the SNMP and the web server services are properly activated beforehand by following the instructions provided by the manufacturer or by using any other S7comm-enabled software, e.g., the Step7 Manager. Failure to perform this step may result in the creation of an incomplete profile.

5.2.5 Profiling Allen-Bradley and ABB PLCs

Additionally, we were interested in exploring the capabilities of HoneyPLC to support PLC manufacturers other than Siemens, so we can provide some general recommendations for practitioners interested in obtaining additional PLC profiles.

5.2.6 Environment Description

For this case study, we procured the Allen-Bradley MicroLogix 1100 and the ABB PM554-TP-ETH PLCs, which are shown in Fig. 5. Additionally, we used Allen-Bradley and ABB software tools to configure their IP addresses.

5.2.7 Methodology

As with our previous case study, we deployed the HoneyPLC Profiler Tool and Python 3 in a laptop host and connected each PLC to a special power supply. Also, we connected each PLC model to our experimental laptop host and used our command line-based HoneyPLC Profiler Tool as before.

5.2.8 Results

We successfully produced a profile for each of the PLCs under analysis and obtained the following recommendations to practitioners. First, for non-Siemens PLCs, it may become necessary to identify the network services they provide, as different vendors may implement a variety of protocols on different ports. As an example, the Allen-Bradley MicroLogix 1100 PLC uses port 80 to implement a light web server, similar to Siemens PLCs, whereas such a feature is not implemented by the ABB PM554-TP-ETH. Second, both non-Siemens PLCs under study also fail to support the SNMP service, which prevents the HoneyPLC Profiler Tool from retrieving a MIB database. Third, the Allen-Bradley MicroLogix 1100 PLC implements the industry standard EtherNet/IP protocol on port 2222 for configuration purposes, which differs from Siemens models that use the proprietary S7comm protocol. These differences may ultimately result in PLC Profiles that are different from the ones obtained for Siemens PLCs and may need to be subsequently addressed on a case-by-case basis. Fourth, whereas the Siemens PLCs use the proprietary S7comm protocol for loading Ladder Logic programs, the Allen-Bradley MicroLogix 1100 uses the Ethernet/IP protocol. In such regard, the ABB PM554-TP-ETH PLC uses the Nucleus Sand Database, which is mostly used for database record keeping, and whose use in PLC devices is not customary. Because both protocols are not currently supported by HoneyPLC, additional modifications may be required. For example, for the M554-TP-ETH PLC Profile, we modified the Honeyd template to open port 1201 as a Nucleus Sand DB simulation that can be used through the subsystem virtualization is not currently supported. For the MicroLogix 1100 PLC Profile, we modified the Profiler Tool port scan range to include not only well-known ports but also registered ports such as port 2222. Finally, Table 3 provides a comparison of the PLC models supported *out of the box* by related honeypots for ICS, which were also shown in Table 1. The positive results obtained in our two case studies give support to answer Q-1 in the affirmative.

5.3 Resilience to Reconnaissance Experiment

The moment the true nature of HoneyPLC (or any other honeypot) is revealed to an attacker, the quantity and value of the gathered interaction data may significantly decrease. Therefore, we aimed to test the resilience of HoneyPLC to Nmap and PLCScan, described in Sect. 2, which are well-known tools for reconnaissance. Additionally, we tested how existing honeypots, namely Gaspot [50], S7commTrace [51], SCADA HoneyNet [39], and Conpot [30], perform in this regard.

5.3.1 Environment Description

Our experimental setup was composed of two physical computers: a *desktop* and a *laptop* host. The desktop host featured Ubuntu 18.04 LTS along with HoneyPLC, as well as the following tools: Honeyd, lighttpd, snmpsim, and S7comm server. We built Honeyd version 1.6d from source; the latest version is available in the official GitHub repository [9]. Also, we installed the lighttpd web server version 1.4.45. Next, we installed snmpsim version 0.4.7 and all its dependencies. Finally, we installed our S7comm server and our custom library. Conversely, the laptop host included the latest version of Nmap 7.80 as well as the three Siemens PLCs fingerprints in Nmap's fingerprint database nmap-os-db that were obtained as a result of the previous experiment. Additionally, we installed the latest version of PLCScan obtained from GitHub [43]. Both hosts were directly connected via an Ethernet cable. Subsequently, we downloaded and deployed the related honeypots mentioned before and connected them to the scanning host so that all of them would be in the local network.

5.3.2 Methodology

To create a baseline to compare the results of our experiments, the Nmap confidence data of the *real* PLCs featured in the previous experiment was obtained. With that in mind, a second test environment was composed of an additional host with Ubuntu 18.04 LTS and Nmap 7.80. Later, the additional host was directly connected to one of the three different PLCs (S7-300, S7-1200, and S7-1500) using an Ethernet cable. We installed the Step7 Manager in order to configure the network settings of the PLCs. Next, two different sets of Nmap scans were conducted with OS detection enabled: one set for HoneyPLC and another set for the *real* PLCs. Each PLC model was scanned 10 times. For the HoneyPLC experiment, the corresponding HoneyPLC Profile was installed so that the aforementioned applications were correctly configured. Next, we used PLCScan to scan each PLC Profile in similar

```
Device type: specialized|printer
Running (JUST GUESSING): Siemens embedded (97%),
Brother embedded (90%),
Toshiba embedded (88%)
OS CPE: cpe:/h:siemens:simatic_300
cpe:/h:brother:mfc-7820n
cpe:/h:toshibatec:e-studio-280
Aggressive OS guesses: Siemens Simatic 300
programmable logic controller (97%),
Siemens SPS programmable logic controller (91%),
Brother MFC-7820N printer (90%)
%\end{verbatim}
```

Fig. 6 Nmap scan results for the S7-300 PLC profile

fashion as the Nmap methodology. Afterwards, we turned to Gaspot, S7commTrace, SCADA HoneyNet, and Conpot. Each honeypot was scanned with Nmap's OS detection enabled 10 times. Finally, we used PLCScan on S7commTrace, SCADA HoneyNet, and Conpot. Gaspot was omitted as it does not support the S7comm protocol.

5.3.3 Results

The results of our Nmap experiment can be seen in Fig. 7 and show that for all three PLC models, the *real* PLCs gets the best confidence by a small margin. However, our PLC Profiles as provided by HoneyPLC were really close behind, thus providing positive evidence that our approach can provide effective covertness, as required by our question Q-2. When Nmap cannot detect a perfect OS match, it suggests near-matches. The match has to be very close for Nmap to do this by default. Nmap will tell you when an imperfect match is printed and display its confidence level (percentage) for each guess [32]. As an example, Fig. 6 shows the Nmap Scan results for our S7-300 PLC Profile. These results are encouraging since for all scans across all sets Nmap identified the correct PLC model with the highest confidence. Our PLCScan experiments were also successful, as we were able to obtain and provide real PLC data using PLCScan against HoneyPLC for all three PLC Profiles. In addition, SCADA HoneyNet was identified as a Siemens CP 343-1 PLC, and however, Gaspot, S7commTrace, and Conpot were fingerprinted as Linux OS with a 100% confidence, with no mention of any PLC device. Regarding PLCScan, Conpot was identified as an S7-200 PLC and SCADA HoneyNet and S7commTrace provided connection information but displayed an empty PLCScan report. Our results are even more significant due to the fact that a Linux kernel simulation of the TCP/IP Stack, as implemented by several related approaches, including Gaspot and Conpot, will not deceive Nmap [5].

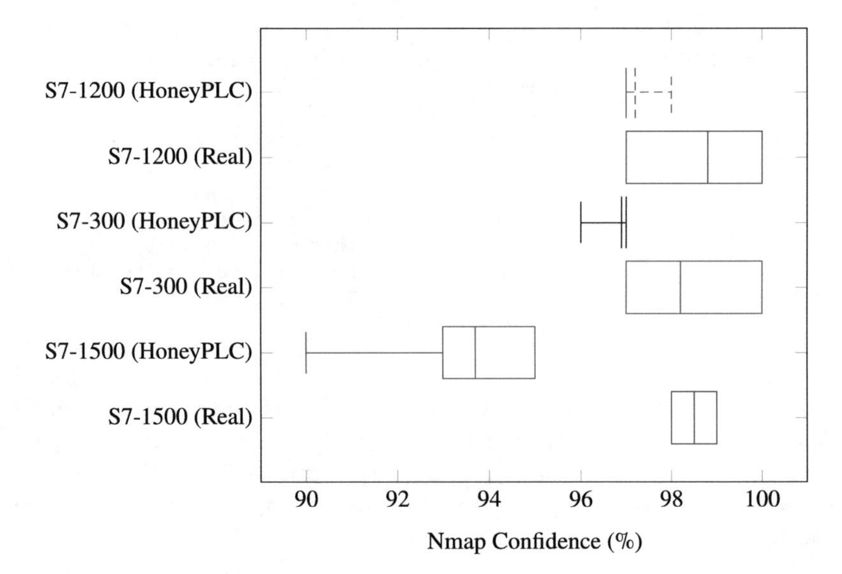

Fig. 7 Nmap scan results. All three profiles obtained at least a 90% confidence rate. The S7-300 and S7-1200 profile obtained rates comparable with their real counterparts. Gaspot and Conpot are fingerprinted as a Linux OS host with a 100% confidence, so they are excluded from this chart

5.4 Shodan's Honeyscore Experiment

As with the previous experiment, Shodan, described in Sect. 2.2, is actively leveraged in practice, along with its corresponding Shodan API to detect honeypots exposed to the Internet with a high degree of accuracy. Therefore, we were interested in the capabilities of HoneyPLC to deal with this state-of-the-art tool.

5.4.1 Environment Description

For this experiment, we deployed three AWS EC2 instances accessible from the Internet with the following specifications: 2 vCPUs, 4GB RAM, and Ubuntu 18.04 LTS OS, exposing TCP ports 80 and 102 and UDP port 161. Then, we deployed HoneyPLC on each one of them featuring all of our three PLC profiles, following the configuration steps detailed in the previous experiment. We also deployed four additional AWS instances hosting Conpot, Gaspot, S7commTrace, and SCADA HoneyNet.

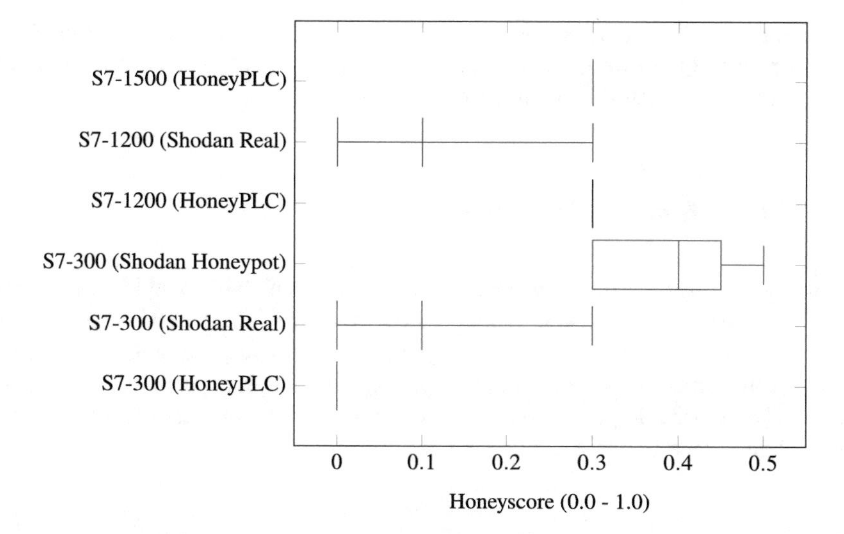

Fig. 8 Shodan Honeyscore results. Our HoneyPLC PLC profiles perform better than other honeypots found in Shodan and at the same level as *real* PLCs

5.4.2 Methodology

We obtained the Shodan Honeyscores, whose methodology is described in Sect. 2.2, of each of our HoneyPLC PLC Profiles, other honeypots for the same PLC models that were publicly exposed to the Internet and Gaspot, Conpot, S7commTrace, and SCADA HoneyNet. For such a purpose, we leveraged Shodan to gather data of Internet-facing *real* PLCs and PLCs flagged as honeypots. We looked at open ports, geolocation, Honeyscore, PLC model and IP addresses. Later, we compared these data to the one obtained for our HoneyPLC PLC Profiles. Once deployed to the Internet, it took about a week for Shodan to index our honeypots and identify the S7comm and HTTP services on ports 102 and 80.

5.4.3 Results

The results of our Shodan experiment, depicted in Fig. 8, show that Shodan assigns a Honeyscore of 0.0 to our S7-300 profile and how this Honeyscore compares to *real* S7-300 PLCs and other S7-300 honeypots found in the wild. Moreover, our S7-1200 and S7-1500 profiles got a 0.3 Honeyscore, which is comparable with the one obtained by *real* S7-1200 PLCs as indexed by Shodan. Unfortunately, at the time this experiment was performed, we were not able to find any S7-1200 honeypots in Shodan for comparison. Regarding the other four AWS instances, S7commTrace, Gaspot, and SCADA HoneyNet were not indexed by Shodan as they crashed when Shodan's crawler tried to interact with them. Thus, they could not be assigned a Honeyscore. Conpot, however, was successfully indexed and was assigned a 0.3 Honeyscore. Overall, these results add compelling evidence with respect to Question Q-2, showing that HoneyPLC is effective at maintaining covertness against state-of-the-art reconnaissance tools.

5.5 Step7 Manager Experiment

We designed an experiment to test the capabilities of the HoneyPLC S7Comm Server, discussed in Sect. 4.3, against Step7 Manager [4], a Siemens proprietary software used to configure, write, and upload ladder logic programs to PLCs. For comparison purposes, we attempted to perform the same experiment on Conpot, the SCADA HoneyNet, and S7commTrace, which claim support for the S7comm protocol, as shown in Table 2.

5.5.1 Environment Description

For this experiment, we used a Windows XP virtual environment installed on a *desktop* host. Additionally, we installed HoneyPLC, the related work honeypots shown in Table 2, and all three Siemens PLC Profiles in different Ubuntu 18 LTS VMs and connected them to the Windows XP host.

5.5.2 Methodology

To test the compatibility of a particular honeypot with Step7 Manager, we performed the following: first, we attempted a direct, initial connection to the tool by using the 'Go Online' GUI feature. Second, we used Step7 Manager to list all the memory blocks contained within a given honeypot. Third, we also tried to upload a memory block to each honeypot, and finally, in a reciprocal action, we tried to download the contents of a memory block, which was previously stored by each honeypot under test.

5.5.3 Results

Our results show that HoneyPLC is the only implementation capable of handling all of the functionality previously mentioned, as is shown in Table 2. Conpot, S7commTrace, and SCADA HoneyNet were able to establish the initial connection, and however, the Step7 Manager threw a connection timeout error, preventing any further interaction and resulting in an aborted execution. Moreover, as S7commTrace is a high-interaction honeypot that implements features similar to the ones provided by HoneyPLC's S7comm Server, we strove to provide an extended comparison between them. The HoneyPLC S7comm Server improves over

Table 4 Comparison of S7comm function codes

S7comm implementation	Functions	Subfunctions
HoneyPLC	13	18
S7commTrace	12	14

S7commTrace by providing more functions and subfunctions as shown in Table 4. Specifically, it adds an error response function and insert block, delete block, blink LED, and cancel password subfunctions. The error response function and the delete and insert block functions, in particular, are important when injecting ladder logic programs and connecting with Step7 Manager. Overall, besides providing compatibility with Step7 Manager, HoneyPLC also provides enhanced capabilities for capturing ladder logic, e.g., reading and writing memory blocks, which are not supported by S7commTrace.

5.6 Internet Interaction Experiment

In order to explore the capabilities of HoneyPLC to interact with external, non-controlled agents, e.g., attackers, we designed an experiment intended to expose the PLC Profiles discussed in previous experiments to remote connections via Internet.

5.6.1 Environment Description

We leveraged the environmental setup we designed for our previous Shodan-based experiment in Sect. 5.4. Also, we used the same AWS EC2 instances equipped with PLC Profiles for the S7-300, S7-1200, and S7-1500 PLCs.

5.6.2 Methodology

We exposed the EC2 instances to the Internet for a period of 5 months. Using the HoneyPLC logging capabilities discussed in Sect. 4.5, we logged all received interactions. Later on, we analyzed such logs and obtained the results we discuss next.

5.6.3 Results

As a result of this experiment, more than 5GB of data were recorded. Table 5 shows the different S7comm function commands received by each PLC Profile. The fact that we recorded these functions means that external agents interacted with HoneyPLC beyond a simple connection performing reconnaissance tasks. Additionally, we received 4 PLC Stop functions on our S7-300 Profile, which stops

Table 5 S7comm function commands received

PLC profile	Setup communication	Read SZL	PLC stop	List blocks
S7-300	600	1013	4	80
S7-1200	202	324	0	0
S7-1500	292	343	0	0

Table 6 HTTP and SNMP interactions received

PLC profile	HTTP conversations	HTTP login attempts	SNMP get requests
S7-300	2060	205	1925
S7-1200	1791	30	567
S7-1500	13	0	1271

the current ladder logic program execution, suggesting that external agents tried to disrupt the PLCs' operation. Table 6 shows that our honeypots also received thousands of HTTP conversations and logged multiple HTTP authentication attempts on their administration websites, including the usernames and passwords used by the external parties. These authentication attempts could have been made by web crawlers or malicious actors trying different well-known or default passwords to log into the PLCs admin website. Additionally, we also recorded thousands of SNMP get requests that downloaded our PLC Profile's MIBs several times. Table 7 shows the distribution of S7comm connections based on geographical location. It can be noted that countries with most connections have historically been either the target or the initiators of attacks against ICS [14] recorded in the literature. Finally, at the time of writing this chapter, no attempts to inject malicious ladder logic into our honeypots were recorded. Such an attack would have been signaled by an attempt to write a memory block inside a PLC. Despite this limitation, the amount and nature of the interactions obtained provide additional support for affirmatively answering Question Q-2, showing that HoneyPLC can effectively engage external agents and tools.

5.7 Ladder Logic Capture Experiment

Finally, we were interested in exploring the capabilities of HoneyPLC to properly collect Ladder Logic malware that is injected by attackers, following the Case Scenario described in Sect. 4.1.

5.7.1 Environment Description

For this experiment, we leveraged the same HoneyPLC AWS test environment described in Sect. 5.4 for our Shodan experiment. Additionally, we locally deployed

Table 7 S7comm connections received by geolocation

Geo-location	S7-300	S7-1200	S7-1500	Geo-location	S7-300	S7-1200	S7-1500
United States	359	142	250	Netherlands	22	13	11
United Kingdom	2	1	3	Japan	8	2	2
Turkey	2	0	1	Italy	1	0	0
Switzerland	3	1	1	Iceland	1	1	2
Sweden	1	1	1	Hong Kong	2	1	1
South Korea	1	0	0	Germany	18	9	12
Slovakia	0	1	1	France	10	5	7
Singapore	4	3	5	Denmark	1	1	0
Russia	28	12	14	China	42	16	26
Romania	6	2	4	Canada	3	2	3
Poland	1	0	0	Bulgaria	2	1	0
Panama	2	1	3	Belize	3	3	3

Fig. 9 Ladder Logic payload example found in the Stuxnet malware

```
UC FC1865
POP
L DW#16#DEADF007
==D
BEC
L DW#16#0
L DW#16#0
```

Conpot, Gaspot, S7commTrace, and SCADA HoneyNet. For Gaspot, we downloaded the latest version from GitHub [15] and installed it in an Ubuntu 18 LTS host. Next, for Conpot, we also downloaded the latest version from GitHub [30] and installed it from source in an Ubuntu 18 LTS host. Finally, we deployed the latest version of the SCADA HoneyNet [39] also in an Ubuntu 18 LTS. We faced some problems deploying the SCADA HoneyNet as it is currently not maintained at all (the latest version was released in 2004), and however, we were able to deploy the S7comm portion of the honeypot, enabling us to conduct this experiment. To test our implementation, we employed PLCinject [18], a research tool published by the SCADACS team, which is capable of injecting arbitrary compiled ladder logic programs into a PLC memory block. Figure 9 shows a sample of the ladder logic code dropped by the Stuxnet malware. We also set up a laptop host with Ubuntu 18.04 LTS installed with the latest version of PLCinject available on GitHub [41]. Since PLCinject also leverages the Snap7 framework, we installed a custom library and compiled PLCinject from source. We also used the Windows XP host described in Sect. 5.5 with Step7 Manager.

5.7.2 Methodology

Figure 10 illustrates our setup and methodology. The PLCinject host contains the ladder logic program sample that PLCinject will upload into HoneyPLC, which

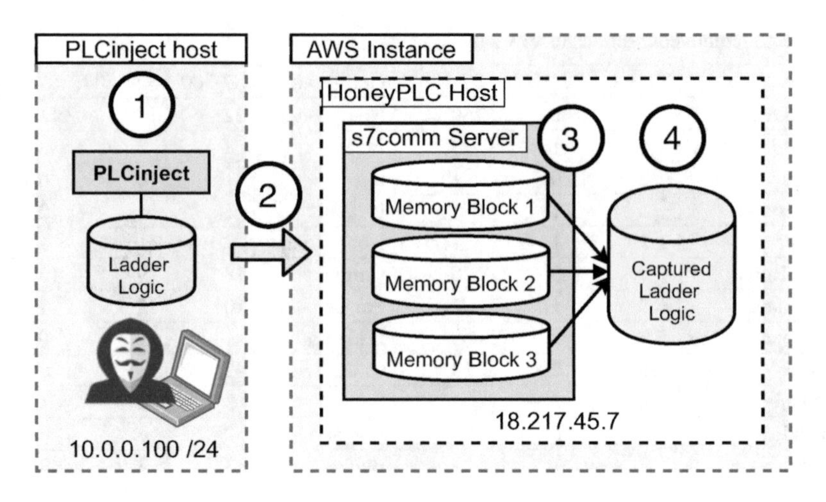

Fig. 10 Capturing Ladder Logic: initially, the attacker selects a malicious program and leverages PLCinject (1), which then establishes communication with an AWS instance running HoneyPLC (2). Malicious code is injected into a previously selected memory block exposed by the S7comm server (3) and finally written into a file repository (4)

resides inside an AWS instance exposing a set of standard PLC memory blocks. We leveraged the capabilities of PLCinject to connect and interact with the HoneyPLC host, eventually injecting the desired Ladder Logic program by using the command line. Later, using the Step7 Manager GUI, we created a new project and wrote a sample ladder logic to be injected into HoneyPLC. Next, we used the Step7 Manager to list the available memory blocks and then use the upload function to inject the sample ladder logic program into HoneyPLC. Later, we conducted another set of experiments focused on Gaspot, Conpot, S7commTrace, and the SCADA HoneyNet. We configured each of the honeypots with the correct IP addresses and ports and used PLCinject and the Step7 Manager to write the sample program into them, following the same process used for HoneyPLC.

5.7.3 Results

Our experiments were successful as we were able to inject a sample ladder logic program into HoneyPLC using both, PLCinject and the Step7 Manager. After the injection was completed, we logged into our honeypot file system and found the ladder logic file with its corresponding timestamp, which matched the contents of the blocks previously updated to PLCinject, as described in the previous paragraph. More to the point, after the Step7 Manager injection was completed, we downloaded our own sample program from HoneyPLC's S7comm server and used the ladder logic editor (included with Step7 Manager) to corroborate that our sample program was in fact saved in HoneyPLC's S7comm server. It is worth mentioning that the Step7 Manager did not crash or threw any errors while interacting with HoneyPLC's

S7comm server. This adds evidence as to the level of interaction that HoneyPLC provides. Regarding the Gaspot honeypot, our results show that it is not possible to inject any program into it. In fact, the TCP connection times out, and there is no reply. The results from Conpot show that it can, in fact, open a connection to TCP port 102, and however, it is reset, and the program upload cannot continue. S7commTrace results in the S7comm connection not being established. Finally, the S7comm portion of the SCADA HoneyNet accepts the TCP port 102 connection and starts the upload function needed to upload the ladder logic program, and however, after the upload function ends, there is no data saved or even transmitted. There results provide evidence for answering Question Q-3 affirmatively.

6 Discussion and Future Work

Before rounding up this chapter, we now present an extended discussion on the novelty, the features, and the experimental results obtained using HoneyPLC, as presented in previous sections. Also, we engage in a short discussion on the observed shortcomings of our approach and discuss interesting topics for future work that may benefit from using HoneyPLC as a supporting framework.

6.1 Comparing HoneyPLC with Previous Approaches

Following the comparison shown in Table 1, HoneyPLC provides significant improvements over the current state of the art of honeypots for PLCs. First, HoneyPLC provides better covertness capabilities than the ones provided by related works in the literature, as shown in the experimental procedures summarized in Table 2. Moreover, as detailed in Sect. 4.3, HoneyPLC provides advanced TCP/IP simulation based on Honeyd, plus the careful simulation of different domain-specific protocols. Whereas the simulation of various protocols is shared by many approaches in the literature, only HoneyPLC and SCADA HoneyNet [39] leverage the rich simulation features provided by the Honeyd framework. Second, the extensibility features of HoneyPLC, discussed in Sect. 4.2, allow for the effective simulation of different PLCs deployed in practice, as it was shown in the experimental procedures detailed in Sect. 5.2. Such a feature is not shared by any other approach in the literature, as shown in Table 1. Only a few approaches provide limited extensibility features, but those are mostly based on manually changing some configuration settings for the PLCs they support. As shown in Sect. 4.2, the HoneyPLC's Profiler Tool supports the collection and configuration settings for different *real* PLCs, which may allow for practitioners to create and distribute PLC Profiles for HoneyPLC for many different brands and models used in practice. Finally, HoneyPLC's Ladder Logic Capture feature is optimal for the understanding and analysis of malicious programs tailored for PLCs, which is not provided by any other related work, as shown in Table 2.

6.2 Limitations

Despite the innovative features of HoneyPLC and the promising evaluation results shown in Sect. 5, we identified the following limitations to our approach. First, as shown in Table 1, HoneyPLC does not provide support for modeling physical interactions as depicted by PLCs in practice. To solve this, future versions of HoneyPLC may be enhanced with a generic, general-purpose framework that facilitates the collection and subsequent modeling of physical interactions that can further engage and deceive attackers. Second, despite numerous attempts, we were unable to test HoneyPLC against Stuxnet, shown in Sect. 2.3, up to the point in which PLCs are injected with Ladder Logic code. This problem was also encountered by seasoned partners in industry, as it was revealed to us in private conversations. As an alternative, we strove to replicate a similar code injection scenario as shown in Sect. 5.7. Finally, as discussed in Sect. 5.6, we were not able to capture any Ladder Logic code injection attempts while exposing HoneyPLC to the internet during an extended period of time. We believe that such a thing may not necessarily represent a limitation in the capabilities of our approach, as shown in Sect. 5.7. However, we agree that future work focused on capturing instances of malicious code may obtain significant evidence of the suitability of HoneyPLC for engaging and deceiving external agents.

6.3 Future Work

First, we plan to add support to other ICS specific network protocols such as Modbus, which is widely implemented by other approaches in the literature. Second, we plan to expand the PLC Profile Repository of HoneyPLC, which is graphically depicted in Fig. 2 as an important part of our approach, to include several different PLC Profiles simulating other *real* PLCs widely used in practice, which may have been produced by different manufacturers and may include a diverse set of configuration options. We believe such a feature will likely increase the impact of HoneyPLC in many different projects in the research community, as well as in real-life ICS environments. Third, we plan to use HoneyPLC as a basis for simulating rich ICS infrastructures completely in software, modeling components like SCADA and other devices. Current ICSs are proprietary, closed, and composed of a plethora of costly devices, which clearly complicates the effective development and testing of new protection tools by researchers. In such regard, we believe that HoneyPLC can be combined with other emerging technologies such as *software-defined networks* (SDN) [22], to produce an automated, highly configurable, and automated approach effectively simulating ICS environments. Finally, we plan to turn HoneyPLC into a comprehensive suite for malware analysis for ICS by incorporating Ladder Logic analysis tools such as ICSREF [17], as well as other works such as PLCinject, featured in Sect. 5.7.

7 Conclusions

Attacks targeting ICS are now more real than ever and their consequences may be catastrophic. In such regard, honeypots help us understand and prepare for these attacks, and however, current implementations do not allow us to analyze and tackle brand new threats as desired. To overcome this situation, we have introduced HoneyPLC, a convenient and flexible honeypot, which significantly pushes the *state of the art* of the field forward. Additionally, we have provided experimental evidence that demonstrates that HoneyPLC outperforms existing honeypots in the literature, achieving a performance level comparable to *real* PLC devices. Finally, the HoneyPLC advanced extensibility features, which may allow HoneyPLC to better serve the heterogeneous world of ICS. As an example, we expect for practitioners to create and openly distribute many new PLC Profiles for a variety of PLCs used in practice, thus positioning HoneyPLC not only as a helpful tool for preventing and deterring ongoing attacks but also as the starting point for designing and evaluating new protection technologies for mission-critical cyber-physical systems and infrastructure.

Acknowledgments This work was supported in part by the National Science Foundation (NSF) under grant 1651661, the Department of Energy (DoE) under grant DE-OE0000780, the Army Research Office under grant W911NF-17-1-0370, the Defense Advanced Research Projects Agency (DARPA) under the agreements HR001118C0060 and FA875019C0003, the Institute for Information & communications Technology Promotion (IITP) under grant 2017-0-00168 funded by the Korea government (MSIT), a grant from the Center for Cybersecurity and Digital Forensics (CDF) at Arizona State University, and a grant from Texas A&M University—Corpus Christi. Any opinions, findings, conclusions, or recommendations expressed in this material are those of the author(s) and do not necessarily reflect the views of the United States Government or any agency thereof.

References

1. ABB: Plc automation. https://new.abb.com/plc. Accessed: 2020-02-24
2. Allen-Bradley: Programmable controllers. https://ab.rockwellautomation.com/Programmable-Controllers. Accessed: 2020-02-24
3. Antonioli, D., Agrawal, A., Tippenhauer, N.O.: Towards high-interaction virtual ics honeypots-in-a-box. In: Proc. of the 2nd ACM Workshop on Cyber-Physical Systems Security and Privacy, pp. 13–22 (2016)
4. Berger, H.: Automating with STEP7 in STL and SCL: programmable controllers Simatic S7-300/400. Publicis (2006)
5. Buza, D.I., Juhász, F., Miru, G., Félegyházi, M., Holczer, T.: Cryplh: Protecting smart energy systems from targeted attacks with a plc honeypot. In: Int. Workshop on Smart Grid Security, pp. 181–192. Springer (2014)
6. Cao, J., Li, W., Li, J., Li, B.: Dipot: A distributed industrial honeypot system. In: Int. Conference on Smart Computing and Communication, pp. 300–309. Springer (2017)
7. Case, D.U.: Analysis of the cyber attack on the Ukrainian power grid. Electricity Information Sharing and Analysis Center (E-ISAC) vol. 388 (2016)
8. Cybersecurity (CISA) I.S.A.: Apt cyber tools targeting ics/scada devices (2022). https://www.cisa.gov/uscert/ncas/alerts/aa22-103a

9. DataSoft/Honeyd: (2020). https://github.com/DataSoft/Honeyd. Original-date: 2011-12-09T22:40:03Z
10. Dragos, I.: Crashoverride: Analysis of the threat to electric grid operations (2017). Online: https://dragos.com/blog/crashoverride/CrashOverride-01.pdf
11. Dragos, I.: Chernovite's pipedream malware targeting industrial control systems (ics) (2022). https://www.dragos.com/blog/industry-news/chernovite-pipedream-malware-targeting-industrial-control-systems/
12. Falliere, N., Murchu, L.O., Chien, E.: W32. stuxnet dossier. White paper, Symantec Corp., Security Response **5**(6), 29 (2011)
13. Hahn, A.: Operational technology and information technology in industrial control systems. In: Cyber-security of SCADA and Other Industrial Control Systems, pp. 51–68. Springer (2016)
14. Hemsley, K.E., Fisher, E., et al.: History of industrial control system cyber incidents. Tech. rep., Idaho National Lab.(INL), Idaho Falls, ID (United States) (2018)
15. Hilt, S.: Gaspot released at blackhat 2015 (2016). https://github.com/sjhilt/GasPot
16. Jicha, A., Patton, M., Chen, H.: Scada honeypots: An in-depth analysis of conpot. In: 2016 IEEE Conference on Intelligence and Security Informatics (ISI), pp. 196–198. IEEE (2016)
17. Keliris, A., Maniatakos, M.: ICSREF: A framework for automated reverse engineering of industrial control systems binaries. In: Network and Distributed System Security Symposium, (NDSS). The Internet Society (2019)
18. Klick, J., Lau, S., Marzin, D., Malchow, J.O., Roth, V.: Internet-facing plcs-a new back orifice. Blackhat USA, pp. 22–26 (2015)
19. Kneschke, J.: Lighttpd-fly light. https://www.lighttpd.net/ (2020)
20. Kołtyś, K., Gajewski, R.: SHaPe: A honeypot for electric power substation. J. Telecomm. Inf. Technol. (4), 37–43 (2015)
21. Langner, R.: Stuxnet: Dissecting a cyberwarfare weapon. IEEE Secur. Priv. **9**(3), 49–51 (2011)
22. Lantz, B., Heller, B., McKeown, N.: A network in a laptop: Rapid prototyping for software-defined networks. In: Proc. of the 9th ACM SIGCOMM Workshop on Hot Topics in Networks, Hotnets-IX. Association for Computing Machinery, New York, NY, USA (2010)
23. Lian, F.L., Moyne, J., Tilbury, D.: Network design consideration for distributed control systems. IEEE Trans. Control Syst. Technol. **10**(2), 297–307 (2002)
24. Litchfield, S., Formby, D., Rogers, J., Meliopoulos, S., Beyah, R.: Rethinking the honeypot for cyber-physical systems. IEEE Internet Comput. **20**(5), 9–17 (2016)
25. López-Morales, E., Rubio-Medrano, C., Doupé, A., Shoshitaishvili, Y., Wang, R., Bao, T., Ahn, G.J.: HoneyPLC: A next-generation honeypot for industrial control systems. In: Proceedings of the 2020 ACM SIGSAC Conference on Computer and Communications Security, CCS '20, pp. 279–291. Association for Computing Machinery, New York, NY, USA (2020). https://doi.org/10.1145/3372297.3423356
26. Lyon, G.F.: Nmap network scanning: The official Nmap project guide to network discovery and security scanning. Insecure (2009)
27. Matherly, J.: Complete guide to shodan. Shodan, LLC (2016-02-25), vol. 1 (2015)
28. Matherly, J.: Personal communication (2019)
29. Mokube, I., Adams, M.: Honeypots: concepts, approaches, and challenges. In: Proc. of the 45th Annual Southeast Regional Conference, pp. 321–326 (2007)
30. MushMush: Conpot (2020). https://github.com/mushorg/conpot
31. Nardella, D.: Snap7 (2018). http://snap7.sourceforge.net/
32. nmap.org: Os detection (2022). https://nmap.org/book/man-os-detection.html
33. Provos, N.: Honeyd-a virtual honeypot daemon. In: 10th DFN-CERT Workshop, Hamburg, Germany, vol. 2, p. 4 (2003)
34. Provos, N., Holz, T.: Virtual Honeypots: From Botnet Tracking to Intrusion Detection. Pearson Education (2007)
35. Repository, U.M.: snmpwalk - retrieve a subtree of management values using snmp getnext requests (2019). http://manpages.ubuntu.com/manpages/bionic/man1/snmpwalk.1.html
36. Repository, U.M.: Wget - the non-interactive network downloader (2019). http://manpages.ubuntu.com/manpages/disco/en/man1/wget.1.html

37. Response, S.I.: Dragonfly: Cyberespionage attacks against energy suppliers. Tech. Rep., July (2014)
38. S7comm - The Wireshark Wiki (2016). https://wiki.wireshark.org/S7comm
39. SCADA HoneyNet Project: Building Honeypots for Industrial Networks (2020). http://scadahoneynet.sourceforge.net/
40. SCADACS: Snap7 (2017). https://github.com/SCADACS/snap7
41. SCADACS/PLCinject (2020). https://github.com/SCADACS/PLCinject. Original-date: 2015-07-13T09:38:19Z
42. Schwartz, M.D., Mulder, J., Trent, J., Atkins, W.D.: Control system devices: Architectures and supply channels overview. Sandia Report SAND2010-5183, Sandia National Laboratories, Albuquerque, New Mexico, vol. 102, 103 (2010)
43. Searle, J.: plcscan (2015). https://github.com/meeas/plcscan
44. Shi, J., Wan, J., Yan, H., Suo, H.: A survey of cyber-physical systems. In: 2011 International Conference on Wireless Communications and Signal Processing (WCSP), pp. 1–6 (2011). https://doi.org/10.1109/WCSP.2011.6096958
45. Siemens: The intelligent choice for your automation task: Simatic controllers. https://new.siemens.com/global/en/products/automation/systems/industrial/plc.html. Accessed: 2020-02-24
46. Slowik, J.: Anatomy of an attack: Detecting and defeating crashoverride. VB2018, October (2018)
47. Stouffer, K., Falco, J., Scarfone, K.: Nist special publication 800-82: Guide to industrial control systems (ics) security. National Institute of Standards and Technology (NIST), Gaithersburg, MD (2008)
48. Stouffer, K., Lightman, S., Pillitteri, V., Abrams, M., Hahn, A.: Nist special publication 800-82, revision 2: Guide to industrial control systems (ics) security. National Institute of Standards and Technology (2014)
49. Wade, S.M.: Scada honeynets: The attractiveness of honeypots as critical infrastructure security tools for the detection and analysis of advanced threats (2011)
50. Wilhoit, K., Hilt, S.: The gaspot experiment : Unexamined perils in using gas-tank-monitoring systems. GitHub repository (2020)
51. Xiao, F., Chen, E., Xu, Q.: S7commtrace: A high interactive honeypot for industrial control system based on s7 protocol. In: Int. Conference on Information and Communications Security, pp. 412–423. Springer (2017)

Using Amnesia to Detect Credential Database Breaches

Ke Coby Wang and Michael K. Reiter

1 Introduction

Credential database breaches have become a widespread security problem. Verizon confirmed 3950 database breaches globally between Nov. 2018 and Oct. 2019 inclusive; of those 1665 breaches for which they identified victims, 60% leaked credentials [43].[1] Credential database breaches are the largest source of compromised passwords used in credential stuffing campaigns [42], which themselves are the cause of the vast majority of account takeovers [41]. Unfortunately, there is usually a significant delay between the breach of a credential database and the discovery of that breach; estimates of the average delay range from 7 [23] to 15 [41] months. The resulting window of vulnerability gives attackers the opportunity to crack the passwords offline (if the stolen credential database stores only password hashes), to determine their value by probing accounts using them [41], and then to either use them directly to extract value or sell them through illicit forums for trafficking stolen credentials [41, 42].

Decoy passwords have been proposed in various forms to interfere with the attacker's use of a stolen credential database. In these proposals (see Sect. 2), a

This paper is originally appeared in the *Proceedings of the 30th USENIX Security Symposium*, August 2021.

[1] This number excludes 14 breaches of victims in Latin America and the Caribbean for which the rate of credential leakage was not reported.

K. C. Wang · M. K. Reiter (✉)
Duke University, Durham, NC, USA
e-mail: kwang@cs.unc.edu; coby.wang@duke.edu; michael.reiter@duke.edu

T. Bao et al. (eds.), *Cyber Deception*, Advances in Information Security 89,
https://doi.org/10.1007/978-3-031-16613-6_9

site (the target) stores decoy passwords alongside real passwords in its credential database, so that if the attacker breaches the database, the correct passwords are hidden among the decoys. The attacker's entry of a decoy password can alert the target to its breach; the term *honeywords* has been coined for decoys used in this way [25].

While potentially effective, honeywords suffer from two related shortcomings that, we believe, have limited their use in practice. First, previous proposals that leverage honeywords require a trusted component to detect the entry of a honeyword, i.e., a component that retains secret state even *after* the target has been breached. Such a trusted component is a strong assumption, however, and begs the question of whether one could have been relied upon to prevent the breach of the target's database in the first place. Second, the effectiveness of honeywords depends on the indistinguishability of the user-chosen password from the decoys when they are exposed to an attacker. However, because so many users reuse their chosen passwords across multiple accounts [11, 36, 44], an attacker can simply test (or *stuff*) all passwords for an account leaked from the target at accounts for the same user at other sites. Any password that works at another site is almost certainly the user-chosen password at the target.

In this paper, we resolve both of these difficulties and realize their solutions in a framework called Amnesia. First, we show that honeywords can be used to detect a target's database breach with no persistent secret state at the target, a surprising result in light of previous work. Specifically, we consider a threat model in which the target is breached passively but completely and potentially repeatedly. Without needing to keep secrets from the attacker, Amnesia nevertheless enables the target to detect its own breach probabilistically, with benefits that we quantify through probabilistic model checking. Our results show, for example, that Amnesia substantially reduces the time an attacker can use breached credentials to access accounts without alerting the target to its breach.

To address credential stuffing elsewhere to distinguish the user-chosen password from the honeywords, Amnesia enables the target to monitor for the entry of passwords stolen from it at other sites, called monitors. Via this framework, incorrect passwords entered for the same user at monitors are treated (for the purposes of breach detection) as if they had been entered locally at the target. One innovation to accomplish this is a cryptographic protocol by which a monitor transfers the password attempted in an unsuccessful login there to the target, but only if the attempted password is one of the passwords (honey or user-chosen) for the same account at the target; otherwise, the target learns nothing. We refer to this protocol as a *private containment retrieval* (PCR) protocol, for which we detail a design and show it secure. Leveraging this PCR protocol, we show that Amnesia requires no trust in the monitors for the target to accept a breach notification. In other words, even if a monitor is malicious, it cannot convince an unbreached target that it has been breached.

We finally describe the performance of our Amnesia implementation. Our performance results suggest that the computation, communication, and storage costs of distributed monitoring are minimal. For example, generating a monitoring response takes constant time and produces a constant-size result, as a function of the number of honeywords, and is practical (e.g., no more than 10ms and about 1KB, respectively).

To summarize, our contributions are as follows:

- We develop the first algorithm leveraging honeywords by which a target site can detect the breach of its password database, while relying on *no* secret persistent state. We evaluate this design using probabilistic model checking to quantify the security it provides.
- We extend this algorithm with a protocol to monitor accounts at monitors to detect the use of the target's honeywords there. Our algorithm is the first such proposal to ensure no false detections of a database breach, despite even malicious behavior by monitors.
- A core component of this algorithm is a new cryptographic protocol we term a *private containment retrieval* protocol, which we detail and prove correct.
- We describe the performance of our algorithm using an implementation and show that it is practical.

2 Related Work

Within research on decoy passwords, we are aware of only two proposals by which a target can detect its *own* breach using them. Juels and Rivest [25] coined the term *honeywords* for decoy passwords submitted in login attempts to signal to a site that it was breached by an attacker. In their proposal and works building on it (e.g., [14]), the target is augmented with a trusted *honeychecker* that stores which of the passwords listed with the account is the user-chosen one; login attempts with others alert the site to its breach. Almeshekah et al. [2] use a machine-dependent function (e.g., hardware security module) in the password hash at the target site to prevent offline cracking of its credential database if breached. Of more relevance here, an attacker who is unaware of this defense and so attempts to crack its database offline will produce plausible decoy passwords (*ersatzpasswords*) that, when submitted, alert the target site to its breach. The primary distinction between these proposals and ours is that ours permits a target to detect its own breach *without any secret persistent state*. In contrast, these proposals require a trusted component—the honeychecker or the machine-dependent function—whose state is assumed to remain secret even after the attacker breaches the site. In addition, we reiterate that ersatzpasswords are effective in alerting the target to its breach only if the attacker is unaware of the use of this scheme, as otherwise the attacker will know that passwords generated through offline cracking without access to the machine-dependent function are ersatzpasswords.

Other uses of decoy passwords leverage defenses at *other, unbreached* sites—either their online guessing defenses generically [5, 28] or their cooperation to check for decoy passwords specifically [5, 48]—to defend accounts whose credentials have been stolen, whether by phishing [48], user device compromise [5], or the target site's database breach [28]. While we extend our design in Sect. 5 to monitor for a target's honeywords being submitted in login attempts at monitors, to our knowledge our design is the first to eliminate the need for the target to trust another site in order to accept that a detected breach actually occurred. Specifically, in our design a monitor, even if malicious, cannot convince an unbreached site that it has been breached.

Various other works have leveraged decoy *accounts* to detect credential database breaches, i.e., accounts with no owner that, if ever accessed, reveal the breach of the account's site or a site where a replica of the account was created (e.g., [14, 21]). In Tripwire [13], each decoy account is registered with a distinct email address and password, for which the password at the email provider is the same. Any login to the email account (provided that the email provider is itself not compromised) suggests the breach of the website where that email address was used to register an account. Like the previously discussed proposals, this design places trust in the detecting party (the email provider or, in this case, the researchers working with it) to be truthful when reporting the breach of a target. Indeed, DeBlasio et al. report that sites' unwillingness to trust the evidence they provided of the sites' breaches was an obstacle to getting them to act.[2] Moreover, the utility of artificial accounts hinges critically on their indistinguishability from real ones, and if methods using them became effective in hindering attacker activity, ensuring the indistinguishability of these accounts would presumably become its own arms race. Our design is agnostic to whether it is deployed on real or decoy accounts, sidestepping the need for convincing decoy accounts but also demanding attention to the risks to real accounts that it might introduce.

To be fair, generation of honeywords that are sufficiently indistinguishable from real ones is itself a topic of active investigation (e.g., [1, 14, 45]). Here we will simply assume that a site can generate honeywords in isolation to satisfy certain properties, detailed in Sect. 3. The development of methods to achieve these properties is a separate concern.

An alternative to decoy passwords or accounts for defending against a breach of a site's credential database is for the site to instead leverage a breach-hardening service. Even after having breached the target's credential database, the attacker must succeed in an *online* dictionary attack with the breach-hardening service per stolen credential he wishes to use, provided that the breach-hardening service is itself not simultaneously breached (e.g., [15, 30–32, 40]). While differing in their details, these schemes integrate the breach-hardening service tightly into the target's operation, in the sense that, e.g., the benign failure of a breach-hardening service would interfere with login attempts at the target. In contrast, while the

[2] The paper concludes, "A major open question, however, is how much (probative, but not particularly illustrative) evidence produced by an external monitoring system like Tripwire is needed to convince operators to act, such as notifying their users and forcing a password reset" [13, Section 8].

benign failure of our $\mathsf{monitors}$ would render them useless for helping to detect the target's breach, the operation of the target would be otherwise unaffected.

3 Honeywords

We assume the existence of a randomized honeyword generator $\mathsf{HoneyGen}$ that, given an account identifier a, user-chosen password π_a, and integer k, produces a set Π_a containing π_a and k other strings and having the following properties. We use "$\leftarrow$" to denote assignment of the result of evaluating the expression on its right to the variable on its left, and "$\stackrel{\$}{\leftarrow}$" to denote sampling an element uniformly at random from the set on its right and assigning the result to the variable on its left.

First, the essential purpose of honeywords is to make it difficult for an adversary who breaches a credential database to determine which of the passwords listed for an account a is the user-chosen one. In other words, for any attacker algorithm A that is given the account identifier a and its set of passwords Π_a, we assume

$$\mathbb{P}\left(\pi = \pi_a \;\middle|\; \Pi_a \leftarrow \mathsf{HoneyGen}(a, \pi_a, k); \pi \leftarrow A(a, \Pi_a)\right) \approx \frac{1}{k+1} \tag{1}$$

Second, because honeywords are intended to alert the target to a breach of its credential database, avoiding false alarms requires that an adversary be unable to generate a honeyword for an account without having actually breached the target. In particular, this property would ideally be achieved even if the user-chosen password π_a is known, e.g., because the user was phished or because she reused π_a as her password at another site that was compromised. While these place the user's account at the target at risk, neither equates to the target's wholesale breach and so should not suffice to induce a breach detection at the target. That is, for any attacker algorithm B that knows only the account identifier a and user-chosen password π_a, we assume:

$$\mathbb{P}\left(\pi \in \Pi_a \setminus \{\pi_a\} \;\middle|\; \Pi_a \leftarrow \mathsf{HoneyGen}(a, \pi_a, k); \pi \leftarrow B(a, \pi_a)\right) \approx 0 \tag{2}$$

This assumption implies that any two invocations of $\mathsf{HoneyGen}(a, \pi_a, k)$ produce sets Π_a, Π'_a that intersect *only* in Π_a with near certainty. Otherwise, an adversary $B(a, \Pi_a)$ that invokes $\Pi'_a \leftarrow \mathsf{HoneyGen}(a, \pi_a, k)$ and returns a random $\pi \in \pi'_a \setminus \{\pi_a\}$ would violate (2). In other words, (2) implies that the honeywords generated at two different sites for the same user's accounts are distinct, even if the user reuses the same password for both accounts.

4 Detecting Honeyword Entry Locally

The first contribution of this paper is in demonstrating how the target site can detect its own breach while relying on no secret persistent state. We detail the threat model for this section in Sect. 4.1 and provide the detection algorithm in Sect. 4.2. We demonstrate the efficacy of this algorithm in Sect. 4.3.

4.1 Threat Model

Our goal is to enable a site, called the target, to detect that its credential database has been stolen. We assume that the target uses standard password-based authentication, i.e., in which the password is submitted to the target under the protection of a cryptographic protocol such as TLS.

We allow for an attacker to breach the target *passively only*, in which case it captures all persistent storage at the site associated with validating or managing account logins. Throughout this paper, this persistent storage is denoted DB, and information associated specifically with account a is denoted DB_a. In particular, the information captured includes the passwords listed for each of the site's user accounts (DB_a.auths); if stored as salted hashes, the attacker can crack the passwords offline. The attacker also captures any long-term cryptographic keys of the site. As will become relevant below, we allow the attacker to capture the site's persistent storage multiple times, periodically.

We stress that the information captured by the attacker includes only information *stored persistently* at the site. Recall that the principle behind honeywords is to leverage their use in login attempts to alert the target that its credential database has been stolen. As such, we must assume that transient information that arrives in a login attempt but is not stored persistently at the site is unavailable to the attacker. Otherwise, the attacker would simply capture the correct password for an account once the legitimate owner of that account logs in. Since the site's breach leaks any long-term secrets, this assumption implies that the cryptographic protocol protecting user logins provides perfect forward secrecy [20][3] or that the attacker simply cannot observe login traffic. Similarly, we assume that despite breaching the target site, the attacker cannot predict future randomness generated at the site.

We also highlight that, like in Juels and Rivest's honeyword design [25], we do not consider the active compromise of the target. In particular, the integrity of the target's persistent storage is maintained despite the attacker's breach, and the site always executes its prescribed algorithms. Without this assumption, having the

[3] Cohn-Gordon et al. [9] observe that for a passive attacker, perfect forward secrecy implies protection not only against the future compromise of the long-term key but also its past compromise.

target detect its own breach is not possible. We do, however, permit the attacker to submit login attempts to the target via its provided login interface.

Finally, while the adversary might steal passwords chosen by some legitimate users of the target (e.g., by phishing, keylogging, or social engineering) and be a user of the site himself, Amnesia leverages the activity of other account owners, each of whose chosen password is indistinguishable to the attacker in the set of passwords listed for her account. As such, when we refer to account owners below, we generally mean ones who have not been phished or otherwise compromised.

4.2 *Algorithm*

In this section we detail our algorithm for a target to leverage honeywords for each of its accounts to detect its own breach. Somewhat counterintuitively, in our design the honeywords the target site creates for each account are indistinguishable from the correct password, even to itself (and so to an attacker who breaches it)—hence the name *Amnesia*. However, the passwords for an account (i.e., both user-chosen and honey) are *marked* probabilistically with binary values. Marking ensures that the password last used to access the account is always marked (i.e., its associated binary value is 1). Specifically, upon each successful login to an account, the set of passwords is remarked with probability p_{remark}, in which case the entered password is marked (with probability 1.0) and each of the other passwords is marked independently with probability p_{mark}. As such, if an attacker accesses the account using a honeyword, then the user-chosen password becomes unmarked with probability $p_{\mathsf{remark}}(1 - p_{\mathsf{mark}})$. In that case, the breach will be detected when the user next accesses the account, since the password she supplies is unmarked.

More specifically, the algorithm for the target to detect its own breach works as follows. The algorithm is parameterized by probabilities p_{mark} and p_{remark}, and an integer $k > 0$. It leverages a procedure mark shown in Fig. 1, which marks the given element e with probability 1.0, marks other elements of $\mathsf{DB}_a.\mathsf{auths}$ for the given account a with probability p_{mark}, and stores these markings in the credential database for account a as the function $\mathsf{DB}_a.\mathsf{marks}$.

mark(a, e): /* Assumption: $e \in \mathsf{DB}_a.\mathsf{auths}$ */

- $\mathcal{X} \leftarrow \mathsf{DB}_a.\mathsf{auths}$
- Choose *marked* : $\mathcal{X} \rightarrow \{0, 1\}$ subject to:
  - $\mathit{marked}(e) = 1$
  - $\forall e' \in \mathcal{X} \setminus \{e\} : \mathit{marked}(e') \sim \mathsf{Bernoulli}\ (p_{\mathsf{mark}})$
- $\mathsf{DB}_a.\mathsf{marks} \leftarrow \mathit{marked}$

Fig. 1 Procedure mark, used in Sects. 4–5

Password Registration When the user sets (or resets) the password for her account a, she provides a user-chosen password π. The password registration system generates $\mathsf{DB}_a.\mathsf{auths} \leftarrow \mathsf{HoneyGen}(a, \pi, k)$ and then invokes $\mathsf{mark}(a, \pi)$.

Login When a login is attempted to account a with password π, the outcome is determined as follows:

- If $\pi \notin \mathsf{DB}_a.\mathsf{auths}$, then the login attempt is unsuccessful.
- If $\pi \in \mathsf{DB}_a.\mathsf{auths}$ and $\mathsf{DB}_a.\mathsf{marks}(\pi) = 0$, then the login attempt is unsuccessful *and a credential database breach is detected.*
- Otherwise (i.e., $\pi \in \mathsf{DB}_a.\mathsf{auths}$ and $\mathsf{DB}_a.\mathsf{marks}(\pi) = 1$) the login attempt is successful.[4] In this case, $\mathsf{mark}(a, \pi)$ is executed with probability p_{remark}.

This algorithm requires that a number of considerations be balanced if an attacker can breach the site repeatedly to capture its credential database many times. Consider that:

- Repeatedly observing the passwords left marked by user logins permits the attacker to narrow in on the user-chosen password as the one that is always marked. This suggests that legitimate logins should remark the passwords as rarely as possible (i.e., p_{remark} should be small) or that, when remarking occurs, doing so results in passwords already marked staying that way (i.e., p_{mark} should be large).
- If the attacker accesses an account between two logins by the user, a remarking *must* occur between the legitimate logins if there is to be any hope of the second legitimate login triggering a detection (i.e., p_{remark} should be large).
- If the attacker is permitted to trigger remarkings many times between consecutive legitimate logins, however, then it can do so repeatedly until markings are restored on most of the passwords that were marked when it first accessed the account. The attacker could thereby reduce the likelihood that the next legitimate login detects the breach. This suggests that it must be difficult for the attacker to trigger many remarkings on an account (i.e., p_{remark} should be small) or that when remarkings occur, significantly many passwords are left unmarked (i.e., p_{mark} should be small).

All of this is complicated by the fact that the target site cannot distinguish between legitimate and attacker logins, of course. While an anomaly detection system (ADS) using features of each login attempt *other* than the password entered (e.g., [18]) could provide a noisy indication, unfortunately our threat model permits the attacker to learn *all persistent state* that the target site uses to manage logins; this

[4] Or more precisely, the stage of the login pipeline dealing with the password is deemed successful. Additional steps, such as a second-factor authentication challenge, could still be required for the login to succeed.

would presumably include the ADS model for each account, thereby enabling the adversary to potentially evade it. For this reason, we eschew this possibility, instead settling for a probability p_{remark} of remarking passwords on a successful login and, if so, a probability p_{mark} with which each password is marked (independently), that together balance the above concerns. We explore such settings in Sect. 4.3.

4.3 Security

Methodology To evaluate the security of our algorithm, we model an attack as a Markov decision process (MDP) consisting of a set of states and possible transitions among them. When the MDP is in a particular state, the attacker can choose from a set of available actions, which determines a probability distribution over the possible next states as a function of the current state and the action chosen. Using probabilistic model checking, we can evaluate the success of the adversary in achieving a certain goal (see below) under his *best* possible strategy for doing so. In our evaluations below, we use the Prism model checker [29].

The basic distributions for modeling our algorithm for a single account are straightforward. Let $\mathbb{N}_\ell$ denote the number of passwords that the attacker always observes as marked in ℓ breaches of the target, with each pair of breaches separated by at least one remarking in a legitimate user login. (Breaches with no remarking between them will observe the same marks.) Then, $\mathbb{N}_\ell \sim \text{binomial}\left(k, (p_{\text{mark}})^\ell\right) + 1$, where the "+ 1" represents the user-chosen password, which remains marked across these ℓ remarkings. Now, letting $\mathbb{A}_n$ denote the number of these passwords that are marked after an adversary-induced remarking, conditioned on $\mathbb{N}_\ell = n + 1$, we know $\mathbb{A}_n \sim \text{binomial}(n, p_{\text{mark}}) + 1$, where the "+ 1" represents the marked password that the adversary submitted to log into the account, which remains marked with certainty. If $\mathbb{A}_n = \alpha + 1$ after the adversary's login, then the probability of the target detecting its own breach upon the legitimate user's next login to this account is $1 - \frac{\alpha+1}{n+1}$.

To turn these distributions into a meaningful MDP, however, we need to specify some additional limits.

- The number of attacker breaches until it achieves ℓ that each follows a distinct remarking induced by a legitimate user login is dependent not only on p_{remark} but also on the rate of user logins. In our experiments, we model user logins as Poisson arrivals with an expected number $\lambda = 1$ login per time unit. We permit the attacker to breach the site and capture all stored state at the end of each time unit.
- Even with this limit on the rate of legitimate user logins, an attacker that breaches the site arbitrarily many times will eventually achieve $\mathbb{N}_\ell = 1$ and so will know the legitimate user's password. In practice, however, the attacker cannot wait arbitrarily long to access an account, since there is a risk that his breaches will be detected by other means (i.e., not by our algorithm). To model this limited window of vulnerability, we assume that the time unit in which the breach is

discovered by other means (at the end of the time unit), and so the experiment *stops*, is represented as a random variable $\mathbb{S}$ distributed normally with mean μ_{stop} and relative standard deviation $\chi_{\mathsf{stop}} = 0.2$. For example, assuming a seven-month average breach discovery delay [23], an account whose user accesses it once per week on average, would have $\mu_{\mathsf{stop}} \approx 30$ time units (weeks).

- Once the attacker logs into the account with one of the $n + 1$ passwords that it observed as always marked in its breaches, it can log in repeatedly (i.e., resample $\mathbb{A}_n$) to leave the account with marks that minimize its probability of detection on the next legitimate user login. If allowed an unbounded number of logins, it can drive its probability of detection to zero. Therefore, we assume that the site monitors accounts for an unusually high rate of *successful* logins, limiting the adversary to at most Λ per time unit.

Let random variable $\mathbb{L}$ denote the time unit at which the attacker logs into the account for the first time, and let random variable $\mathbb{D} \le \mathbb{S}$ denote the time unit at which the attacker is detected. That is, $\mathbb{D} < \mathbb{S}$ means that our algorithm detected the attacker before he was detected by other means. Moreover, note that $\mathbb{L} < \mathbb{D}$, since our algorithm can detect the attacker only after he logs into the account. We define the $\mathsf{benefit}$ of our algorithm to be the expected number of time units that our algorithm deprives the attacker of undetectably accessing the account, expressed as a fraction of the number of time units it could have done so in the absence of our algorithm. In symbols:

$$\mathsf{benefit} = \frac{\mathbb{E}(\mathbb{S} - \mathbb{L}) - \mathbb{E}(\mathbb{D} - \mathbb{L})}{\mathbb{E}(\mathbb{S} - \mathbb{L})} = 1 - \frac{\mathbb{E}(\mathbb{D} - \mathbb{L})}{\mathbb{E}(\mathbb{S} - \mathbb{L})} \tag{3}$$

When computing $\mathsf{benefit}$, we do so for an attacker strategy maximizing $\mathbb{E}(\mathbb{D} - \mathbb{L})$, i.e., against an attacker that maximizes the time for which it accesses the account before it is detected.

Results The computational cost of model checking this MDP is such that we could complete it for only relatively small (but still meaningful) parameters. The results we achieved are reported in Figs. 2, 3, and 4. To explore how increasing each of k, Λ, and μ_{stop} affects $\mathsf{benefit}$, each of the tables in Fig. 2 corresponds to modifying one parameter from the baseline table shown in Fig. 2a, where $k = 48$, $\Lambda = 4$, and $\mu_{\mathsf{stop}} = 8$. Each number in each table is the $\mathsf{benefit}$ of a corresponding $\langle p_{\mathsf{remark}}, p_{\mathsf{mark}} \rangle$ parameter pair, where higher numbers are better. When k is increased from 48 to 64 (Fig. 2b), we can see a slight boost to the $\mathsf{benefit}$. However, increasing Λ or μ_{stop}, shown in Fig. 2c and d, respectively, causes $\mathsf{benefit}$ to drop slightly. The reasons behind these drops are that larger Λ (i.e., more repeated logins by the attacker) give him a better chance to leave with a reduced probability of detection, and a larger μ_{stop} allows the attacker to observe more user logins and so more remarkings (to minimize $\mathbb{N}_\ell$) before he is detected by other means.

This latter effect is illustrated in Fig. 3, which shows $\mathsf{benefit}$ as a function of μ_{stop}. When $\mu_{\mathsf{stop}} \le 7$, the settings $p_{\mathsf{mark}} = 0.2$, $p_{\mathsf{remark}} = 0.9$ yield the best $\mathsf{benefit}$ among the combinations pictured in Fig. 3. However, as μ_{stop} grows, the longer time (i.e., larger ℓ) the attacker can wait to access the account affords him a

	p_{mark}								
p_{remark}	.10	.20	.30	.40	.50	.60	.70	.80	.90
.10	.06	.06	.05	.04	.04	.03	.02	.02	.01
.20	.11	.11	.10	.09	.07	.06	.04	.03	.02
.30	.16	.15	.14	.12	.10	.08	.06	.04	.02
.40	.21	.21	.19	.16	.14	.11	.08	.05	.02
.50	.27	.26	.24	.20	.17	.13	.10	.07	.03
.60	.31	.30	.27	.23	.19	.15	.11	.07	.03
.70	.34	.35	.32	.27	.23	.18	.13	.09	.04
.80	.32	.38	.35	.30	.25	.19	.14	.09	.04
.90	.32	.41	.40	.34	.28	.22	.15	.11	.05
1.0	.33	.40	.42	.38	.31	.24	.17	.12	.05

(a)

p_{mark}								
.10	.20	.30	.40	.50	.60	.70	.80	.90
.06	.06	.05	.04	.04	.03	.02	.02	.01
.12	.11	.10	.09	.07	.06	.04	.03	.02
.17	.16	.15	.12	.10	.08	.06	.04	.02
.23	.22	.19	.16	.14	.11	.08	.05	.03
.29	.27	.24	.21	.17	.14	.10	.07	.03
.33	.31	.28	.24	.20	.16	.11	.08	.03
.37	.36	.33	.28	.23	.19	.13	.09	.04
.35	.40	.36	.31	.26	.21	.15	.10	.04
.34	.43	.41	.35	.29	.23	.16	.11	.05
.34	.42	.45	.38	.32	.26	.18	.12	.05

(b)

	p_{mark}								
p_{remark}	.10	.20	.30	.40	.50	.60	.70	.80	.90
.10	.06	.06	.05	.04	.04	.03	.02	.02	.01
.20	.11	.10	.10	.08	.07	.05	.04	.03	.01
.30	.15	.15	.14	.12	.10	.08	.05	.04	.02
.40	.20	.19	.18	.15	.12	.10	.07	.05	.02
.50	.25	.24	.22	.19	.15	.12	.08	.06	.03
.60	.29	.28	.26	.22	.18	.14	.10	.07	.03
.70	.29	.32	.30	.25	.21	.16	.11	.08	.03
.80	.29	.35	.33	.28	.23	.18	.12	.08	.03
.90	.28	.38	.37	.31	.25	.19	.13	.09	.04
1.0	.28	.36	.41	.35	.28	.22	.15	.10	.04

(c)

p_{mark}								
.10	.20	.30	.40	.50	.60	.70	.80	.90
.06	.06	.06	.05	.04	.03	.02	.02	.01
.12	.12	.11	.10	.08	.06	.05	.03	.02
.17	.17	.15	.13	.11	.09	.06	.04	.02
.23	.22	.21	.18	.15	.11	.08	.06	.03
.29	.28	.26	.22	.18	.14	.10	.07	.03
.31	.32	.30	.25	.21	.16	.12	.08	.04
.30	.36	.35	.30	.24	.19	.14	.09	.04
.28	.34	.38	.33	.27	.21	.15	.10	.04
.28	.34	.39	.37	.30	.23	.16	.11	.05
.31	.38	.40	.40	.33	.26	.18	.13	.05

(d)

Fig. 2 benefit of local detection, as k (**b**), Λ (**c**), and μ_{stop} (**d**) are increased individually from the "baseline" (**a**) of $k = 48$, $\Lambda = 4$, and $\mu_{stop} = 8$. (**a**) Baseline. (**b**) $k = 64$. (**c**) $\Lambda = 8$. (**d**) $\mu_{stop} = 12$

lower $\mathbb{N}_\ell$ and so a lower probability of being detected when the legitimate user subsequently logs in. This effect can be offset by decreasing p_{remark} (Fig. 3a), increasing p_{mark} (Fig. 3b), or both.

The impact of Λ is shown in Fig. 4, which plots benefit as a function of k for various Λ. Figure 4 shows that even when the attacker logs in more frequently than the user by a factor of $\Lambda = 10$, our algorithm still remains effective with benefit ≈ 0.5 for moderately large k. That said, while Fig. 4 suggests that increasing k into the hundreds should suffice, we will see in Sect. 5 that an even larger k might be warranted when credential stuffing is considered.

Interpreting benefit As we define it, benefit is a conservative measure, in two senses. First, benefit is calculated (via probabilistic model checking) against the strongest attacker possible in our threat model. Second, benefit is computed only for one account, but detection on *any* account is enough to inform the target of

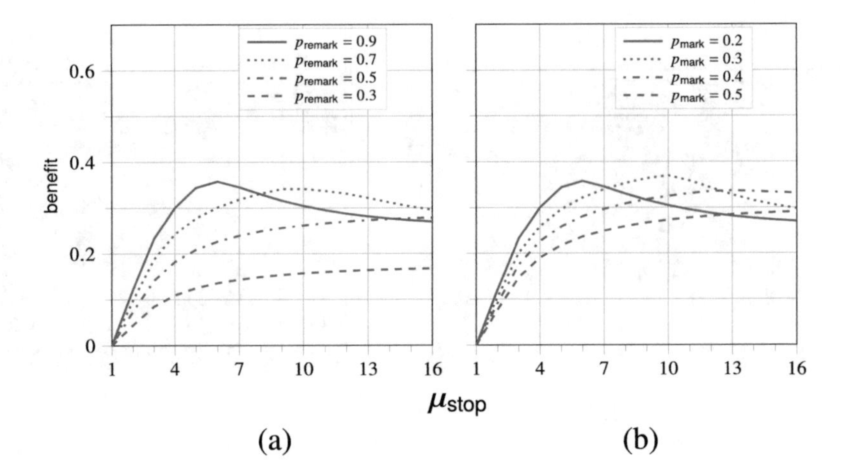

Fig. 3 benefit as a function of μ_{stop} with varying p_{remark} and varying p_{mark} ($k = 32$, $\Lambda = 4$). (**a**) $p_{\text{mark}} = 0.2$. (**b**) $p_{\text{remark}} = 0.9$

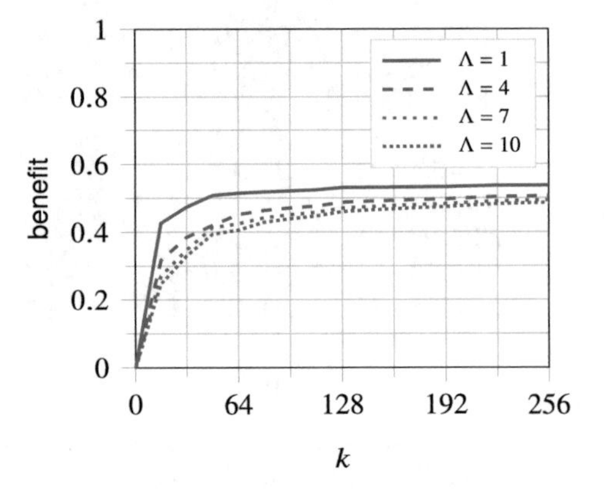

Fig. 4 benefit as a function of k with varying Λ ($p_{\text{mark}} = 0.3$, $p_{\text{remark}} = 1.0$, $\mu_{\text{stop}} = 8$)

its breach. For an attacker whose goal is to assume control of a large number of accounts at the target (vs. one account specifically), the detection power of our algorithm will be much higher.

That said, quantifying that detection power holistically for the target is not straightforward. Recall that benefit is defined in terms of time units wherein the legitimate user is expected to login $\lambda = 1$ time. As such, the real-time length of this unit for a frequently accessed account will be different than for an infrequently accessed one. And, since μ_{stop} is expressed in this time unit, μ_{stop} will be larger for a frequently accessed account than for an infrequently accessed one, even though the real-time interval that passes before a site detects its own breach by means other than Amnesia might be independent of the legitimate login rates to accounts. Thus, extrapolating the per-account benefit to the security improvement for a target holistically requires knowledge of the legitimate login rates across all the sites' accounts as a function of real time, adjusting μ_{stop} (and χ_{stop}) accordingly per account, and translating the per-account benefits back into a real-time measure.

5 Detecting Remotely Stuffed Honeywords

When a credential database is breached, it is common for attackers to submit the login credentials therein (i.e., usernames and passwords) to *other* sites, in an effort to access accounts whose user sets the same password as she did at the breached site. These attacks, called *credential stuffing*, are already the primary attack yielding account takeovers today [41]. But even worse for our purposes here, credential stuffing enables an attacker to circumvent the honeywords at a breached target site: If a user reused her password at another site, then stuffing the breached passwords there will reveal which is the user-chosen password, i.e., as the one that gains access. The attacker can then return to the target site with the correct password to access the user's account at the target.

The design in this section mitigates credential stuffing as a method to identify the user's chosen password, by ensuring that stuffing honeywords at other sites probabilistically still alerts the target site to its breach. At a high level, the target maintains a set of monitor sites and can choose to monitor an account at any of those monitors. To monitor the account at a monitor, the target sends the monitor a *private containment retrieval* (PCR) query for this account identifier, to which the monitor responds after any unsuccessful login attempt to this account (potentially even if the account does not exist at the monitor). In the abstract, a PCR query is a private (encrypted) representation of a set $\mathcal{X}$ of elements known to the target, and a response computed with element e reveals to the target the element e if $e \in \mathcal{X}$ and nothing otherwise. In this case, the target's set $\mathcal{X}$ contains the local password hashes for the user's account. If a monitor then sends a response computed using some $e \in \mathcal{X}$, the target can treat e as if it were attempted locally, permitting the detection of a breach just as in Sect. 4.

5.1 Threat Model

As in Sect. 4.1, we allow the adversary to breach the target passively, thereby learning all information persistently stored by the site for the purpose of determining the success of its users' login attempts. We highlight that in this section, the breached information includes a private key that is part of the target's stored state for managing login attempts in our algorithm. So, if the target is breached, then this private key is included in the data that the attacker learns.

We permit the attacker that breaches the target to also *actively* compromise monitors, in which case these monitors can behave arbitrarily maliciously. Malicious monitors can refuse to help the target detect its own breach via our design, e.g., by simply refusing to respond. However, our scheme must ensure that even malicious monitors cannot convince a target that it has been breached when it has not. Moreover, malicious monitors should not be able to leverage their participation in this protocol to attack passwords at a target that is never breached.

We do not permit the attacker to interfere with communication between a (breached or unbreached) target and an uncompromised monitor. Otherwise, the attacker could prevent the target from discovering its breach by simply refusing to let it communicate with uncompromised monitors.

Our design assumes that different sites can ascertain a common identifier a for the same user's accounts at their sites, at least as well as an attacker could. In practice, this would typically be the email address (or some canonical version thereof, see [46]) registered by the user for account identification or password reset purposes.

5.2 *Private Containment Retrieval*

The main building block for our design is a private containment retrieval (PCR) protocol with the following algorithms.

- pcrQueryGen is an algorithm that, on input a public key pk and a set $\mathcal{X}$, generates a PCR query $Y \leftarrow \mathsf{pcrQueryGen}_{pk}(\mathcal{X})$.
- pcrRespGen is an algorithm that, on input a public key pk, an element e, and a query $Y \leftarrow \mathsf{pcrQueryGen}_{pk}(\mathcal{X})$, outputs a PCR response $Z \leftarrow \mathsf{pcrRespGen}_{pk}(e, Y)$.
- pcrReveal is an algorithm that on input the private key sk corresponding to pk, an element $e' \in \mathcal{X}$, and a response $Z \leftarrow \mathsf{pcrRespGen}_{pk}(e, Y)$ where $Y \leftarrow \mathsf{pcrQueryGen}_{pk}(\mathcal{X})$, outputs a Boolean $z \leftarrow \mathsf{pcrReveal}_{sk}(e', Z)$ where $z = true$ iff $e' = e$.

Informally, this protocol ensures that Y reveals nothing about $\mathcal{X}$ (except its size) to anyone not holding sk; that Z computed on $e \notin \mathcal{X}$ reveals nothing about e (except $e \notin \mathcal{X}$); and that if $\mathsf{pcrReveal}_{sk}(e', Z) = true$, then the party that computed Z knows e'. We make these properties more precise and provide an implementation in Sect. 6.

5.3 *Algorithm*

We first provide greater detail about how the target maintains its credential database. Whereas in Sect. 4 we left hashing of the honey and user-chosen passwords in DB_a.auths implicit, in this section we need to expose this hashing explicitly for functional purposes. Consistent with current best practices, the target represents DB_a.auths as a set of hashes salted with a random κ-bit salt DB_a.salt, including one hash $f(s, \pi)$ of the user-chosen password π where $s \leftarrow \mathsf{DB}_a$.salt and a salted hash $f(s, \pi')$ for each of k honeywords π'. Then, testing whether π is either a honey or user-chosen password amounts to testing $f(s, \pi) \in \mathsf{DB}_a$.auths. In addition to these refinements, for this algorithm the target is also initialized with a public-key/private-key pair $\langle pk, sk \rangle$ for use in the PCR protocol, and a set $\mathcal{S}$ of possible monitors (URLs). If the target R is breached, then all of DB, $\mathcal{S}$, and $\langle pk, sk\rangle$ are captured by the attacker.

The algorithm below treats local logins at the target R similar to how they were treated in Sect. 4, with the exception of exposing the hashing explicitly. In addition, the algorithm permits R to ask monitor S to monitor a. To do so, R sends a PCR query Y to S computed on $\mathsf{DB}_a.\mathsf{auths}$. Upon receiving this request, S simply saves it for use on each incorrect login to a at S, to generate a PCR response to R. The hash encoded in this response is then treated at R (for the purposes of detecting a breach) as if it has been entered in a local login attempt. In sum, the protocol works as described below.

Password Registration at R When the user (re)sets the password for her account a at the target site R, she provides her chosen password π. The password registration system at R executes:

- $\Pi_a \leftarrow \mathsf{HoneyGen}(a, \pi, k)$
- $\mathsf{DB}_a.\mathsf{salt} \xleftarrow{\$} \{0, 1\}^\kappa$
- $\mathsf{DB}_a.\mathsf{auths} \leftarrow \{f(\mathsf{DB}_a.\mathsf{salt}, \pi')\}_{\pi' \in \Pi_a}$
- $\mathsf{mark}(a, f(\mathsf{DB}_a.\mathsf{salt}, \pi))$

Login Attempt at R For a login attempted to account a with password π at R, the outcome is determined as follows, where $h \leftarrow f(\mathsf{DB}_a.\mathsf{salt}, \pi))$:

- If $h \notin \mathsf{DB}_a.\mathsf{auths}$, the login attempt is unsuccessful.
- If $h \in \mathsf{DB}_a.\mathsf{auths}$ and $\mathsf{DB}_a.\mathsf{marks} = 0$, then the login attempt is unsuccessful *and a credential database breach is detected.*
- Otherwise (i.e., $h \in \mathsf{DB}_a.\mathsf{auths}$ and $\mathsf{DB}_a.\mathsf{marks} = 1$), the login attempt is successful and R executes $\mathsf{mark}(a, h)$ with probability p_{remark}.

R **Monitors** a **at** S At an arbitrary time, R can ask $S \in \mathcal{S}$ to monitor account a by generating $Y \leftarrow \mathsf{pcrQueryGen}_{pk}(\mathsf{DB}_a.\mathsf{auths})$ and sending $\langle a, \mathsf{DB}_a.\mathsf{salt}, pk, Y\rangle$ to S.

S **Receives a Monitoring Request** $\langle a, s, pk, Y\rangle$ **from** R S saves $\langle R, a, s, pk, Y\rangle$ locally.

Login Attempt at S For an *unsuccessful* login attempt to an account a using (incorrect) password π, if S holds a monitoring request $\langle R, a, s, pk, Y\rangle$, then it computes $Z \leftarrow \mathsf{pcrRespGen}_{pk}(f(s, \pi), Y)$ and sends $\langle a, Z\rangle$ to R.

R **Receives a Monitoring Response** $\langle a, Z\rangle$ If $\mathsf{pcrReveal}_{sk}(h, Z)$ is *false* for all $h \in \mathsf{DB}_a.\mathsf{auths}$, then R discards $\langle a, Z\rangle$ and returns. Otherwise, let $h \in \mathsf{DB}_a.\mathsf{auths}$ be some hash for which $\mathsf{pcrReveal}_{sk}(h, Z)$ is *true*. *R detects a breach* if $\mathsf{DB}_a.\mathsf{marks}(h) = 0$ and otherwise executes $\mathsf{mark}(a, h)$ with probability p_{remark}.

In the above protocol, the only items received by the monitor S in $\langle a, s, pk, Y \rangle$ are all available to an attacker who breaches R. In this sense, a malicious S gains nothing that an attacker who breaches the target R does not also gain, and in fact gains less, since it learns none of sk, DB_a.auths, or $\mathcal{S}$. Indeed, the *only* advantage an attacker gains by compromising S in attacking passwords at R is learning the salt $s = \mathsf{DB}_a$.salt, with which it can precompute information (e.g., rainbow tables [35]) to accelerate its offline attack on DB_a.auths if it eventually breaches R. If this possibility is deemed too risky, R can refuse to send s to S in its request but instead permit S to compute $f(s, \pi')$ when needed by interacting with R, i.e., with f being implemented as an oblivious pseudo-random function (OPRF) [17] keyed with s, for which there are efficient implementations (e.g., the DH-OPRF implementation leveraged by OPAQUE [24]). This approach would require extra interaction between S and R per response from S, however, and so we do not consider this alternative further here.

S should authenticate a request $\langle a, s, pk, Y \rangle$ as coming from R, e.g., by requiring that R digitally sign it. Presuming that this digital signing key (different from sk) is vulnerable to capture when R is breached, S should echo each monitoring request back to R upon receiving it. If R receives an echoed request bearing its own signature but that it did not create, it can again detect its own breach. (Recall that we cannot permit the attacker to interfere with communications between R and an uncompromised S and still have R detect its breach.)

In practice, a monitor will not retain a monitoring record forever, as its list of monitoring records—and the resulting cost incurred due to generating responses to them—would only grow. Moreover, it cannot count on R to withdraw its monitoring requests, since R does not retain records of where it has deposited what requests, lest these records be captured when it is breached and the attacker simply avoid monitored accounts. Therefore, presumably a monitor should unilaterally expire each monitoring record after a period of time or in a randomized fashion. We do not investigate specific expiration strategies here, nor do we explore particular strategies for a target to issue monitoring requests over time.

5.4 Security

Several security properties are supported directly by the PCR protocol, which will be detailed in Sect. 6. Here we leverage those properties to argue the security of our design.

No Breach Detected by Unbreached target If the target R has not been breached, then the PCR protocol will ensure that S must know h for it to generate a Z for which $\mathsf{pcrReveal}_{sk}(h, Z)$ returns *true* at R. Assuming S cannot guess a $h \in \mathsf{DB}_a$.auths without guessing a password π such that $h = f(s, \pi)$ and that (ignoring collisions in f) guessing such a π is infeasible (see (2)), generating such a Z is infeasible for S unless the user provides such a π to S herself. Since the only such π she knows is the one she chose during password registration at R, π is the user-chosen password at a. And, since R has not been breached, the hash of π will still be marked there. As such, R will not detect its own breach.

No Risk to Security of Account at Unbreached target If the target R has not been breached, then the PCR request Y reveals nothing about $\mathsf{DB}_a.\mathsf{auths}$ (except its size) to S. As such, sending a monitoring request poses no risk to the target's account.

No Risk to Security of Account at Uncompromised monitor We now consider the security of the password π for account a at the monitor S (if this account exists at S). First recall that S generates PCR responses only for *incorrect* passwords attempted in local login attempts for account a; the correct password at S will not be used to generate a response. Moreover, S could even refuse to generate responses for passwords very close to the correct password for a, e.g., the correct password with typos [7]. Second, the PCR protocol ensures that the target R learns nothing about the attempted (and again, incorrect) password π if S is not compromised, unless R included $h = f(s, \pi)$ in the set from which it generated its PCR query Y. In this case, $\mathsf{pcrReveal}_{sk}(h, Z)$ returns *true* but, again, R already guessed it.

Detection of the target's **Breach** We now consider the ability of R to detect its own breach by monitoring an account a at an uncompromised monitor S, which is the most nuanced aspect of our protocol's security. Specifically, an attacker who can both repeatedly breach R and simultaneously submit login attempts at an uncompromised S poses the following challenge: Because this attacker can see what hashes for a are presently marked at R, it can be sure to submit to S a password for one of the marked hashes at R, so that the induced PCR response Z will not cause R to detect its own breach. Moreover, if the user reused her password at both R and S, then the attacker will know when it submits this password to S, since S will accept the login attempt.

As such, for R to detect its own breach in these (admittedly extreme) circumstances, the attacker must be unable to submit enough stolen passwords for a to S to submit the user-chosen one with high probability, in the time during which it can repeatedly breach R and before the next legitimate login to a at R or S. To slow the attacker somewhat, R can reduce p_{mark} and p_{remark} to limit the pace of remarkings and, when remarkings occur, the number of hashes that are marked (which are the ones that the attacker can then submit to S).

Two other defenses will likely be necessary, however. First, R can greatly increase the attacker's workload by increasing the number of honeywords per account, say to the thousands or tens of thousands (cf., [28]). Second, since honeywords from R submitted to S will be incorrect for the account a at S, online guessing defenses (account lockout or rate limiting) at S can (and should) be used to slow the attacker's submissions at S. In particular, NIST recommends that a site "limit consecutive failed authentication attempts on a single account to no more than 100" [19, Section 5.2.2], in which case an attacker would be able to eliminate, say, at most 2% of the honeywords for an account with 5000 honeywords stolen from

R by submitting them in login attempts at S. Our design shares the need for these defenses with most other methods for using decoy passwords [5, 14, 25, 28, 48]. In particular, if the user reused her password at other sites that permit the attacker to submit passwords stolen from the target without limitation, then the attacker discovering the user's reuse of that password is simply a matter of time, after which the attacker can undetectably take over the account.

5.5 Alternative Designs

The algorithm presented above is the result of numerous iterations, in which we considered and discarded other algorithm variants for remote detection of stuffed honeywords. Here we briefly describe several variants and why we rejected them.

- The target could exclude the known (entered at password reset) or likely (entered in a successful login) user-chosen password π from the monitor request, i.e., $Y \leftarrow \mathsf{pcrQueryGen}_{pk}(\mathsf{DB}_a.\mathsf{auths} \setminus \{f(s, \pi)\})$. In this case, *any* "non-empty" PCR response Z (i.e., $\mathsf{pcrReveal}_{sk}(h, Z)$ returns *true* for some $h \in \mathsf{DB}_a.\mathsf{auths}$) would indicate a breach. However, combining the data breached at the target with Y at a malicious monitor would reveal the password not included in Y as the likely user-chosen one.
- Since a monitor returns a PCR response only for an *incorrect* password attempted locally, the target could plausibly treat any non-empty PCR response as indicating its breach. That is, if the user reused her password, it would not be used to generate a response anyway, and so the response would seemingly have to represent a honeyword attempt. However, if the user did *not* reuse her target password at the monitor, then her mistakenly entering it at the monitor would cause the target to falsely detect its own breach.
- The monitor could return a PCR response for *any* login attempt, correct or not, potentially hastening the target detecting its own breach. However, a PCR request would then present an opportunity for a malicious target to guess $k + 1$ passwords for the account at the monitor, and be informed if the user enters one there.
- Any two PCR responses for which $\mathsf{pcrReveal}_{sk}$ returns true with distinct $h, h' \in \mathsf{DB}_a.\mathsf{auths}$ is a reliable breach indicator; one must represent a honeyword. This suggests processing responses in batches, batched either at the monitor or target. However, ensuring that the attacker cannot artificially "fill" batches with repeated password attempts can be complex; batching can delay detection; and batching risks disclosure of a user-chosen password if one might be included in a response and responses are saved in persistent storage (to implement batching).

6 Private Containment Retrieval

Recall that in the algorithm of Sect. 5, upon receiving a monitoring request for an account a from a target, a monitor stores the request locally and uses it to generate a PCR response per failed login attempt to a. Since a response is generated per failed login attempt, it is essential that pcrRespGen be efficient and that the response Z be small. Moreover, considering that a database breach is an uncommon event for a site, we expect that most of the time, the response would be generated using a password that is not in the set used by the target to generate the monitoring request. (Indeed, barring a database breach at the target, this should never happen unless the user enters at the monitor her password for her account at the target.) So, in designing a PCR, we place a premium on ensuring that pcrReveal is very efficient in this case.

6.1 Comparison to Related Protocols

Since the monitor's input to pcrRespGen is a singleton set (i.e., a hash), a natural way to achieve the functionality of a private containment retrieval is to leverage existing private set intersection (PSI) protocols, especially *unbalanced* PSIs that are designed for the use case where two parties have sets of significantly different sizes [8, 26, 27, 39, 42]. Among these protocols, those based on oblivious pseudo-random functions (OPRFs) [26, 27, 39, 42] require both parties to obliviously agree on a privacy-preserving but deterministic way of representing their input sets so at least one party can compare and output elements in the intersection, if any. To achieve this, both parties participate in at least one round of interaction (each of at least two messages) during an online phase, and so would require more interaction in our context than our framework as defined in Sect. 5. Chen et al. [8] proposed a PSI protocol with reduced communication, but at the expense of leveraging fully homomorphic encryption. And, interestingly, these unbalanced PSI protocols, as well as private membership tests (e.g., [34, 38, 46, 47]), are all designed for the case where the target has the smaller set and the monitor has the larger one, which is the opposite of our use case.

Among other PSI protocols that require no more than one round of interaction, that of Davidson and Cid [12] almost meets the requirements of our framework on the monitor side: its monitor's computation complexity and response message size are manageable and, more importantly, constant in the target's set size. However, in their design, the query message size depends on the false-positive probability (of the containment test) due to their use of Bloom filters and bit-by-bit encryption, while ours is also constant in the false-positive probability. If applied in our context, their design would generate a significantly larger query and so significantly greater storage overhead at the monitor than ours, especially when a relatively low false-positive probability is enforced. For example, to achieve a 2^{-96} false-positive probability, their query message would include $\approx 131\times$ more ciphertexts than ours.

Our PCR protocol, on the other hand, is designed specifically for the needs of our framework, where the target has a relatively large set and the monitor's set is smaller (in fact, of size 1) that keeps changing over time. Our protocol requires only one message from the monitor to the target. In addition, the response message computation time and output size is constant in the target's set size. We also constructed our algorithm so that determining that $\mathsf{pcrReveal}_{sk}(h, Z)$ is *false* for all $h \in \mathsf{DB}_a.\mathsf{auths}$, which should be the common case, costs much less time than finding the $h \in \mathsf{DB}_a.\mathsf{auths}$ for which $\mathsf{pcrReveal}_{sk}(h, Z)$ is *true*. We demonstrate these properties empirically in Sect. 6.5. While our protocol leverages tools (e.g., partially homomorphic encryption, cuckoo filters) utilized in other protocols (e.g., [47]), ours does so in a novel way and with an eye toward our specific goals here.

6.2 Building Blocks

Partially Homomorphic Encryption Our protocol builds on a partially homomorphic encryption scheme $\mathcal{E}$ consisting of algorithms Gen, Enc, isEq, and $+_{[\cdot]}$.

- Gen is a randomized algorithm that on input 1^κ outputs a public-key/private-key pair $\langle pk, sk\rangle \leftarrow \mathsf{Gen}(1^\kappa)$. The value of pk determines a prime r for which the plaintext space for encrypting with pk is the finite field $\langle \mathbb{Z}_r, +, \times\rangle$ where $+$ and $\times$ are addition and multiplication modulo r, respectively. For clarity below, we denote the additive identity by $\mathbf{0}$, the multiplicative identity by $\mathbf{1}$, and the additive inverse of $m \in \mathbb{Z}_r$ by $-m$. The value of pk also determines a ciphertext space $C_{pk} = \bigcup_m C_{pk}(m)$, where $C_{pk}(m)$ denotes the ciphertexts for plaintext m.
- Enc is a randomized algorithm that on input public key pk and a plaintext m, outputs a ciphertext $c \leftarrow \mathsf{Enc}_{pk}(m)$ chosen uniformly at random from $C_{pk}(m)$.
- isEq is a deterministic algorithm that on input a private key sk, plaintext m, and ciphertext $c \in C_{pk}$, outputs a Boolean $z \leftarrow \mathsf{isEq}_{sk}(m, c)$ where $z = \mathit{true}$ iff $c \in C_{pk}(m)$.
- $+_{[\cdot]}$ is a randomized algorithm that, on input a public key pk and ciphertexts $c_1 \in C_{pk}(m_1)$ and $c_2 \in C_{pk}(m_2)$, outputs a ciphertext $c \leftarrow c_1 +_{pk} c_2$ chosen uniformly at random from $C_{pk}(m_1 + m_2)$.

Note that our protocol does not require an efficient decryption capability. Nor does the encryption scheme on which we base our empirical evaluation in Sect. 6.5, namely "exponential ElGamal" (e.g., [10]), support one. It does, however, support an efficient isEq calculation.

Given this functionality, it will be convenient to define a few additional operators involving ciphertexts. Below, "$\mathbf{Y} \stackrel{d}{=} \mathbf{Y}'$" denotes that random variables $\mathbf{Y}$ and $\mathbf{Y}'$ are distributed identically; "$\mathbf{Z} \in (\mathcal{X})^{\alpha\times\alpha'}$" means that $\mathbf{Z}$ is an α-row, α'-column matrix of elements in the set $\mathcal{X}$; and "$(\mathbf{Z})_{i,j}$" denotes the row-i, column-j element of the matrix $\mathbf{Z}$.

- $\sum_{pk}$ denotes summing a sequence using $+_{pk}$, i.e.,

$$\sum_{k=1}^{z}{}_{pk} \quad c_k \stackrel{d}{=} c_1 +_{pk} c_2 +_{pk} \ldots +_{pk} c_z$$

- If $\mathbf{C} \in (C_{pk})^{\alpha \times \alpha'}$ and $\mathbf{C}' \in (C_{pk})^{\alpha \times \alpha'}$, then $\mathbf{C} +_{pk} \mathbf{C}' \in (C_{pk})^{\alpha \times \alpha'}$ is the result of component-wise addition using $+_{pk}$, i.e., so that

$$\left(\mathbf{C} +_{pk} \mathbf{C}'\right)_{i,j} \stackrel{d}{=} (\mathbf{C})_{i,j} +_{pk} \left(\mathbf{C}'\right)_{i,j}$$

- If $\mathbf{M} \in (\mathbb{Z}_r)^{\alpha \times \alpha'}$ and $\mathbf{C} \in (C_{pk})^{\alpha \times \alpha'}$, then $\mathbf{M} \circ_{pk} \mathbf{C} \in (C_{pk})^{\alpha \times \alpha'}$ is the result of Hadamard (i.e., component-wise) "scalar multiplication" using repeated application of $+_{pk}$, i.e., so that

$$\left(\mathbf{M} \circ_{pk} \mathbf{C}\right)_{i,j} \stackrel{d}{=} \sum_{k=1}^{(\mathbf{M})_{i,j}}{}_{pk} \ (\mathbf{C})_{i,j}$$

- If $\mathbf{M} \in (\mathbb{Z}_r)^{\alpha \times \alpha'}$ and $\mathbf{C} \in (C_{pk})^{\alpha' \times \alpha''}$, then $\mathbf{M} *_{pk} \mathbf{C} \in (C_{pk})^{\alpha \times \alpha''}$ is the result of standard matrix multiplication using $+_{pk}$ and "scalar multiplication" using repeated application of $+_{pk}$, i.e., so that

$$\left(\mathbf{M} *_{pk} \mathbf{C}\right)_{i,j} \stackrel{d}{=} \sum_{k=1}^{\alpha'}{}_{pk} \ \sum_{k'=1}^{(\mathbf{M})_{i,k}}{}_{pk} \ (\mathbf{C})_{k,j}$$

Cuckoo Filters A cuckoo filter [16] is a set representation that supports insertion and deletion of elements, as well as testing membership. The cuckoo filter uses a "fingerprint" function $\mathsf{fp} : \{0,1\}^* \to F$ and a hash function $\mathsf{hash} : \{0,1\}^* \to [\beta]$, where for an integer z, the notation "$[z]$" denotes $\{1, \ldots, z\}$, and where β is a number of "buckets". We require that $F \subseteq \mathbb{Z}_r \setminus \{\mathbf{0}\}$ for any r determined by $\langle pk, sk \rangle \leftarrow \mathsf{Gen}(1^\kappa)$. For an integer bucket "capacity" χ, the cuckoo filter data structure is a β-row, χ-column matrix $\mathbf{X}$ of elements in $\mathbb{Z}_r$, i.e., $\mathbf{X} \in (\mathbb{Z}_r)^{\beta \times \chi}$. Then, the membership test $e \stackrel{?}{\in} \mathbf{X}$ returns *true* if and only if there exists $j \in [\chi]$ such that either

$$(\mathbf{X})_{\mathsf{hash}(e),j} = \mathsf{fp}(e) \quad \text{or} \tag{4}$$

$$(\mathbf{X})_{\mathsf{hash}(e) \oplus \mathsf{hash}(\mathsf{fp}(e)),j} = \mathsf{fp}(e) \tag{5}$$

Cuckoo filters permit false positives (membership tests that return *true* for elements not previously added or already removed) with a probability that, for fixed χ, can be decreased by increasing the size of F [16].

6.3 Protocol Description

Our PCR protocol is detailed in Fig. 5. Figure 5a shows the message flow, which conforms with the protocol's use in our algorithm of Sect. 5, and Fig. 5b shows the procedures. In this protocol, the target R has a public-key pair $\langle pk, sk\rangle$ for the encryption scheme defined in Sect. 6.2 and a cuckoo filter $\mathbf{X}$. In the context of Sect. 5, $\mathbf{X}$ holds the password hashes (for k honeywords and one user-chosen password) for an account. $\mathsf{pcrQueryGen}_{pk}$ simply encrypts each element of the cuckoo filter individually and returns this matrix $\mathbf{Y}$ as the PCR query. R sends pk and $\mathbf{Y}$ to the monitor S in message m1.

S has an input e—which is the hash of a password entered in a failed login attempt, in the algorithm of Sect. 5—and invokes $\mathsf{pcrRespGen}_{pk}(e, \mathbf{Y})$ to produce a response $\langle \mathbf{Z}, \mathbf{Z}'\rangle$. $\mathsf{pcrRespGen}$ first generates a $2 \times \beta$ matrix $\mathbf{Q}$ with $\mathbf{1}$ at the indices i_1 and i_2 in the first and second rows, respectively (lines s2–s4), and $\mathbf{0}$ elsewhere, and a $2 \times \chi$ matrix $\mathbf{F}$ that contains encryptions of $-\mathsf{fp}(e)$ (lines s5–s6). Referring to line s8, the operation $\mathbf{Q} *_{pk} \mathbf{Y}$ thus produces the two buckets (rows) of $\mathbf{Y}$ that could include a ciphertext of $\mathsf{fp}(e)$ (ignoring collisions in fp), and $\left(\mathbf{Q} *_{pk} \mathbf{Y}\right) +_{pk} \mathbf{F}$ produces a matrix where that ciphertext (if any) has been changed to a ciphertext of $\mathbf{0}$. This ciphertext of $\mathbf{0}$ remains after multiplying this matrix component-wise by the random matrix $\mathbf{M}$ to produce $\mathbf{Z}$. The remaining steps (lines s9–s12) simply rerandomize $\mathbf{Z}$ and transform this ciphertext of $\mathbf{0}$ to a ciphertext of $\mathsf{fp}'(e)$ in $\mathbf{Z}'$, for a fingerprint function $\mathsf{fp}' : \{0, 1\}^* \to F$ that is "unrelated" to fp. (We will model fp' as a random oracle [4] for the security argument in Sect. 6.4.) Rerandomization using $\mathbf{M}'$ in the creation of $\mathbf{Z}'$ is essential to protect the privacy of e if $e \notin \mathbf{X}$, since without rerandomizing, the component-wise differences of the plaintexts of $\mathbf{Z}$ and $\mathbf{Z}'$ would reveal $\mathsf{fp}'(e)$ to R.

For (an artificially small) example, suppose $\beta = 3$, $\chi = 1$, and that the monitor S invokes $\mathsf{pcrRespGen}_{pk}(e, \mathbf{Y})$ where $i_1 = \mathsf{hash}(e) = 3$ and $i_2 = \mathsf{hash}(e) \oplus \mathsf{hash}(\mathsf{fp}(e)) = 2$. Furthermore, suppose that $(\mathbf{X})_{i_1,1} \stackrel{d}{=} \mathsf{Enc}_{pk}(e)$. Then,

$$\mathbf{Q} *_{pk} \mathbf{Y} \stackrel{d}{=} \begin{bmatrix} \mathbf{0}\ \mathbf{0}\ \mathbf{1} \\ \mathbf{0}\ \mathbf{1}\ \mathbf{0} \end{bmatrix} *_{pk} \begin{bmatrix} c_1 \\ c_2 \\ \mathsf{Enc}_{pk}(e) \end{bmatrix} \stackrel{d}{=} \begin{bmatrix} \mathsf{Enc}_{pk}(e) \\ c_2 \end{bmatrix}$$

and so

$$\left(\mathbf{Q} *_{pk} \mathbf{Y}\right) +_{pk} \mathbf{F} \stackrel{d}{=} \begin{bmatrix} \mathsf{Enc}_{pk}(e) \\ c_2 \end{bmatrix} +_{pk} \begin{bmatrix} \mathsf{Enc}_{pk}(-e) \\ \mathsf{Enc}_{pk}(-e) \end{bmatrix} \stackrel{d}{=} \begin{bmatrix} \mathsf{Enc}_{pk}(\mathbf{0}) \\ \mathsf{Enc}_{pk}(m_2 - e) \end{bmatrix}$$

$R(\langle pk, sk\rangle, \mathbf{X})$ $S(e)$

$\mathbf{Y} \leftarrow \mathsf{pcrQueryGen}_{pk}(\mathbf{X})$

m1. $\xrightarrow{\langle pk, \mathbf{Y}\rangle}$

$\langle \mathbf{Z}, \mathbf{Z}'\rangle \leftarrow \mathsf{pcrRespGen}_{pk}(e, \mathbf{Y})$

m2. $\xleftarrow{\langle \mathbf{Z}, \mathbf{Z}'\rangle}$

return $\arg_{e' \in \mathbf{X}} \mathsf{pcrReveal}_{sk}(e', \langle \mathbf{Z}, \mathbf{Z}'\rangle)$

(a)

$\mathsf{pcrQueryGen}_{pk}(\mathbf{X})$:

r1. abort if $\mathbf{X} \notin (\mathbb{Z}_r)^{\beta\times\chi}$

r2. $\forall i \in [\beta], j \in [\chi] : (\mathbf{Y})_{i,j} \leftarrow \mathsf{Enc}_{pk}((\mathbf{X})_{i,j})$

r3. return $\mathbf{Y}$

$\mathsf{pcrRespGen}_{pk}(e, \mathbf{Y})$:

s1. abort if $\mathbf{Y} \notin (C_{pk})^{\beta\times\chi}$

s2. $i_1 \leftarrow \mathsf{hash}(e)$

s3. $i_2 \leftarrow \mathsf{hash}(e) \oplus \mathsf{hash}(\mathsf{fp}(e))$

s4. $\forall i \in [2], j \in [\beta] : (\mathbf{Q})_{i,j} \leftarrow \begin{cases} \mathbf{1} \text{ if } \langle i, j\rangle \in \{\langle 1, i_1\rangle, \langle 2, i_2\rangle\} \\ \mathbf{0} \text{ otherwise} \end{cases}$

s5. $f \leftarrow \mathsf{Enc}_{pk}(-\mathsf{fp}(e))$

s6. $\forall i \in [2], j \in [\chi] : (\mathbf{F})_{i,j} \leftarrow f$

s7. $\forall i \in [2], j \in [\chi] : (\mathbf{M})_{i,j} \xleftarrow{\$} \mathbb{Z}_r \setminus \{\mathbf{0}\}$

s8. $\mathbf{Z} \leftarrow \mathbf{M} \circ_{pk} \left((\mathbf{Q} *_{pk} \mathbf{Y}) +_{pk} \mathbf{F}\right)$

s9. $f' \leftarrow \mathsf{Enc}_{pk}(\mathsf{fp}'(e))$

s10. $\forall i \in [2], j \in [\chi] : (\mathbf{F}')_{i,j} \leftarrow f'$

s11. $\forall i \in [2], j \in [\chi] : (\mathbf{M}')_{i,j} \xleftarrow{\$} \mathbb{Z}_r$

s12. $\mathbf{Z}' \leftarrow (\mathbf{M}' \circ_{pk} \mathbf{Z}) +_{pk} \mathbf{F}'$

s13. return $\langle \mathbf{Z}, \mathbf{Z}'\rangle$

$\mathsf{pcrReveal}_{sk}(e', \langle \mathbf{Z}, \mathbf{Z}'\rangle)$:

r4. return *false* if $\mathbf{Z} \notin (C_{pk})^{2\times\chi} \vee \mathbf{Z}' \notin (C_{pk})^{2\times\chi}$

r5. $\langle \hat{\imath}, \hat{\jmath}\rangle \leftarrow \arg_{\langle i,j\rangle} \mathsf{isEq}_{sk}(\mathbf{0}, (\mathbf{Z})_{i,j})$

r6. return *false* if $\langle \hat{\imath}, \hat{\jmath}\rangle = \langle \bot, \bot\rangle$

r7. return $\mathsf{isEq}_{sk}(\mathsf{fp}'(e'), (\mathbf{Z}')_{\hat{\imath},\hat{\jmath}})$

(b)

Fig. 5 Private containment retrieval protocol, with matrices $\mathbf{X} \in (\mathbb{Z}_r)^{\beta\times\chi}$; $\mathbf{Y} \in (C_{pk})^{\beta\times\chi}$; $\mathbf{Q} \in (\mathbb{Z}_r)^{2\times\beta}$; $\mathbf{M}, \mathbf{M}' \in (\mathbb{Z}_r)^{2\times\chi}$; $\mathbf{F}, \mathbf{F}', \mathbf{Z}, \mathbf{Z}' \in (C_{pk})^{2\times\chi}$. (**a**) Message flow. (**b**) Procedures

where $c_2 \in C_{pk}(m_2)$. Assuming $m_2 \neq e$, we then have

$$\mathbf{Z} \stackrel{d}{=} \mathbf{M} \circ_{pk} \left((\mathbf{Q} *_{pk} \mathbf{Y}) +_{pk} \mathbf{F} \right)$$
$$\stackrel{d}{=} \begin{bmatrix} m_3 \\ m_4 \end{bmatrix} \circ_{pk} \begin{bmatrix} \mathsf{Enc}_{pk}(\mathbf{0}) \\ \mathsf{Enc}_{pk}(m_2 - e) \end{bmatrix} \stackrel{d}{=} \begin{bmatrix} \mathsf{Enc}_{pk}(\mathbf{0}) \\ \mathsf{Enc}_{pk}(m_5) \end{bmatrix}$$

where $m_3, m_4 \stackrel{\$}{\leftarrow} \mathbb{Z}_r \setminus \{\mathbf{0}\}$ and so $m_5 \neq \mathbf{0}$. Finally,

$$\mathbf{Z}' \stackrel{d}{=} \left(\mathbf{M}' \circ_{pk} \mathbf{Z} \right) +_{pk} \mathbf{F}'$$
$$\stackrel{d}{=} \left(\begin{bmatrix} m_6 \\ m_7 \end{bmatrix} \circ_{pk} \begin{bmatrix} \mathsf{Enc}_{pk}(\mathbf{0}) \\ \mathsf{Enc}_{pk}(m_5) \end{bmatrix} \right) +_{pk} \begin{bmatrix} \mathsf{Enc}_{pk}(\mathsf{fp}'(e)) \\ \mathsf{Enc}_{pk}(\mathsf{fp}'(e)) \end{bmatrix} \stackrel{d}{=} \begin{bmatrix} \mathsf{Enc}_{pk}(\mathsf{fp}'(e)) \\ \mathsf{Enc}_{pk}(m_8) \end{bmatrix}$$

where $m_6, m_7 \stackrel{\$}{\leftarrow} \mathbb{Z}_r$ and so m_8 is uniformly random in $\mathbb{Z}_r$.

Given this structure of $\langle \mathbf{Z}, \mathbf{Z}' \rangle$, $\mathsf{pcrReveal}_{sk}(e', \langle \mathbf{Z}, \mathbf{Z}' \rangle)$ must simply find the location $\langle \hat{i}, \hat{j} \rangle$ where $\mathbf{Z}$ holds a ciphertext of $\mathbf{0}$ (line r5) and, unless there is none (line r6), return whether the corresponding location in $\mathbf{Z}'$ is a ciphertext of $\mathsf{fp}'(e')$ (line r7).

6.4 Security

The use of this protocol to achieve the security arguments of Sect. 5.4 depends on the PCR protocol achieving certain key properties. We present these properties below.

Security Against a Malicious monitor When the target R is not breached, our primary goals are twofold. First, we need to show that monitoring requests do not weaken the security of R's accounts or, in other words, that the request $\mathbf{Y}$ does not leak information about $\mathbf{X}$ (except its size). This is straightforward, however, since in this protocol S observes only ciphertexts $\mathbf{Y}$ and the public key pk with which these ciphertexts were created. (The target R need not, and should not, divulge the result of the protocol to the monitor S.) As such, the privacy of $\mathbf{X}$ reduces trivially to the IND-CPA security [3] of the encryption scheme.

The second property that we require of this protocol is that a malicious monitor be unable to induce the target to evaluate $\mathsf{pcrReveal}_{sk}(e', \langle \mathbf{Z}, \mathbf{Z}' \rangle)$ to *true* for any $e' \in \mathbf{X}$ unless the monitor knows e'. That is, in the context of Sect. 5, we want to ensure that the monitor must have received (a password that hashes to) e' in a login attempt, as otherwise the monitor might cause the target to falsely detect its own breach. This is straightforward to argue in the random oracle model [4], however, since if fp' is modeled as a random oracle, then to create a ciphertext $(\mathbf{Z}')_{i,j} \in C_{pk}(\mathsf{fp}'(e'))$ with non-negligible probability in the output length of fp', S must invoke the fp' oracle with e' and so must "know" it.

Security Against a Malicious target Though our threat model in Sect. 5.1 does not permit a malicious target for the purposes of designing an algorithm for it to detect its own breach, a monitor will participate in this protocol only if doing so does not impinge on the security of its own accounts, even in the case where the target is malicious. The security of the monitor's account a is preserved since if the monitor correctly computes $\mathsf{pcrRespGen}_{pk}(e, \mathbf{Y})$, then the output $\langle \mathbf{Z}, \mathbf{Z}' \rangle$ carries information about e only if some $(\mathbf{Y})_{i,j} \in C_{pk}(\mathsf{fp}(e))$, i.e., only if the target already enumerated this password among the $k + 1$ in $\mathbf{Y}$ (ignoring collisions in fp). That is, even a malicious target learns nothing about e from the response computed by an honest monitor unless the target already guessed e (or more precisely, $\mathsf{fp}(e)$).

This reasoning requires that pk is a valid public key for the cryptosystem, and so implicit in the algorithm description in Fig. 5 is that the monitor verifies this. This verification is trivial for the cryptosystem with which we instantiate this protocol in Sect. 6.5.

Proposition *Given* $\langle pk, \mathbf{Y} \rangle$ *and* e *where* $(\mathbf{Y})_{i,j} \notin C_{pk}(\mathsf{fp}(e))$ *for each* $i \in [\beta]$, $j \in [\chi]$, *if the* monitor *correctly computes* $\langle \mathbf{Z}, \mathbf{Z}' \rangle \leftarrow \mathsf{pcrRespGen}_{pk}(e, \mathbf{Y})$, *then*

$$\mathbb{P}\left((\mathbf{Z})_{i,j} \in C_{pk}(m) \wedge \left(\mathbf{Z}'\right)_{i,j} \in C_{pk}(m')\right) = \frac{1}{r(r-1)}$$

for any $i \in [2]$, $j \in [\chi]$, $m \in \mathbb{Z}_r \setminus \{\mathbf{0}\}$, *and* $m' \in \mathbb{Z}_r$. □

Proof Since each $(\mathbf{Y})_{i,j} \notin C_{pk}(\mathsf{fp}(e))$ by assumption, the constructions of $\mathbf{Q}$ and $\mathbf{F}$ imply that $\left(\mathbf{Q} *_{pk} \mathbf{Y}\right)_{i,j} \notin C_{pk}(\mathsf{fp}(e))$ and so $\left(\left(\mathbf{Q} *_{pk} \mathbf{Y}\right) +_{pk} \mathbf{F}\right)_{i,j} \notin C_{pk}(\mathbf{0})$ for any $i \in [2]$, $j \in [\chi]$. Then, since $(\mathbf{M})_{i,j}$ is independently and uniformly distributed in $\mathbb{Z}_r \setminus \{\mathbf{0}\}$, it follows that $(\mathbf{Z})_{i,j} = \left(\mathbf{M} \circ_{pk} \left(\left(\mathbf{Q} *_{pk} \mathbf{Y}\right) +_{pk} \mathbf{F}\right)\right)_{i,j} \in C_{pk}(m)$ for m distributed uniformly in $\mathbb{Z}_r \setminus \{\mathbf{0}\}$, as well. Finally, since $\left(\mathbf{M}'\right)_{i,j}$ is independently and uniformly distributed in $\mathbb{Z}_r$, we know that $\left(\left(\mathbf{M}' \circ_{pk} \mathbf{Z}\right) +_{pk} \mathbf{F}'\right)_{i,j} \in C_{pk}(m')$ for m' distributed uniformly in $\mathbb{Z}_r$. □

The proposition above shows that the *plaintexts* in the response are uniformly distributed if $(\mathbf{Y})_{i,j} \notin C_{pk}(\mathsf{fp}(e))$. The following proposition also points out that the *ciphertexts* are uniformly distributed.

Proposition *If the* monitor *follows the protocol, then*

$$\mathbb{P}\left((\mathbf{Z})_{i,j} = c \;\middle|\; (\mathbf{Z})_{i,j} \in C_{pk}(m)\right) = \frac{1}{|C_{pk}(m)|}$$

$$\mathbb{P}\left(\left(\mathbf{Z}'\right)_{i,j} = c \;\middle|\; \left(\mathbf{Z}'\right)_{i,j} \in C_{pk}(m)\right) = \frac{1}{|C_{pk}(m)|}$$

for any $i \in [2]$, $j \in [\chi]$, $m \in \mathbb{Z}_r$, *and* $c \in C_{pk}(m)$. □

Proof This is immediate since $+_{pk}$ ensures that for $c_1 \in C_{pk}(m_1)$ and $c_2 \in C_{pk}(m_2)$, $c_1 +_{pk} c_2$ outputs a ciphertext c chosen uniformly at random from $C_{pk}(m_1+m_2)$. □

6.5 Performance

We implemented the protocol of Fig. 5 to empirically evaluate its computation and communication costs. The implementation is available at https://github.com/k3coby/pcr-go.

Parameters In our implementation, we instantiated the underlying cuckoo filter with bucket size $\chi = 4$, as recommended by Fan et al. [16]. We chose fingerprints of length 224 bits to achieve a low false-positive probability, i.e., about 2^{-221}. For the underlying partially homomorphic encryption scheme, we chose exponential ElGamal (e.g., see [10]) implemented in the elliptic-curve group secp256r1 [6] to balance performance and security (roughly equivalent to 3072-bit RSA security or 128-bit symmetric security).

Experiment Setup Our prototype including cuckoo filters and cryptography, were implemented in Go. We ran the experiments reported below on two machines with the same operating system and hardware specification: Ubuntu 20.04.1, AMD 8-core processor (2.67GHz), and 72GiB RAM. These machines played the role of the target and the monitor. We report all results as the means of 50 runs of each experiment and report relative standard deviations (rsd) in the figure captions.

Results We report the computation time of pcrQueryGen, pcrRespGen, and pcrReveal in Fig. 6. As shown in Fig. 6a, the computation time of pcrQueryGen is linear in the target's set size (i.e., $k + 1$). One takeaway here is that even if the number of honeywords is relatively large, e.g., $k = 1000$, it only takes the target about 100ms to generate a query with four logical CPU cores. Moreover, since a query is generated only when choosing to monitor an account at a monitor, the target can choose when to incur this cost. Figure 6b shows that the computational cost of PCR response generation is essentially unchanged regardless of k. This is important so that the computational burden on the monitors does not increase even if the target grows its number of honeywords per account. Another observation from Fig. 6b is that it only takes less than 9ms for the monitor, with even a single logical core, to produce a response when a failed login attempt occurs.

The computation time of $\arg_{e' \in \mathbf{X}} \mathsf{pcrReveal}_{sk}(e', \langle \mathbf{Z}, \mathbf{Z}' \rangle)$ is shown in Fig. 6c–d in two separate cases: when for all $e' \in \mathbf{X}$ is $\mathsf{pcrReveal}_{sk}(e', \langle \mathbf{Z}, \mathbf{Z}' \rangle) = \mathit{false}$ (and so the result $= \perp$, Fig. 6c) and when for some $e' \in \mathbf{X}$, $\mathsf{pcrReveal}_{sk}(e', \langle \mathbf{Z}, \mathbf{Z}' \rangle) = \mathit{true}$ (i.e., the result $\neq \perp$, Fig. 6d). We report these cases separately since they have significantly different performance characteristics. Again, we expect the former to be the common case. This operation takes constant time in the former case, since the target needs only to test if any of the 2χ ciphertexts (e.g., 8 ciphertexts with

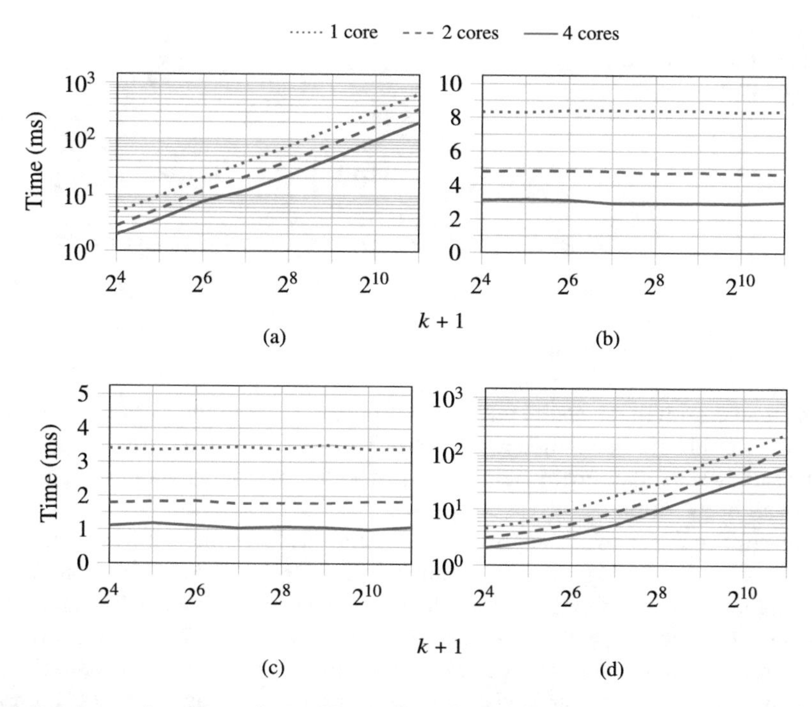

Fig. 6 Runtimes of $\mathsf{pcrQueryGen}_{pk}(\mathbf{X})$, $\mathsf{pcrRespGen}_{pk}(e, \mathbf{Y})$, and $\arg_{e' \in \mathbf{X}} \mathsf{pcrReveal}_{sk}(e', \langle \mathbf{Z}, \mathbf{Z}' \rangle)$ when $= \perp$ and when $\neq \perp$, as functions of $k + 1$ with varying numbers of logical CPU cores. (**a**) $\mathsf{pcrQueryGen}_{pk}(\mathbf{X})$ ($\mathsf{rsd} < 0.10$). (**b**) $\mathsf{pcrRespGen}_{pk}(e, \mathbf{Y})$ ($\mathsf{rsd} < 0.10$). (**c**) $\arg_{e' \in \mathbf{X}} \mathsf{pcrReveal}_{sk}(e', \langle \mathbf{Z}, \mathbf{Z}' \rangle) = \perp$ ($\mathsf{rsd} < 0.20$). (**d**) $\arg_{e' \in \mathbf{X}} \mathsf{pcrReveal}_{sk}(e', \langle \mathbf{Z}, \mathbf{Z}' \rangle) \neq \perp$ ($\mathsf{rsd} < 0.65$)

$\chi = 4$) are encryptions of zeros. In our experiments for Fig. 6d, the element e' for which $\mathsf{pcrReveal}_{sk}(e', \langle \mathbf{Z}, \mathbf{Z}' \rangle) = \mathit{true}$ was randomly picked from $\mathbf{X}$, and the target immediately returned once e' was identified. So the position of e' in $\mathbf{X}$ has a large impact on the computation time for each run, yielding an increased relative standard deviation. Since the target on average performs approximately $\frac{k+1}{2}$ isEq operations to identify e' in this case, the cost is linear in the target's set size, as shown in Fig. 6d.

As shown in Fig. 7, the query (message m1) is of size linear in the target's set size, while the response (m2) size is constant ($\approx$1 KB). These communication and storage costs are quite manageable. For example, even 100,000 monitoring requests would require only about 32GB of storage at the monitor when $k + 1 = 4096$.

Performance Example To put these performance results in context, consider the STRONTIUM credential harvesting attacks launched against over 200 organizations from September 2019 to June 2020. Microsoft [33] reported that their most aggressive attacks averaged 335 login attempts per hour per account for hours or days at a time, and that organizations targeted in these attacks saw login attempts on an average of 20% of their total accounts. So, if all of a target's monitors had

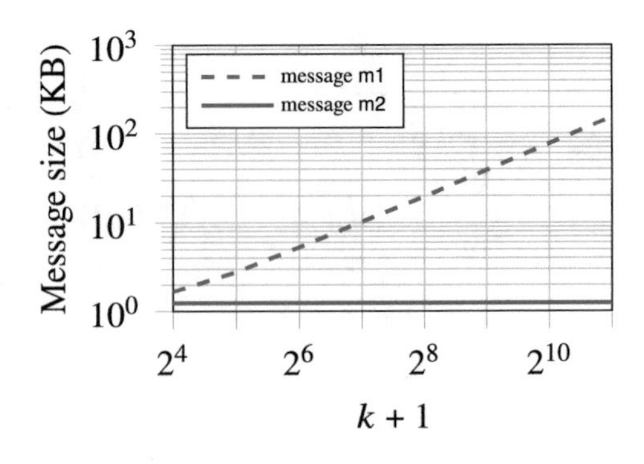

Fig. 7 Message size as a function of R's password set size for a (rsd < 0.01)

been attacked simultaneously by STRONTIUM, then 20% of the target's monitoring requests would have been triggered to generate responses to the target. Suppose that in the steady state, the target had maintained a total of x active monitoring requests across all of its monitors.

We now consider two scenarios. First, if monitors would not have limited the number of incorrect logins per account that induced monitoring responses, then each triggered monitoring request would have induced an average of 335 monitoring responses per hour. As such, the target would have averaged $(20\%)(335)(x) = 67x$ monitoring responses per hour, or $\frac{67}{3600}x$ monitoring responses per second. Since in our experiments, processing each monitoring response averaged ≈ 0.002s on a 2-core computer (Fig. 6c), this computer could have sustained the processing load that would have been induced on the target provided that $x < \frac{3600}{(0.002)(67)} \approx 26{,}865$ monitoring requests. Even if all x monitoring requests had been active at the same monitor, this monitor (using the same type of computer) could have sustained generating responses as long as $x < \frac{3600}{(0.005)(67)} \approx 10{,}746$, since generating responses on a 2-core computer averaged $\approx$0.005s (Fig. 6b). If the x monitoring requests had been spread across even only three monitors, however, the bottleneck would have been the target.

The second scenario we consider is one in which monitors would have limited the number of incorrect logins per account that induced a monitoring response, as recommended in Sect. 5.4. If each monitor would have limited the number of consecutive incorrect logins (and so monitoring responses) to 100 per account [19, Section 5.2.2], then the target would have averaged $(20\%)(100)(x) = 20x$ monitoring responses per hour and, using reasoning similar to that above, could have absorbed the induced processing load provided that $x < \frac{3600}{(0.002)(20)} = 90{,}000$ monitoring requests. And, in the extreme case that the same monitor held all x monitoring requests, the monitor (using the same type of computer) could have sustained generating responses for $x < \frac{3600}{(0.005)(20)} = 36{,}000$ monitoring requests.

7 Discussion

In this section we discuss various risks associated with Amnesia. The first is a general risk associated with Amnesia, and the others are specific to the distributed defenses against credential stuffing proposed in Sect. 5.

Password Reset Because detection happens in Amnesia when the legitimate user logs into her account at the target after the attacker has, the attacker can try to interfere with breach detection by changing the account password upon gaining access to the account. The legitimate user will be locked out of her account and so will presumably be forced to reset her password, but this will not serve as unequivocal evidence of the breach; after all, users reset their passwords all the time, due to simply forgetting them [22]. As such, target sites should utilize a backup authentication method (e.g., a code sent to a contact email or phone for the account) before enabling password reset.

Denial-of-Service Attacks There are mainly two potential ways of launching denial-of-service (DoS) attacks against a target: one in which the attacker submits login attempts at a high rate to a benign monitor to induce monitor responses to the target, and one in which a malicious monitor directly sends responses to the target at a high rate. The former DoS should be difficult for an attacker to perform effectively, since it requires the attacker to know or predict where the target will send monitoring requests and for what accounts. While we have not prescribed a specific strategy by which a target deploys monitor requests, such a strategy would need to be unpredictable; otherwise, rather than using this knowledge to conduct DoS, the attacker could instead use it to sidestep the accounts at sites while they are monitored, to avoid alerting the target to its breach. Another reason the former DoS will likely be ineffective is that, as discussed in Sect. 5.4, a target that can be breached repeatedly must rely on monitors to slow stuffing attacks to identify a user's reused password. These defenses will correspondingly help defend the target from this type of DoS. The latter DoS against a target, i.e., by a malicious monitor, would alert the target that this monitor is either conducting DoS or not implementing these slowing defenses. In either case, the target can remove this monitor from its list of monitors and drop responses from it.

As any site, a monitor should deploy state-of-the-art defenses against online guessing attacks which, in turn, can benefit targets as discussed above and in Sect. 5.4. The primary DoS risk introduced by Amnesia to monitors is the storage overhead of monitoring requests, though as discussed in Sect. 6.5, this need not be substantial. Moreover, the monitor has discretion to expire or discard monitoring requests as needed, and so can manage these costs accordingly.

User Privacy Privacy risks associated with remote monitoring of a user account include revealing to monitors the targets at which a user has an account and revealing to a target when a user attempts to log into a monitor. To obscure the former information, a target could send (ineffective) monitoring requests for accounts that have not been registered locally, e.g., using inputs $\mathcal{X}$ to pcrQueryGen consisting of uniformly random values. The latter information will likely be naturally obscured since failed login attempts to an account at a monitor due to automated attacks (online guessing, credential stuffing, etc.) would trigger PCR responses even if the account does not exist at the monitor and can outnumber failed login attempts by a legitimate user even if it does [41]. In addition, a monitor could further obscure user login activity on accounts for which it holds monitoring requests by generating monitoring responses at arbitrary times using uniformly random passwords.

Incentives to Monitor Accounts Given the overheads that monitoring requests induce on monitors, it is natural to question whether monitors have adequate incentives to perform monitoring for targets and, if so, at what rates. Moreover, these questions are complicated by site-specific factors.

On the one hand, large disparities in the numbers of accounts at various sites that might participate in a monitoring ecosystem could result in massive imbalances in the monitoring loads induced on sites. For example, issuing monitoring requests at a rate to induce expected steady-state monitoring of, say, even 10% of Gmail users' accounts, each at only a single monitor, would impose $\approx$180 million monitoring requests across monitors on an ongoing basis [37]. This could easily induce more load on monitors than they would find "worth it" for participating in this ecosystem.

On the other hand, dependencies among sites might justify substantial monitoring investment by the web community as a whole. For example, the benefit to internet security in the large for detecting a breach of Google's credential database quickly is considerable: as one of the world's largest email providers, it is trusted for backup authentication and account recovery (via email challenges) for numerous accounts at other sites. Indeed, as discussed above, some form of backup authentication *needs* to be a gatekeeper to resetting account passwords at a site who wishes to itself participate as target in our design, to ensure it will detect its own breach reliably. Such a site might thus be willing to participate as a monitor for numerous accounts of a target site on which many of its accounts depend for backup authentication.

Balancing these considerations to produce a viable monitoring ecosystem is a topic of ongoing research. We recognize, however, that establishing and sustaining such an ecosystem might benefit from additional inducements, e.g., monetary payments from targets to monitors or savings in the form of reduced insurance premiums for sites that agree to monitor for one another.

8 Conclusion

In this paper, we have proposed Amnesia, a methodology for using honeywords to detect the breach of a site *without relying on any secret persistent state*. Our algorithm remains effective to detect breaches even against attackers who repeatedly access the target site's persistent storage, including any long-term cryptographic keys. We extended this algorithm to allow the target site to detect breaches when the attacker tries to differentiate a (potentially reused) real password from honeywords by stuffing them at other sites. We realized this remote detection capability using a new private containment retrieval protocol with rounds, computation, communication, and storage costs that work well for our algorithm. We expect that, if deployed, Amnesia could effectively shorten the time between credential database breaches and their discovery.

Acknowledgments This research was supported in part by grant numbers 2040675 from the National Science Foundation and W911NF-17-1-0370 from the Army Research Office. The views and conclusions in this document are those of the authors and should not be interpreted as representing the official policies, either expressed or implied, of the National Science Foundation, Army Research Office, or the U.S. Government. The U.S. Government is authorized to reproduce and distribute reprints for Government purposes notwithstanding any copyright notices herein.

References

1. Akshima, Chang, D., Goel, A., Mishra, S., Sanadhya, S.K.: Generation of secure and reliable honeywords, preventing false detection. IEEE Trans. Depend. Secure Comput. **16**(5), 757–769 (2019).
2. Almeshekah, M.H., Gutierrez, C.N., Atallah, M.J., Spafford, E.H.: ErsatzPasswords: ending password cracking and detecting password leakage. In: 31st Annual Computer Security Applications Conference, pp. 311–320 (2015)
3. Bellare, M., Desai, A., Pointcheval, D., Rogaway, P.: Relations among notions of security for public-key encryption schemes. In: Advances in Cryptology—CRYPTO 1998, volume 1462 of Lecture Notes in Computer Science (1998)
4. Bellare, M., Rogaway, P.: Random oracles are practical: a paradigm for designing efficient protocols. In: 1st ACM Conference on Computer and Communications Security (1993)
5. Bojinov, H., Bursztein, E., Boyen, X., Boneh, D.: Kamouflage: Loss-resistant password management. In: European Symposium on Research in Computer Security, volume 6345 of Lecture Notes in Computer Science (2010)
6. Certicom Research: SEC 2: recommended elliptic curve domain parameters (2000). http://www.secg.org/SEC2-Ver-1.0.pdf. Standards for Efficient Cryptography
7. Chatterjee, R., Athayle, A., Akhawe, D., Juels, A., Ristenpart, T.: pASSWORD tYPOS and how to correct them securely. In: 37th IEEE Symposium on Security and Privacy, pp. 799–818 (2016)
8. Chen, H., Laine, K., Rindal, P.: Fast private set intersection from homomorphic encryption. In: 24nd ACM Conference on Computer and Communications Security (2017)
9. Cohn-Gordon, K., Cremers, C., Garratt, L.: On post-compromise security. In: 29th IEEE Computer Security Foundations Symposium (2016)
10. Cramer, R., Gennaro, R., Schoenmakers, B.: A secure and optimally efficient multi-authority election scheme. In: Advances in Cryptology—EUROCRYPT '97, volume 1233 of Lecture Notes in Computer Science, pp. 103–118 (1997)
11. Das, A., Bonneau, J., Caesar, M., Borisov, N., Wang, X.: The tangled web of password reuse. In: ISOC Network and Distributed System Security Symposium (2014)
12. Davidson, A., Cid, C.: An efficient toolkit for computing private set operations. In: 22nd Australasian Conference on Information Security and Privacy, volume 10343 of Lecture Notes in Computer Science (2017)

13. DeBlasio, J., Savage, S., Voelker, G.M., Snoeren, A.C.: Tripwire: Inferring internet site compromise. In: 17th Internet Measurement Conference (2017)
14. Erguler, I.: Achieving flatness: selecting the honeywords from existing user passwords. IEEE Trans. Parallel Distrib. Syst. **13**(2), 284–295 (2015)
15. Everspaugh, A., Chaterjee, R., Scott, S., Juels, A., Ristenpart, T.: The Pythia PRF service. In: 24th USENIX Security Symposium, pp. 547–562 (2015)
16. Fan, B., Andersen, D.G., Kaminsky, M., Mitzenmacher, M.D.: Cuckoo filter: practically better than Bloom. In: 10th ACM Conference on Emerging Networking Experiments and Technologies, pp. 75–88 (2014)
17. Freedman, M.J., Ishai, Y., Pinkas, B., Reingold, O.: Keyword search and oblivious pseudorandom functions. In: 2nd Theory of Cryptography Conference, volume 3378 of Lecture Notes in Computer Science (2005)
18. Freeman, D., Jain, S., Dürmuth, M., Biggio, B., Giacinto, G.: Who are you? A statistical approach to measuring user authenticity. In: 23rd ISOC Network and Distributed System Security Symposium (2016)
19. Grassi, P.A., et al.: Digital Identity Guidelines: Authentication and Lifecycle Management (2017). https://doi.org/10.6028/NIST.SP.800-63b. NIST Special Publication 800-63B.
20. Günther, C.G.: An identity-based key-exchange protocol. In: Advances in Cryptology—EUROCRYPT '89, volume 434 of Lecture Notes in Computer Science, pp. 29–37 (1989).
21. Herley, C., Florêncio, D.: Protecting financial institutions from brute-force attacks. In: 23rd International Conference on Information Security, volume 278 of IFIP Advances in Information and Communication Technology, pp. 681–685 (2008)
22. HYPR: New password study by HYPR finds 78% of people had to reset a password they forgot in past 90 days (2019). https://www.hypr.com/hypr-password-study-findings/
23. IBM Security: Cost of a data breach report 2020 (2020). https://www.ibm.com/security/digital-assets/cost-data-breach-report/
24. Jarecki, S., Krawczyk, H., Xu, J.: OPAQUE: an asymmetric PAKE protocol secure against pre-computation attacks. In: Advances in Cryptology—EUROCRYPT 2018, volume 10822 of Lecture Notes in Computer Science, pp. 456–486 (2018)
25. Juels, A., Rivest, R.L.: Honeywords: making password-cracking detectable. In: 20th ACM Conference on Computer and Communications Security (2013)
26. Kales, D., Rechberger, C., Schneider, T., Senker, M., Weinert, C.: Mobile private contact discovery at scale. In: 28th USENIX Security Symposium (2019)
27. Kiss, Á, Liu, J., Schneider, T., Asokan, N., Pinkas, B.: Private set intersection for unequal set sizes with mobile applications. In: 17th Privacy Enhancing Technologies Symposium, vol. 4, pp. 177–197 (2017)
28. Kontaxis, G., Athanasopoulos, E., Portokalidis, G., Keromytis, A.D.: SAuth: protecting user accounts from password database leaks. In: 20th ACM Conference on Computer and Communications Security (2013)
29. Kwiatkowska, M., Norman, G., Parker, D.: PRISM 4.0: verification of probabilistic real-time systems. In: International Conference on Computer Aided Verification, volume 6806 of Lecture Notes in Computer Science (2011)
30. Lai, R.W.F., Egger, C., Schröder, D., Chow, S.S.M.: Phoenix: rebirth of a cryptographic password-hardening service. In: 26th USENIX Security Symposium, pp. 899–916 (2017)
31. MacKenzie, P., Reiter, M.K.: Delegation of cryptographic servers for capture-resilient devices. Distrib. Comput. **16**(4), 307–327 (2003)
32. MacKenzie, P., Reiter, M.K.: Networked cryptographic devices resilient to capture. International J. Inform. Secur. **2**(1), 1–20 (2003)
33. Microsoft Threat Intelligence Center: STRONTIUM: Detecting new patterns in credential harvesting (2020). https://www.microsoft.com/security/blog/2020/09/10/strontium-detecting-new-patters-credential-harvesting/
34. Nojima, R., Kadobayashi, Y.: Cryptographically secure Bloom-filters. Trans. Data Privacy **2**(2), 131–139 (2009)

35. Oechslin, P.: Making a faster cryptanalytic time-memory trade-off. In: Advances in Cryptology—CRYPTO 2003, volume 2729 of Lecture Notes in Computer Science, pp. 617–630 (2003)
36. Pearman, S., Thomas, J., Naeini, P.E., Habib, H., Bauer, L., Christin, N., Cranor, L.F., Egelman, S., Forget, A.: Let's go in for a closer look: Observing passwords in their natural habitat. In: 24th ACM Conference on Computer and Communications Security (2017)
37. Petrov, C.: 50 Gmail statistics to show how big it is in 2020 (2020). https://techjury.net/blog/gmail-statistics/
38. Ramezanian, S., Meskanen, T., Naderpour, M., Junnila, V., Niemi, V.: Private membership test protocol with low communication complexity. In: 11th International Conference on Network and System Security, volume 10394 of Lecture Notes in Computer Science (2017)
39. Resende, A.C.D., Aranha, D.F.: Faster unbalanced private set intersection. In: 22nd International Conference on Financial Cryptography and Data Security, pp. 203–221 (2018)
40. Schneider, J., Fleischhacker, N., Schröder, D., Backes, M.: Efficient cryptographic password hardening services from partially oblivious commitments. In: 23rd ACM Conference on Computer and Communications Security, pp. 1192–1203 (2016)
41. Shape Security: 2018 credential spill report (2018). https://info.shapesecurity.com/rs/935-ZAM-778/images/Shape_Credential_Spill_Report_2018.pdf
42. Thomas, K., Li, F., Zand, A., Barrett, J., Ranieri, J., Invernizzi, L., Markov, Y., Comanescu, O., Eranti, V., Moscicki, A., Margolis, D., Paxson, V., Bursztein, E.: Data breaches, phishing, or malware? Understanding the risks of stolen credentials. In: 24th ACM Conference on Computer and Communications Security (2017)
43. Verizon: 2020 data breach investigations report (2020). https://enterprise.verizon.com/resources/reports/dbir/
44. Wang, C., Jan, S.T.K., Hu, H., Bossart, D., Wang, G.: The next domino to fall: empirical analysis of user passwords across online services. In: 8th ACM Conference on Data and Application Security and Privacy (2018)
45. Wang, D., Cheng, H., Wang, P., Yan, J., Huang, X.: A security analysis of honeywords. In: 25th ISOC Network and Distributed System Security Symposium (2018)
46. Wang, K.C., Reiter, M.K.: How to end password reuse on the web. In: 26th ISOC Network and Distributed System Security Symposium (2019)
47. Wang, K.C., Reiter, M.K.: Detecting stuffing of a user's credentials at her own accounts. In: 29th USENIX Security Symposium (2020)
48. Yue, C., Wang, H.: BogusBiter: a transparent protection against phishing attacks. ACM Trans. Internet Technol. **10**(2), 1–31 (2010)

Deceiving ML-Based Friend-or-Foe Identification for Executables

Keane Lucas, Mahmood Sharif, Lujo Bauer, Michael K. Reiter, and Saurabh Shintre

1 Introduction

Deceiving an adversary who may, e.g., attempt to reconnoiter a system before launching an attack, typically involves changing the system's behavior such that it deceives the attacker while still permitting the system to perform its intended function. For example, if a system hosting a database is using deception to defend against attack, it may employ measures that cause the attacker to believe that the system is running a different version of a database or that it is running other services. At the same time, legitimate clients of the system should continue to be able to interact with the database.

This chapter is based on the following paper: Keane Lucas, Mahmood Sharif, Lujo Bauer, Michael K. Reiter, and Saurabh Shintre. Malware makeover: Breaking ML-based static analysis by modifying executable bytes. In *Proceedings of the ACM Asia Conference on Computer and Communications Security*, 2021. https://doi.org/10.1145/3433210.3453086.

K. Lucas · L. Bauer (✉)
Carnegie Mellon University, Pittsburgh, PA, USA
e-mail: keanelucas@cmu.edu; lbauer@cmu.edu

M. Sharif
Tel Aviv University and VMware, Tel Aviv, Israel
e-mail: mahmoods@vmware.com

M. K. Reiter
Duke University, Durham, NC, USA
e-mail: michael.reiter@duke.edu

S. Shintre
NortonLifeLock Research Group, Sunnyvale, CA, USA
e-mail: saurabh.shintre@nortonlifelock.com

T. Bao et al. (eds.), *Cyber Deception*, Advances in Information Security 89,
https://doi.org/10.1007/978-3-031-16613-6_10

We seek to create deceptive behaviors by leveraging *evasion attacks* against deep neural networks (DNNs). In particular, we propose to model an attacker as a DNN whose input is a trace of the observable behavior of a defended system. We then attempt evasion attacks that *modify* the observed behavior of the defended system such that the modified behavior obeys the above constraints: deceiving the attacker (into taking some action other than the action that would compromise the defended system), while remaining compatible with the original intended behavior of the system.

A central challenge in developing strategies for deception is the difficulty of evaluating them: attackers' behavior is often not well enough understood to evaluate how it would change in response to changes in the behavior of the system under attack. Hence, we develop and evaluate techniques for implementing deception by studying a *proxy problem*: malware detection.

Modern malware detectors, both academic (e.g., [4, 44]) and commercial (e.g., [25, 90]), increasingly rely on machine learning (ML) to classify executables as benign or malicious based on features such as imported libraries and API calls. In the space of static malware detection, where an executable is classified prior to its execution, recent efforts have proposed deep neural networks (DNNs) that detect malware from binaries' raw byte-level representation, with effectiveness similar to that of detectors based on hand-crafted features selected through tedious manual processing [54, 76].

As old techniques for obfuscating and packing malware (see Sect. 4) are rendered ineffective in the face of static ML-based detection, recent advances in adversarial ML might provide a new opening for attackers to bypass detectors. Specifically, ML algorithms, including DNNs, have been shown vulnerable to adversarial examples—modified inputs that resemble normal inputs but are intentionally designed to be misclassified. For instance, adversarial examples can enable attackers to impersonate users that are enrolled in face-recognition systems [85, 86], fool street-sign recognition algorithms into misclassifying street signs [30], and trick voice-controlled interfaces to misinterpret commands [21, 74, 83].

In the malware-detection domain, the attackers' goal is to alter programs to mislead ML-based malware detectors to misclassify malicious programs as benign or vice versa. In doing so, attackers face a non-trivial constraint: in addition to misleading the malware detectors, alterations to a program must not change its functionality. For example, a keylogger altered to evade being detected as malware should still carry out its intended function, including invoking necessary APIs, accessing sensitive files, and exfiltrating information. This constraint is arguably more challenging than ones imposed by other domains (e.g., evading image recognition without making changes conspicuous to humans [30, 85, 86]) as it is less amenable to being encoded into traditional frameworks for generating adversarial examples, and most changes to a program's raw binary are likely to break a program's syntax or semantics. Prior work proposed attacks to generate adversarial examples to fool static malware-detection DNNs [27, 49, 55, 72, 89] by adding adversarially crafted byte values in program regions that do not affect execution (e.g., at the end of programs or between sections). These attacks can be defended against by eliminating the added content before classification (e.g., [56]); we confirm this empirically.

In contrast, we develop a new way to modify binaries to both retain their functionality and mislead state-of-the-art DNN-based static malware detectors [54, 76]. We leverage binary-diversification tools—originally proposed to defend against code-reuse attacks by transforming program binaries to create diverse variants [52, 71]—to evade malware-detection DNNs. While these tools preserve the functionality of programs by design (e.g., functionality-preserving randomization), their naïve application is insufficient to evade malware detection. We propose optimization algorithms to guide the transformations of binaries to fool malware-detection DNNs, both in settings where attackers have access to the DNNs' parameters (i.e., white-box) and ones where they have no access (i.e., black-box). The algorithms we propose can produce program variants that often fool DNNs in 100% of evasion attempts and, surprisingly, even evade some commercial malware detectors (likely over-reliant on ML-based static detection), in some cases with success rates as high as 85%. Because our attacks transform functional parts of programs, they are particularly difficult to defend against, especially when augmented with complementary methods to further deter static or dynamic analysis (as our methods alone should have no effect on dynamic analysis). We explore potential mitigations to our attacks (e.g., by normalizing programs before classification [3, 18, 98]) but identify their limitation in thwarting adaptive attackers.

In a nutshell, the contributions of our work are as follows:

- We repair and extend prior binary-diversification implementations to iteratively yield candidate transformations. We also reconstruct them to be composable, more capable, and resource-efficient. The code is available online.[1]
- We propose a novel functionality-preserving attack on DNNs for static malware detection from raw bytes (Sect. 3). The attack precisely composes the updated binary-diversification techniques, evades defenses against prior attacks, and applies to both white- and black-box settings.
- We evaluate and demonstrate the effectiveness of the proposed attack in different settings, including against commercial malware detectors (Sect. 4). We show that our attack effectively undermines ML-based static analysis, a significant component of state-of-the-art malware detection, while being robust to defenses that can thwart prior attacks.
- We explore the effectiveness of prior and new defenses against our proposed attack (Sect. 5). While some defenses seem promising against specific variants of the attack, none explored neutralize our most effective attack, and they are likely vulnerable to adaptive attackers.

[1] https://github.com/pwwl/enhanced-binary-diversification.

2 Background and Related Work

We first discuss previous work on DNNs that detect malware by examining program binaries. We then discuss research on attacking and defending ML algorithms generally, and malware detection specifically. Finally, we provide background on binary-randomization methods, which serve as building blocks for our attacks.

2.1 *DNNs for Static Malware Detection*

We study attacks targeting two DNN architectures for detecting malware from the raw bytes of Windows binaries (i.e., executables in Portable Executable format) [54, 76]. The main appeal of these DNNs is that they achieve state-of-the-art performance using automatically learned features, instead of manually crafted features that require tedious human effort (e.g., [4, 43, 50]). Due to their desirable properties, computer-security companies use DNNs similar to the ones we study (i.e., ones that operate on raw bytes and use a convolution architectures) for malware detection [24].

The DNNs proposed by prior work follow standard convolutional architectures similar to the ones used for image classification [54, 76]. Yet, in contrast to image classifiers that classify continuous inputs, malware-detection DNNs classify discrete inputs—byte values of binaries. To this end, the DNNs were designed with initial embedding layers that map each byte in the input to a vector in $\mathbb{R}^8$. After the embedding, standard convolutional and non-linear operations are performed by subsequent layers.

2.2 *Attacking and Defending ML Algorithms*

Attacks on Image Classification Adversarial examples—inputs that are minimally perturbed to fool ML algorithms—have emerged as challenge to ML. Most prior attacks (e.g., [9, 11, 14, 33, 70, 91]) focused on DNNs for image classification, and on finding adversarial perturbations that have small L_p-norm (p typically $\in \{0, 2, \infty\}$) that lead to misclassification when added to input images. By limiting perturbations to small L_p-norms, attacks aim to ensure that the perturbations are imperceptible to humans. Attacks are often formalized as optimization processes; e.g., Carlini and Wagner [14] proposed the following formulation for finding adversarial perturbations that target a class c_t and have small L_2-norms:

$$\arg\min_{r} \; Loss_{cw}(x + r, c_t) + \kappa \cdot ||r||_2$$

where x is the original image, r is the perturbation, and κ is a parameter to tune the L_2-norm of the perturbation. $Loss_{cw}$ is a function that, when minimized, leads $x+r$ to be (mis)classified as c_t. It is roughly defined as:

$$Loss_{cw}(x+r, c_t) = \max_{c \neq c_t}\{\mathbb{L}_c(x+r)\} - \mathbb{L}_{c_t}(x+r)$$

where $\mathbb{L}_c$ is the output for class c at the logits of the DNN—the output of the one-before-last layer. Our attacks use $Loss_{cw}$ to mislead the malware-detection DNNs.

Attacks on Static Malware Detection Modern malware-detection systems often leverage both dynamic and static analyses to determine maliciousness [8, 25, 44, 90, 93]. While in most cases an attacker would hence need to adopt countermeasures against both of these types of analyses, in other situations, such as potential attacks on end-user systems protected predominantly through static analysis based anti-virus detectors [20, 95], defeating a static malware detector could be sufficient for an attacker to achieve their goals. Even when a combination of static and dynamic analyses is used for detecting malware, fooling static analysis is necessary for an attack to succeed. Here we focus on attacks that target ML-based static analyzers for detecting malware.

Multiple attacks were proposed to evade ML-based malware classifiers while preserving the malware's functionality. Some (e.g., [26, 88, 97, 102]) tweak malware to mimic benign files (e.g., adding benign code-snippets to malicious PDF files). Others (e.g., [1, 27, 35, 41, 49, 55, 72, 89]) tweak malware using gradient-based optimizations or generative methods (e.g., to find which APIs to import). Still others combine mimicry and gradient-based optimizations [79].

Differently from some prior work (e.g., [1, 79, 97]) that studied attacks against dynamic ML-based malware detectors, we explore attacks that target DNNs for malware detection from raw bytes (i.e., static detection methods). Furthermore, the attacks we explore do not introduce adversarially crafted bytes to unreachable regions of the binaries [49, 55, 89] (which may be possible to detect and sanitize statically, see Sect. 4.4), or by mangling bytes in the header of binaries [27] (which can be stripped before classification [78]). Instead, our attacks transform actual instructions of binaries in a functionality-preserving manner to achieve misclassification.

More traditionally, attackers use various obfuscation techniques to evade malware detection. Packing [12, 80, 92, 94]—compressing or encrypting binaries' code and data, and then uncompressing or decrypting them at run time—is commonly used to hide malicious content from static detection methods. As we explain later (Sect. 3.1) we mostly consider unpacked binaries in this work, as is typical for static analysis [12, 54]. Attackers also obfuscate binaries by substituting instructions or altering their control-flow graphs [16, 17, 45, 92]. We demonstrate that such obfuscation methods do not fool malware-detection DNNs when applied naïvely (see Sect. 4.3). To address this, our attacks guide the transformation of binaries via stochastic optimization techniques to mislead malware detection.

Pierazzi et al. formalized the process of adversarial example generation in the problem space and used their formalization to produce malicious Android apps that evade detection [73]. Our attack fits the most challenging setting they describe, where mapping the problem space to features space is non-invertible and non-differentiable.

Most closely related to our work is the recent work on misleading ML algorithms for authorship attribution [65, 75]. Meng et al. proposed an attack to mislead authorship attribution at the binary level [65]. Unlike the attacks we propose, Meng et al. leverage weaknesses in feature extraction and modify debug information and non-loadable sections to fool the ML models. Furthermore, their method leaves a conspicuous footprint that the binary was modified (e.g., by introducing multiple data and code sections to the binaries). While this is potentially acceptable for evading author identification, it may raise suspicion when evading malware detection. Quiring et al. recently proposed an attack to mislead authorship attribution from source code [75]. In a similar spirit to our work, their attack leverages an optimization algorithm to guide code transformations that change syntactic and lexical features of the code (e.g., switching between `printf` and `cout`) to mislead ML algorithms for authorship attribution.

Defending ML Algorithms Researchers are actively seeking ways to defend against adversarial examples. One line of work, called adversarial training, aims to train robust models largely by augmenting the training data with correctly labeled adversarial examples [33, 46, 47, 57, 62, 91]. Another line of work proposes algorithms to train certifiably (i.e., provably) robust defenses against certain attacks [22, 51, 60, 67, 103], though these defenses are limited to specific types of perturbations (e.g., ones with small L_2- or L_∞-norms). Moreover, they often do not scale to large models that are trained on large datasets. As we show in Sect. 5, amongst other limitations, these defenses would also be too expensive to practically mitigate our attacks. Some defenses suggest that certain input transformations (e.g., quantization) can "undo" adversarial perturbations before classification [37, 61, 64, 81, 87, 100, 101]. In practice, however, it has been shown that attackers can adapt to circumvent such defenses [5, 6]. Additionally, the input transformations that have been explored in the image-classification domain cannot be applied in the context of malware detection. Prior work has also shown that attackers [13] can circumvent methods for detecting the presence of attacks (e.g., [31, 34, 64, 66]). We expect that such attackers can circumvent attempts to detect our attacks too.

Prior work proposed ML-based malware-classification methods designed to be robust against evasion [28, 43]. However, these methods either have low accuracy [43] or target linear classifiers [28], which are unsuitable for detecting malware from raw bytes.

Fleshman et al. proposed to harden malware-detection DNNs by constraining parameter weights in the last layer to non-negative values [32]. Their approach aims to prevent attackers from introducing additional features to malware to decrease its likelihood of being classified correctly. While this rationale holds for single-layer neural networks (i.e., linear classifiers), DNNs with multiple layers constitute complex functions where feature addition at the input may correspond to feature deletion in deep layers. As a result of the misalignment between the threat model and the defense, we found that DNNs trained with this defense are as vulnerable to prior attacks [55] as undefended DNNs.

2.3 Binary Rewriting and Randomization

Software diversification is an approach to produce diverse binary versions of programs, all with the same functionality, to resist different kinds of attacks, such as memory corruption, code injection, and code reuse [58]. Diversification can be performed on source code, during compilation, or by rewriting and randomizing programs' binaries. In this work, we build on binary-level diversification techniques, as they have wider applicability (e.g., self-spreading malware can use them to evade detection without source-code access [68]). Nevertheless, we expect that this work can be extended to work with different diversification methods.

Binary rewriting takes many forms (e.g., [38, 52, 53, 63, 71, 82, 99]). Certain methods aim to speed up code via expensive search through the space of equivalent programs [63, 82]. Other methods significantly increase binaries' sizes, or leave conspicuous signs that rewriting took place [38, 99]. We build on binary-randomization tools that have little-to-no effect on the size or run time of randomized binaries, thus helping our attacks remain stealthy [52, 71]. We present these tools and our extensions thereof in Sect. 3.2.

3 Technical Approach

Here we present the technical approach of our attack. Before delving into the details, we initially describe the threat model.

3.1 Threat Model

We assume that the attacker has white-box or black-box access to DNNs for malware detection that receive raw bytes of program binaries as input. In the white-box setting, the attacker has access to the DNNs' architectures and weights and can efficiently compute the gradients of loss functions with respect to the DNNs' input via forward and backward passes. On the other hand, the attacker in the black-box setting may only query the model with a binary and receive the probability estimate that the binary is malicious.

The DNNs' weights are fixed and *cannot* be controlled by attackers (e.g., by poisoning the training data). The attackers use binary rewriting to manipulate the raw bytes of binaries and cause misclassification while keeping functionality intact. Namely, attackers aim mislead the DNNs while ensuring that the I/O behavior of program and the order of syscalls remain the same after rewriting. In certain practical settings (e.g., when both dynamic and static detection methods are used [92]) evading static detection techniques as the DNNs we study may be insufficient to evade the complete stack of detectors. Nonetheless, evading the static detection techniques in such settings is *necessary* for evading detection overall. In Sect. 4.6, we show that our attacks can evade commercial detectors, some of which may be using multiple detection methods.

Attacks may seek to cause malware to be misclassified as benign or benign binaries to be misclassified as malware. The former may cause malware to circumvent defenses and be executed on a victim's machine. The latter induces false positives, which may lead users to turn off or ignore the defenses [39]. Our methods are applicable to transform binaries in either direction, but we focus on transforming malicious binaries in this chapter.

As is common for static malware detection [12, 54], we assume that the binaries are unpacked. While adversaries may attempt to evade detection via packing, our attack can act as an alternative or a complementary evasion technique (e.g., once packing is undone). Such a technique is particularly useful as packer-detection (e.g., [12]) and unpacking (e.g., [15]) techniques improve. In fact, we found that packing with a popular packer increases the likelihood of detection for malicious binaries (see Sect. 4.6), thus further motivating the need for complementary evasion measures.

As is standard for ML-based malware detection from raw bytes in particular (Sect. 2.1), and for classification of inputs from discrete domains in general (e.g., [59]), we assume that the first layer of the DNN is an embedding layer. This layer maps each discrete token from the input space to a vector of real numbers via a function $\mathbb{E}(\cdot)$. When computing the DNN's output $\mathbb{F}(x)$ on an input binary x, one first computes the embeddings and feeds them to the subsequent layers. Thus, if we denote the composition of the layers following the embedding by $\mathbb{H}(\cdot)$, then $\mathbb{F}(x) = \mathbb{H}(\mathbb{E}(x))$. While the DNNs we attack contain embedding layers, our attacks conceptually apply to DNNs that do not contain such layers. Specifically, for a DNN function $\mathbb{F}(x) = \ell_{n-1}(\ldots \ell_{i+1}(\ell_i(\ldots \ell_0(x)\ldots))\ldots)$ for which the errors can be propagated back to the $(i+1)^{th}$ layer, the attack presented below can be executed by defining $\mathbb{E}(x) = \ell_i(\ldots \ell_0(x)\ldots)$.

3.2 Functionality-Preserving Attack

The attack we propose iteratively transforms a binary x of class y (y=0 for benign binaries and y=1 for malware) until misclassification occurs or a maximum number of iterations is reached. To keep the binary's functionality intact, only functionality-preserving transformations are used. In each iteration, the attack determines the subset of transformations that can be safely used on each function in the binary. The attack then randomly selects a transformation from each function-specific subset and enumerates candidate byte-level changes. Each candidate set of changes is mapped to its corresponding gradient. The changes are only applied if this gradient has positive cosine similarity with the target model's loss gradient.

Algorithm 1: White-box attack

Input : $\mathbb{F} = \mathbb{H}(\mathbb{E}(\cdot))$, $\mathbb{L}_{\mathbb{F}}$, x, y, *niters*
Output: $\hat{x}$

1 $i \leftarrow 0$;
2 $\hat{x} \leftarrow RandomizeAll(x)$;
3 **while** $\mathbb{F}(\hat{x}) = y$ *and* $i < niters$ **do**
4   **for** $f \in \hat{x}$ **do**
5     $\hat{e} \leftarrow \mathbb{E}(\hat{x})$;
6     $g \leftarrow \frac{\partial \mathbb{L}_{\mathbb{F}}(\hat{x}, y)}{\partial \hat{e}}$;
7     $o \leftarrow RandomTransformationType()$;
8     $\tilde{x} \leftarrow RandomizeFunction(\hat{x}, f, o)$;
9     $\tilde{e} \leftarrow \mathbb{E}(\tilde{x})$;
10     $\delta_f = \tilde{e}_f - \hat{e}_f$;
11   **if** $g_f \cdot \delta_f > 0$ **then**
12       $\hat{x} \leftarrow \tilde{x}$;
13   **end**
14   **end**
15   $i \leftarrow i + 1$;
16 **end**
17 **return** $\hat{x}$;

Algorithm 1 presents the pseudocode of the attack in the white-box setting. The algorithm starts with a random initialization. This is manifested by transforming all the functions in the binary in an undirected way. Namely, for each function in the binary, a transformation type is selected at random from the set of available transformations and applied to that function without consulting loss-gradient similarity. When there are multiple ways to apply the transformation to the function, one is chosen at random. The algorithm then proceeds to further transform the binary using our gradient-guided method for up to *niters* iterations.

Each iteration starts by computing the embedding of the binary to a vector space, $\hat{e}$, and the gradient, g, of the DNN's loss function, $\mathbb{L}_{\mathbb{F}}$, with respect to the embedding. Particularly, we use the $Loss_{cw}$, presented in Sect. 2, as loss function. Because the true value of g is affected by any committed function change and could be unreliable after transforming many preceding functions in large files, it is recalculated prior to transforming each function (lines 5–6).

Ideally, to move the binary closer to misclassification, we would manipulate the binary so that the difference of its embedding from $\hat{e} + \alpha g$ (for some scaling factor α) is minimized (see prior work for examples [49, 55]). However, if applied naively, such manipulation would likely cause the binary to be ill-formed or change its functionality. Instead, we transform the binary via functionality-preserving transformations. As the transformations are stochastic and may have many possible outcomes (in some cases, more than can be feasibly enumerated), we cannot precisely estimate their impact on the binary a priori. Therefore, we implement the transformation of each function, f, as the acceptance or denial of candidate functionality-preserving transformations we iteratively generate throughout the function, where we apply a transformation only if it shifts the embedding in a direction similar to g (lines 5–13). More concretely, if g_f is the gradient with respect to the embedding of the bytes corresponding to f, and δ_f is the difference between the embedding of f's bytes after the attempted transformation and its bytes before, then each small candidate transformation is applied only if the cosine similarity (or, equivalently, the dot product) between g_f and δ_f is positive. Other optimization methods (e.g., genetic programming [102]) and similarity measures (e.g., similarity in the Euclidean space) that we tested did not perform as well.

If the input was continuous, it would be possible to perform the same attack in a black-box setting after estimating the gradients by querying the model (e.g., [42]). In our case, however, it is not possible to estimate the gradients of the loss with respect to the input, as the input is discrete. Therefore, the black-box attack we propose follows a general hill-climbing approach (e.g., [88]) rather than gradient ascent. The black-box attack is conceptually similar to the white-box one, and differs only in the method of checking whether to apply attempted transformations: Whereas the white-box attack uses gradient-related information to decide whether to apply a transformation, the black-box attack queries the model after attempting to transform a function and accepts the transformation only if the probability of the target class increases.

Transformation Types We consider two families of transformation types [52, 71]. As the first family, we adopt and extend transformation types proposed for in-place randomization (*IPR*) [71]. Given a binary to randomize, Pappas et al. proposed to disassemble it and identify functions and basic blocks, statically perform four types of transformations that preserve functionality, and then update the binary accordingly from the modified assembly. The four transformation types are: *(1)* replacing instructions with equivalent ones of the same length (e.g., `sub eax,4` → `add eax,-4`); *(2)* reassigning registers within functions or sets of basic blocks (e.g., swapping all instances of `ebx` and `ecx`) if this does not affect code that follows; *(3)* reordering instructions using a dependence graph to ensure that no instruction appears before one it depends on; and *(4)* altering the order in which register values are pushed to and popped from the stack to preserve them across function calls. To maintain the semantics of the code, the disassembly and transformations are performed conservatively (e.g., speculative disassembly, which is likely to misidentify code, is avoided). *IPR* does not alter binaries' sizes and has no measurable effect on their run time [71]. Figure 1 shows examples of transforming code via *IPR*.

The original implementation of Pappas et al. was unable to produce the majority of functionally equivalent binary variants that should be achievable under the four transformation types. Thus, we extended and improved the implementation

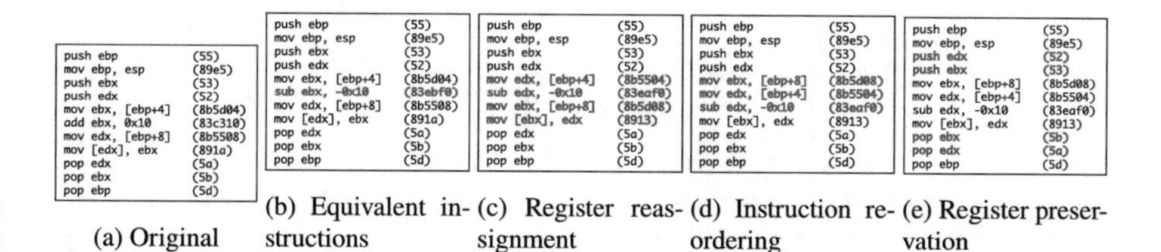

(a) Original (b) Equivalent instructions (c) Register reassignment (d) Instruction reordering (e) Register preservation

Fig. 1 An illustration of *IPR*. We show how the original code (**a**) changes after replacing instructions with equivalent ones (**b**), reassigning registers (**c**), reordering instructions (**d**), and changing the order of instructions that preserve register values (**e**). We provide the hex encoding of each instruction to its right. The affected instructions are boldfaced and colored in red

in various ways. First, we enabled the transformations to compose: unlike Pappas et al.'s implementation, our implementation allows us to iteratively apply different transformation types to the same function. Second, we apply transformations more conservatively to ensure that the functionality of the binaries is preserved (e.g., by not replacing `add` and `sub` instructions if they are followed by instructions that read the flags register). Third, compared to the previous implementation, ours handles a larger number of instructions and function-calling conventions. In particular, our implementation can rewrite binaries containing additional instructions (e.g., `shrd`, `shld`, `ccmove`) and less common calling conventions (e.g., nonstandard returns via increment of `esp` followed by a `jmp` instruction). Last, we fixed significant bugs in the original implementation. These bugs include incorrect checks for writes to memory after reads, as well as memory leaks which required routine experiment restarts.

The second family of transformation types that we build on is based on code displacement (*Disp*) [52]. Similarly to *IPR*, *Disp* begins by conservatively disassembling the binary. The original idea of *Disp* is to break potential gadgets that can be leveraged by code-reuse attacks by moving code to a new executable section. The original code to be displaced has to be at least five bytes in size so that it can be replaced with a `jmp` instruction that passes control to the displaced code. If the displaced code contains more than five bytes, the bytes after the `jmp` are replaced with trap instructions that terminate the program; these would be executed if a code-reuse attack is attempted. In addition, another `jmp` instruction is appended to the displaced code to pass control back to the instruction that should follow. Any displaced instruction that uses an address relative to the instruction pointer (i.e., `IP`) register is also updated to reflect the new address after displacement. *Disp* has a minor effect on binaries' sizes ($\sim$2% increase on average) and causes a small amount of run-time overhead ($<$1% on average) [52].

We extend *Disp* in two main ways. First, we allow it to displace any set of consecutive instructions within a basic block, not only ones that belong to gadgets. Second, instead of replacing the original instructions with traps, we replace them with *semantic nops*—sets of instructions that *cumulatively* do not affect the memory

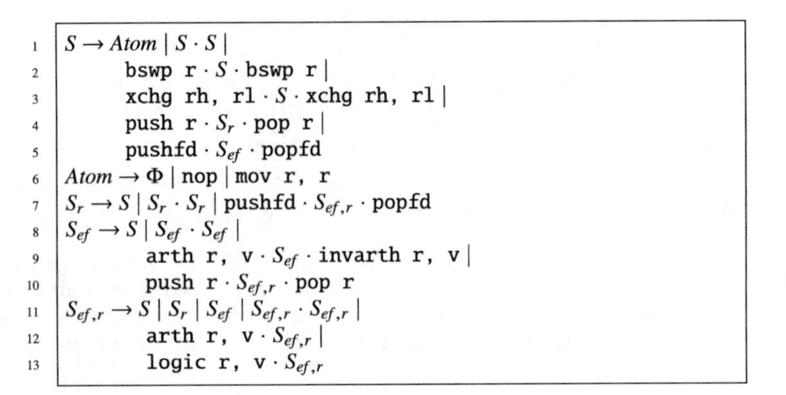

```
1   S → Atom | S · S |
2        bswp r · S · bswp r |
3        xchg rh, rl · S · xchg rh, rl |
4        push r · S_r · pop r |
5        pushfd · S_ef · popfd
6   Atom → Φ | nop | mov r, r
7   S_r → S | S_r · S_r | pushfd · S_ef,r · popfd
8   S_ef → S | S_ef · S_ef |
9         arth r, v · S_ef · invarth r, v |
10        push r · S_ef,r · pop r
11  S_ef,r → S | S_r | S_ef | S_ef,r · S_ef,r |
12        arth r, v · S_ef,r |
13        logic r, v · S_ef,r
```

Fig. 2 A context-free grammar for generating semantic nops. S is the starting symbol; Φ the empty string; `arth` indicates an arithmetic operation (specifically, `add`, `sub`, `adc`, or `sbb`); `invarth` indicates its inverse; `logic` indicates a logical operation (specifically, `and`, `or`, or `xor`); and `r` and `v` indicate a register and a randomly chosen integer, respectively

or register values and have no side effects [17]. These semantic nops get jumped to immediately after the displaced code is done executing.

While nops can be defined atomically (e.g., by a `nop` instruction), initial failures to mislead malware detection indicated that a rich semantic nop language is needed for successful attacks. Such a language enables the attack to search through a large set of functionally equivalent programs to evade DNNs. Therefore, we developed a context-free grammar to create diverse semantic nops (see Fig. 2). At a high level, a semantic nop is an atomic instruction; or an invertible instruction that is followed by a semantic nop and then by the inverse instruction (e.g., `push eax` followed by a semantic nop and then by `pop eax`); or two consecutive semantic nops. When the flags register's value is saved (i.e., between `pushfd` and `popfd` instructions), a semantic nop may contain instructions that affect flags (e.g., add and then subtract a value from a register); and when a register's value is saved (i.e., between `push r` and `pop r`), a semantic nop may contain instructions that affect the register (e.g., decrement it by a random value). Using the grammar for generating semantic nops, for example, one may generate a semantic nop that stores the flags and `ebx` registers on the stack (`pushfd; push ebx`), performs an operation that might affect both registers (e.g., `add ebx, 0xff`), and then restores the registers (`pop ebx; popfd`).

When using *Disp*, our attacks start by displacing code up to a certain budget, to ensure that the resulting binary's size does not increase above a threshold (e.g., 1% above the original size). We divide the budget (expressed as the number of bytes to be displaced) by the number of functions in the binary and attempt to displace exactly that number of bytes per function. If multiple options exist for what code in a function to displace, we choose at random. If a function does not contain enough code to displace, then we attach semantic nops after the displaced code to meet the per-function budget. In the rare case that the function does not have any basic

```
...          ...                ...
0x4587:      add ax, 0x10       (6683c010)
0x458b:      sub bx, 0x10       (6683eb10)
0x458f:      cmp ax, bx         (6639d8)
...          ...                ...
```

(a) Original code

```
...          ...                ...
0x4587:      jmp 0x4800         (e974020000)
0x458c:      mov cx, cx         (6689c9)
0x458f:      cmp ax, bx         (6639d8)
...          ...                ...

...          ...                ...
0x4800:      add ax, 0x10       (6683c010)
0x4804:      sub bx, 0x10       (6683eb10)
0x4808:      nop                (90)
0x4805:      pushfd             (9c)
0x4806:      push ebx           (53)
0x4807:      add ebx, 0x1a      (83c31a)
0x480a:      pop ebx            (5b)
0x480b:      popfd              (9d)
0x480d:      jmp 0x458c         (e97afdffff)
...          ...                ...
```

(b) After *Disp*

Fig. 3 An example of displacement. The two instructions staring at address `0x4587` in the original code (**a**) are displaced to starting address `0x4800`. The original instructions are replaced with a `jmp` instruction and a semantic nop (see (**b**)). To consume the displacement budget, semantic nops are added immediately after the displaced instructions and just before the `jmp` that passes the control back to the original code. Semantic nops are shown in boldface and red

block larger than five bytes, we skip that function. Figure 3 illustrates an example of displacement where semantic nops are inserted to replace original code as well as after displaced code, to consume the budget. Then, in each iteration of modifying the binary to cause it to be misclassified, new semantic nops are chosen at random and used to replace the previously inserted semantic nops if that moves the binary closer to misclassification.

Some of the semantic nops contain integer values that can be set arbitrarily (e.g., see line 12 of Fig. 2). In a white-box setting, the bytes of the binary that correspond to these values can be set to perturb the embedding in the direction that is most similar to the gradient. Namely, if an integer value in the semantic nop corresponds to the ith byte in the binary, we set this i^{th} byte to $b \in \{0, \ldots, 255\}$ such that the cosine similarity between $\mathbb{E}(b) - \mathbb{E}(\hat{x}_i)$ and g_i is maximized. This process is repeated each time a semantic nop is drawn to replace previous semantic nops in white-box attacks.

Known methods [18] for detecting and removing semantic nops from binaries might appear viable for defending against *Disp*-based attacks. However, as we discuss in Sect. 5, attackers can leverage various techniques to evade semantic-nop detection and removal.

Limitations Our implementation leaves room for improvement. For instance, it does not displace code that has been displaced in earlier iterations. A better implementation might apply displacements recursively. Furthermore, the composability of *IPR* and *Disp* transformations could be improved. In particular, when applying both *Disp* and *IPR* transformations to a binary, both types of transformations affect the original instructions of the binary. However, *IPR* does not affect the semantic nops that are introduced by *Disp*. Despite room for improvement, our implementation is already sufficient to generate successful attacks (see below).

4 Evaluation

In this section, we comprehensively evaluate our attack. We first detail the DNNs and data used for evaluation. We then show that naïve, random transformations that are not guided via optimization do not lead to misclassification. Subsequently, we evaluate variants of our attack in the white- and black-box setting and compare with prior work. We then evaluate our attack against commercial anti-viruses and close the section with experiments to validate that the attacks preserve functionality.

4.1 Datasets and Malware-Detection DNNs

4.1.1 Dataset Composition

Our dataset, *VTFeed*, contains raw binaries of malware samples targeting Windows machines. As such, the binaries adhere to the Portable Executable format (*PE*; the standard format for `.dll` and `.exe` files) [48]. Overall, we use significantly more samples than similar prominent prior work (e.g., [4, 50]).

VTFeed was collected by sampling the VirusTotal feed for PE binaries, representing binaries encountered in practice by anti-virus vendors. Collection took around two weeks and was restricted to binaries first seen in 2020, to ensure recency, and smaller than 5 MB. Following prior work [2], binaries were filtered and labeled as benign (resp., malicious) if they were classified as malicious by 0 (resp., over 40) anti-virus vendors as aggregated by VirusTotal. The dataset contains 278,316 binaries with a roughly even distribution between benign and malicious binaries. We sampled training, test, and validation sets at a ratio of 80%, 10%, and 10%, respectively. Exact numbers can be seen in Table 1.

Table 1 The number of benign and malicious binaries used to train, validate, and test the DNNs

VTFeed	Train	Val.	Test
Benign	111,258	13,961	13,926
Malicious	111,395	13,870	13,906

4.1.2 DNN Training

Using the malicious and benign samples, we trained two malware-detection DNNs. All DNNs receive binaries' raw bytes as inputs and output the probability that the binaries are malicious. The first DNN (henceforth, *AvastNet*), proposed by Krčál et al. [54], receives inputs up to 512 KB in size. The second DNN (henceforth, *MalConv*), proposed by Raff et al. [76], receives inputs up to 2 MB in size. Except for the batch size (set to 32 due to memory limitations), we used the same training parameters reported in prior work. When using binaries for training, we excluded their headers so the DNNs would not rely on header values, which are easily manipulable, for classification [27].

Each DNN achieves test accuracy of about 99% (see Table 2). Even when restricting the false positive rates (FPRs) conservatively to 0.1% (as is often done by anti-virus vendors [54]), the true positive rates (TPRs) remain as high as 94–96% (i.e., 94–96% of malicious binaries are detected). These results are superior to those reported in the original papers both for classification from raw bytes and from manually crafted features [54, 76]. This is likely because *VTFeed* was sampled over a narrow time span, and expect the performance would slightly decrease if we increased the diversity of the dataset.

In addition to the two DNNs that we trained, we evaluated our attacks using a publicly available DNN (henceforth, *Endgame*) trained by Anderson and Roth [2]. *Endgame* has a similar architecture to *MalConv*. The salient differences are that: *(1)* *Endgame*'s input dimensionality is 1 MB (compared to 2 MB for *MalConv*); and *(2)* *Endgame* uses the *PE* header for classification. On a dataset curated by a computer-security company, *Endgame* achieved about 92% TPR when the FPR was restricted to 0.1% [2].

To evaluate attacks against the DNNs, we selected binaries according to three criteria. First, the binaries had to be unpacked. We used standard packer detectors, Packerid [84] and Yara [96], and deemed binaries as unpacked only if no detector exhibited a positive detection. This method is similar to the one followed by Biondi et al. [12].[2] We also filtered out binaries labeled as packed in their VirusTotal metadata. While the data used to train and test the DNNs included packed binaries, the high accuracy of the DNNs on the test samples suggests that the DNNs' performance was not impacted by (lack of) packing. Second, the binaries had to be classified correctly and with high confidence by the DNNs that we trained. In

Table 2 The DNNs' accuracy and the TPR at the operating point where the FPR equals 0.1%

	Accuracy			TPR @
	Train	Val.	Test	0.1% FPR
AvastNet	99.89%	98.59%	98.60%	94.78%
MalConv	99.97%	98.67%	98.53%	96.08%

[2] Biondi et al. used three packer-detection tools instead of two. Unfortunately, we were unable to get access to one of the proprietary tools.

particular, malicious binaries had to be classified as malicious and the estimated probability that they are malicious had to be above the threshold where the FPR is 0.1%. Consequently, our evaluation of the attacks' success is conservative: the attacks would be more successful for binaries that are initially classified correctly, but not with high confidence. Third, the binaries' sizes had to be smaller than the DNNs' input dimensionality. We further restricted the binaries' sizes to be smaller than smallest input dimensionality of our DNNs (*AvastNet* at 512 KB). While the DNNs can classify binaries whose size is larger than the input dimensionality (as can be seen from the high classification accuracy on the validation and test sets), we avoided large binaries as a means to prevent evasion by displacing malicious code outside the input range of the DNNs. Using these criteria, we selected 100 malicious binaries from the test set to evaluate the attacks against each of the three DNNs.

The total number of samples we collected is comparable to that used in prior work on evading malware detection [49, 55, 88, 89].

4.2 Attack-Success Criteria

We executed the attacks for up to 200 iterations, stopping early if the binaries were misclassified at the operating point where the FPR equals 0.1%. For malicious binaries, this meant that they were misclassified as benign with a probability higher than a model-specific threshold set to achieve 0.1% FPR. This follows the threshold typically used by anti-virus vendors (e.g., [54]). We also found that attack success on the same binary, given identical experiment parameters, was often stochastic. Therefore, we repeated each attack 10 times to get a reliable measure of attack success.

We compared the overall success of attacks in two ways: by the percentage of binaries that were misclassified in *at least* 1 of the 10 repeated attacks on them (**coverage**); and the overall percentage of attacks that were successful across all attacked binaries (**potency**). Coverageis a measure of what percentage of binaries our attack *can* be successful on whereas potency is a measure of the how often a single attack trial succeeds. As a result of this definition, coverage will always be higher than potency.

4.3 Randomly Applied Transformations

We first evaluated whether naïvely transforming binaries at random would lead to evading the DNNs. For each binary that we used to evaluate the attacks we created 200 variants using the *IPR* and *Disp* transformations and classified them using the DNNs. We transformed the binaries sequentially and at random. Namely, starting from the original variant, we created the next variant by transforming every function using a randomly picked transformation type that was applied at random. If any of the variants were misclassified by a DNN given the 1% FPR threshold, we would consider the evasion attempt successful. We set *Disp* to increase binaries' sizes by 5% (i.e., the displacement budget was set to 5% of the binary's original size). We selected 200 and 5% as parameters for this experiment because our attacks were executed for 200 iterations at most and achieved almost perfect success when increasing binaries' sizes by 5% (see below). This technique was most effective when attempting to misclassify malware as benign on *Endgame*, where four binaries evaded detection. However, for all other attempts to evade, no more than three binaries were successful.

Hence, we conclude the DNNs are robust to naïve transformations and more principled approaches are needed to mislead them.

4.4 White-Box Attacks vs. DNNs

In the white-box setting, we evaluated seven variants of our attack. One variant, to which we refer to as *IPR*, relies on the *IPR* transformations only. Three variants, *Disp*-1, *Disp*-3, and *Disp*-5, rely on the *Disp* transformations only, where the numbers indicate the displacement budget as a percentage of the binaries' sizes (e.g., *Disp*-1 increases binaries' sizes by 1%). The last three attack variants, *IPR+Disp*-{1,3,5}, use the *IPR* and *Disp* transformations combined.

We set 5% as the maximum displacement budget and 200 as the maximum number of iterations, as the attacks were almost always successful with these parameters.

The results of the experiments are provided in Fig. 4 where the lighter part of the bar represents potency and the darker part represents coverage. One can immediately see that attacks using the *Disp* transformations were more successful than *IPR*. While showing some effectiveness in evading *Endgame*, *IPR* at best achieves a coverage of 52% while *Disp* of all budgets on all three models are able to cause at least 92% of binaries to be misclassified.

Moreover, *Disp*-5 achieved misclassification on all binaries except one on *AvastNet*. As one would expect, attacks with higher displacement budgets were more successful than attacks with lower displacement budgets. However, the main

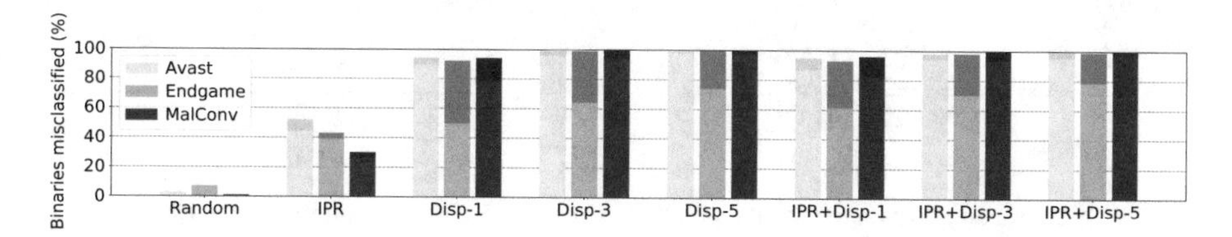

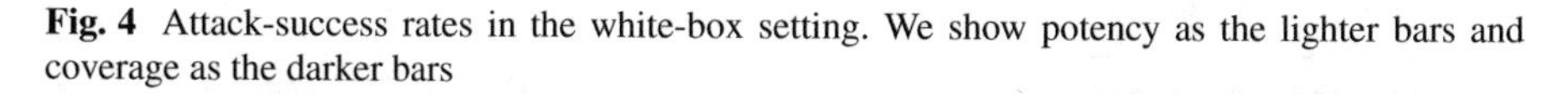

Fig. 4 Attack-success rates in the white-box setting. We show potency as the lighter bars and coverage as the darker bars

difference we see is in the potency of the attack, whereas the coverage only differs by a single missed binary between *Disp*-3 and *Disp*-5.

In addition to achieving higher coverage and potency, another advantage of *Disp*-based attacks over *IPR*-based ones is their time efficiency. While displacing instructions at random from within a function with n instructions has $O(n)$ time complexity, certain *IPR* transformations have $O(n^2)$ time complexity. For example, reordering instructions requires building a dependence graph and extracting instructions one after the other. If every instruction in a function depends on previous ones, this process takes $O(n^2)$ time. In practice, we found that *IPR*-based, *Disp*-based, and *IPR+Disp*-based attacks took 4424, 283, and 961 s on average, respectively.[3]

Combining *IPR* with *Disp* achieved noticeably better results in fewer iterations than respective *Disp*-only attacks when the budget for *Disp* is low. For example, *IPR+Disp*-1 had 11% higher potency than *Disp*-1 when misleading *Endgame* to misclassify a malicious binary as benign (61% vs. 50% potency). Thus, in certain situations, *Disp* and *IPR* can be combined to fool the DNNs while increasing binaries' sizes less than *Disp* alone.

For our most performant attack, *IPR+Disp*-5, we re-executed the attacks with significantly more difficult success criteria. We changed the threshold for attack success to *MalConv* and *AvastNet*'s FNR of 0.01%. Beating this threshold means that a transformed binary must appear less malicious than the least malicious 0.1% of malware in the dataset. For *MalConv*, our potency drops from 97% to 92%, while coverage drops from 100% to 99%. For *AvastNet*, potency drops from 95% to 90% and coverage from 100% to 95%. These results demonstrate our attack's ability to evade more cautious ML detectors, even though this threshold is unlikely to be used as it would flag roughly a third of benign binaries as malware.

In Fig. 5, we averaged and plotted the classification output of the models and the resultant misclassifications of the binaries over the iterations of each attack. As shown, the majority of successful attacks that incorporate *Disp* succeeded in a single iteration, with almost all successful attacks occurring within ten iterations. We also examined the performance of the attacks as a function of the number of modifiable functions in a binary. On average, 89% of functions in a binary were modifiable. As

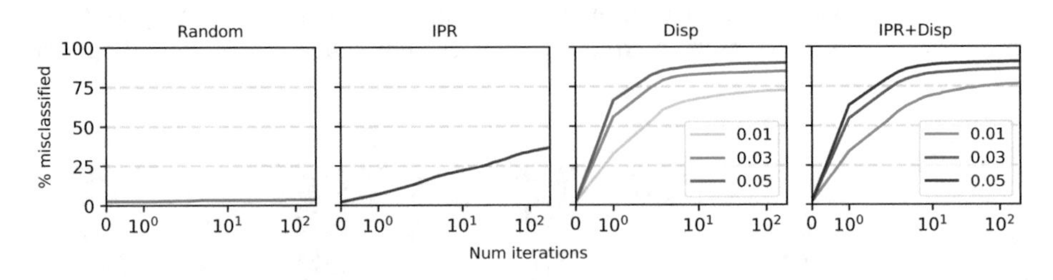

Fig. 5 A contrasting view showing the potency over iteration for the white-box attacks

[3] Times were computed on four machines: one with 2.2GHz AMD Opteron CPU and 128GB RAM, one with 3.4GHz Intel-i7 CPU and 24GB RAM, one with 2.2Ghz AMD Ryzen 3900X and 64GB RAM, and one with 2.7GHz Intel-i5 CPU and 24GB RAM.

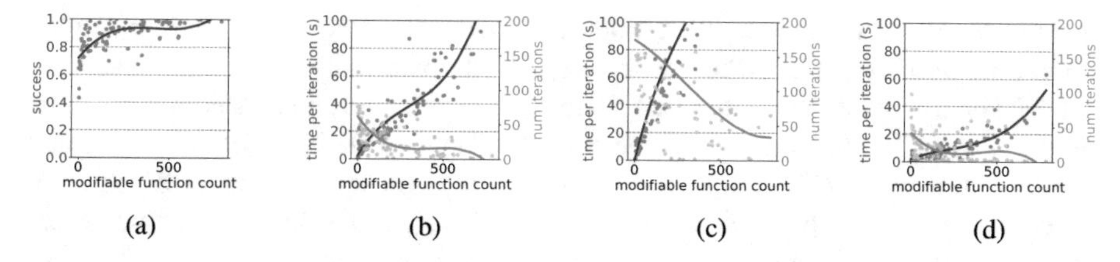

Fig. 6 As the number of modifiable functions increased, the average number of iterations to success decreased, while the time to execute an iteration increased. The lines in each plot are the best fit degree-3 polynomials. (**a**) Success (all attacks). (**b**) Time (all attacks). (**c**) Time (*IPR*). (**d**) Time (*Disp*)

expected, attacks were less likely to succeed when the binaries had few functions to modify (Fig. 6a). Consistent with that finding, as the number of modifiable functions (and number of functions overall) in a binary increased, the average number of iterations required for an attack to succeed decreased (Figs. 6b–d). This trend held across different types of attacks but was more pronounced for less successful attacks (*IPR*) than more successful ones (*Disp*), as the vast majority of the latter completed within a small number of iterations.

Finally, we compared the evasion success rates of our attack with a representative prior attack proposed by Kreuk et al. [55]. To mislead DNNs, this attack appends adversarially crafted bytes to binaries. These bytes are crafted via an iterative algorithm that first computes the gradient g_i of the loss with respect to the embedding $\mathbb{E}(x_i)$ of the binary x_i at the ith iteration, and then sets the adversarial bytes to minimize the L_2 distance of the new embedding $\mathbb{E}(x_{i+1})$ from $\mathbb{E}(x_i) + \epsilon sign(g_i)$, where ϵ is a scaling parameter. We tested three variants of the attack which increase the binaries' sizes by 1%, 3%, and 5%. We used $Loss_{cw}$ as the loss function. As with our attacks, we executed Kreuk et al.'s attacks for up to 200 iterations, stopping sooner if misclassification occurred. We set ϵ=1, as we empirically found it leads to high evasion success.

The variants of Kreuk et al.'s attack achieved success rates comparable to our attack. *Kreuk*-5 was almost always able to mislead the DNNs—it achieved 99% and 98% success rate when attempting to mislead *Endgame* and *MalConv*, respectively, to misclassify malicious binaries, and 100% success rate in all other attempts. Also similar to our attacks, the success rates increased as the attacks increased the binaries' sizes. However, as described in Appendix 1, their attack is easier to defend against by sanitizing bytes (specifically, by masking with zeros) in sections that do not contain instructions.

4.5 *Black-Box Attacks vs. DNNs*

As explained in Sect. 3, because the DNNs' input is discrete, estimating gradient information to mislead them in a black-box setting is not possible. So, the black-box version of Algorithm 1 uses hill climbing to query the DNN after each attempted transformation to decide whether to keep the transformation. Because querying the DNNs after each attempted transformation significantly increased the run time of the attacks ($\sim 30\times$ on a machine with GeForce GTX 980 GPU), we limited our experiments to *Disp* transformations with a displacement budget of 5%. We executed the attacks up to 200 iterations, stopped early if misclassification occurred, and repeated them three times each to account for stochasticity.

The attacks were most successful against *MalConv*, achieving a coverage of 95% and potency of 92%. *AvastNet* and *Endgame* were only slightly more robust with attack coverages of 92% and 59% and potencies of 87% and 56%, respectively. These results show our attack remains effective even in a black-box setting.

4.6 Commercial Anti-Viruses

To assess whether our attacks affect commercial anti-viruses, we tested the malicious transformed binaries that were misclassified by the DNNs in the white-box setting on the anti-viruses available via VirusTotal [19]—a service that aggregates the results of 68 commercial anti-viruses. Since anti-viruses often rely in part on static analysis, with increasing integration of ML, we expected that the malicious binaries generated by our attacks would be detected by fewer anti-viruses than the original binaries.

Due to contractual constraints, we were unable to perform this experiment with our previously described dataset. Thus, we resorted to using binaries taken from other sources. In this alternate dataset, we used 21,741 malicious binaries belonging to seven malware families that were published by Microsoft as part of a malware-classification competition [78]. We complemented these binaries with 19,534 benign binaries collected by installing standard packages (browsers, productivity tools, etc.) on a newly created 32-bit Windows 7 virtual machine.[4] After splitting the binaries for training (21,217), testing (9105), and validation (10,953), we trained variants of *MalConv* and *AvastNet* that achieved 99.15% and 98.92% test accuracy, respectively. Subsequently, we collected 95 malicious binaries from VirusShare [77] that pertain to the seven malware families from the Microsoft competition. We then transformed these malicious binaries using our white-box attack to evade the DNNs we trained as well as *Endgame*, and tested how often the transformed binaries were detected by anti-viruses on VirusTotal.

Original Binaries As a baseline, we first classified the original binaries using the VirusTotal anti-viruses. As one would expect, all the malicious binaries were detected by several anti-viruses. The median number of anti-viruses that detected any particular malware binary as malicious was 55, out of 68 total anti-viruses.

[4] Specifically, we used the Ninite and Chocolatey (https://ninite.com/ and https://chocolatey.org/) package managers to install 179 packages.

Random Transformations To further gauge the efficacy of our guided attack over random diversification, we used commercial anti-viruses to classify binaries that were transformed at random using the *Disp* and *IPR* transformations (as described in Sect. 4.3). We found that certain anti-viruses were susceptible to such simple evasion attempts, presumably due to using fragile detection mechanisms such as signatures. The median number of anti-viruses that correctly detected the malicious binaries decreased from 55 to 43.

Packing We tested whether anti-viruses were susceptible to evasion via packing. We used UPX [69], one of the most popular packers [80], and packed binaries using the highest compression ratios. Interestingly, packing malicious binaries was counter-productive for evading anti-viruses. Packed malicious binaries were more likely to be detected as malware—the median number of anti-viruses that correctly detected malicious binaries increased from 55 for the original binaries to 59 after packing.

Our Attacks Compared to the original malicious binaries and randomly transformed ones, the malicious binaries transformed by our attacks were detected by fewer anti-viruses. The median number of anti-viruses that correctly detected the malicious binaries decreased from 55 for the original binaries and 42 for ones transformed at random to 33–36, depending on the attack variant and the targeted DNN. According to a Kruskal-Wallis test, this reduction is statistically significant ($p < 0.01$ after Bonferroni correction). In other words, the malicious binaries that were transformed by our attacks were detected by only 49%–53% of the VirusTotal anti-viruses in the median case. Table 3 in Appendix 2 summarizes each attack variant's effect on the number of positive detections by anti-viruses.

Because our attack should not affect any dynamic analysis (due to the desired attack property of functional invariance), these results indicate some anti-viruses may be over-reliant on static analyses and/or ML. We also highlight these results cannot only be attributed to breaking signature-based defenses, as the randomly transformed binaries (which were transformed for an equal number of iterations) would have been equally likely to evade anti-viruses as our attacks.

Furthermore, several anti-virus vendors that were misled by our attacks advertise the use of ML detectors. Evading the ML detectors of those vendors was necessary to mislead their anti-viruses. A glance at vendors' websites showed that 15 of the 68 vendors explicitly advertise relying on ML for malware detection. These anti-viruses were especially susceptible to evasion by our attacks. Even more concerning, a popular and highly credible anti-virus whose vendor claims to rely on ML misclassified 85% of the malicious binaries produced by one of our attacks as benign. Generally, malicious binaries that were produced by our attacks were detected by a median number of 7–9 anti-viruses of the 15—down from 12 positive detections for the original binaries. All in all, our results support that binaries that were produced by our attacks were able to evade ML-based static detectors that are used by anti-virus vendors.

4.7 Correctness

A key feature of our attacks is that they transform binaries to mislead DNNs while preserving their functionality. We followed standard practices from the binary-diversification literature [52, 53, 71] to ensure that the functionality of the binaries was kept intact after being processed by our attacks. First, we transformed ten different benign binaries (e.g., `python.exe` of Python version 2.7, and Cygwin's[5] `less.exe` and `grep.exe`) with our attacks and manually validated that they functioned properly after being transformed. For example, we were still able to search files with `grep` after the transformations. Second, we transformed the `.exe` and `.dll` files of a stress-testing tool[6] with our attacks and checked that the tool's tests passed after the transformations. Using stress-testing tools to evaluate binary-transformation correctness is common, as such tools are expected to cover most branches affected by the transformations. Third, and last, we also transformed ten malware binaries and used the Cuckoo Sandbox [36]—a popular sandbox for malware analysis—to check that their behavior remained the same. All ten binaries attempted to access the same hosts, IP addresses, files, APIs, and registry keys before and after being transformed.

5 Potential Mitigations

Our proposed attacks achieved high success rates at fooling DNNs for malware detection in white-box and black-box settings. The attacks were also able to mislead commercial anti-viruses, especially ones that leverage ML algorithms. To protect users and their systems, it is important to develop mitigation measures to make malware detection robust against evasion by our attacks. Our efforts to explore mitigations, however, have met with limited success.

5.1 Prior Defenses

We considered several prior defenses to mitigate our attacks, but, unfortunately, most showed little promise. For instance, adversarial training (e.g., [33, 57]) is currently infeasible, as the attacks are computationally expensive. Depending on the attack variant, it took an average of 283 to 4424 s to run an attack. As a result, running just a single epoch of adversarial training would to take several weeks (using our hardware configuration), as each iteration of training requires running an attack

[5] https://www.cygwin.com/.

[6] https://www.passmark.com/products/performancetest/.

for every sample in the training batch. Moreover, while adversarial training might increase the DNNs' robustness against attackers using certain transformation types, attackers using new transformation types may still succeed at evasion [29]. Defenses that provide formal guarantees (e.g., [51, 67]) are even more computationally expensive than adversarial training. Moreover, those defenses are restricted to adversarial perturbations that, unlike the ones produced by our attacks, have small L_∞- and L_2-norms. Prior defenses that transform the input before classification (e.g., via quantization [101]) are designed mainly for images and do not directly apply to binaries. Lastly, signature-based malware detection would not be effective, as our attacks are stochastic and produce different variants of the binaries after different executions.

Differently from prior attacks on DNNs for malware detection [49, 55, 89], our attacks do not merely append adversarially crafted bytes to binaries, or insert them between sections. Such attacks may be defended against by detecting and sanitizing the inserted bytes via static analysis methods (e.g., similarly to the proof of concept shown in Sect. 4.4, or using other methods [56]). Instead, our attacks transform binaries' original code and extend binaries only by inserting instructions that are executed at run time at various parts of the binaries. As a result, our attacks are difficult to defend against via static or dynamic analyses methods (e.g., by detecting and removing unreachable code), especially when augmented by measures to evade these methods.

Binary normalization [3, 18, 98] is a defense that initially seemed viable for defending against our attacks. The high-level idea of normalization is to employ certain transformations to map binaries to a standard form and thus undo attackers' evasion attempts before classifying the binaries as malicious or benign. For example, Christodorescu et al. proposed a method to detect and remove semantic nops from binaries before classification and showed that it improves the performance of commercial anti-viruses [18]. To mitigate our *Disp*-based attacks, we considered using the semantic nop detection and removal method followed by a method to restore the displaced code to its original location. Unfortunately, we realized that such a defense can be undermined using *opaque predicates* [23, 68]. Opaque predicates are predicates whose value (w.l.g., assume true) is known a priori to the attacker but is hard for the defender to deduce. Often, they are based on *NP*-hard problems [68]. Using opaque predicates, attackers can produce semantic nops that include instructions that affect the memory and registers only if an opaque predicate evaluates to false. Since opaque predicates are hard for defenders to deduce, the defenders are likely to have to assume that the semantic nops impact the behavior of the program. As a result, the semantic nops would survive the defenders' detection and removal attempts. As an alternative to opaque predicates, attackers can also use *evasive predicates*—predicates that evaluate to true or false with an overwhelming probability (e.g., checking if a randomly drawn 32-bit integer is equal to 0) [10]. In this case, the binary will function properly the majority of the time and may function differently or crash once every many executions.

The normalization methods proposed by prior work would not apply to the transformations performed by our *IPR*-based attacks. Therefore, we explored methods to normalize binaries to a standard form to undo the effects of *IPR* before classification. We found that a normalization process that leverages the *IPR* transformations to produce the form with the lowest lexicographic representation (where the alphabet contains all possible 256 byte values) prevented *IPR*-based attacks. Formally, if $[x]$ is the equivalence class of binaries that are functionally equivalent to x and that can be produced via the *IPR* transformation types, then the normalization process produces an output $norm(x) \in [x]$, such that, $norm(x) \leq x_i$ for every $x_i \in [x]$. Appendix 3 presents an algorithm that computes the normalized form of a binary when executed for a large number of iterations and approximates it when executed for a few iterations. At a high level, the algorithm applies the *IPR* transformations iteratively in an effort to reduce the lexicographic representation after every iteration. We found that executing the algorithm for ten iterations was sufficient to defend against *IPR*-based attacks. In particular, we executed the normalization algorithm using the malicious and benign binaries produced by the *IPR*-based attacks to fool *Endgame* in the white-box setting and found that the success rates dropped to 3% and 0%, respectively, compared to 62% and 74% before normalization. At the same time, the classification accuracy over the original binaries was not affected by normalization. As our experiments in Sect. 4 have shown, generating functionally equivalent variants of binaries via random transformations results in correct classifications almost all of the time. Normalization of binaries deterministically led to specific variants that were correctly classified with high likelihood.

5.2 Masking Random Instructions

While normalization was useful for defending against *IPR*-based attacks, it cannot mitigate the more pernicious *Disp*-based attacks that are augmented with opaque or evasive predicates. Moreover, normalization has the general limitations that attackers could use transformations that the normalization algorithm is not aware of or could obfuscate code to inhibit normalization. Therefore, we explored additional defensive measures. In particular, motivated by the fact that randomizing binaries without the guidance of an optimization process is unlikely to lead to misclassification, we explored whether masking instructions at random can mitigate attacks while maintaining high performance on the original binaries. The defense works by selecting a random subset of the bytes that pertain to instructions and masking them with zeros (a commonly used value to pad sections in binaries). While the masking is likely to result in an ill-formed binary that is unlikely to execute properly (if at all), the masking only occurs before classification, which does not require a functional binary. Depending on the classification result, one can decide whether or not to execute the unmasked binary.

We tested the defense on binaries generated via the *IPR+Disp*-5 white-box attack on *Kaggle* and found that it was effective at mitigating attacks. For example, when masking 25% of the bytes pertaining to instructions, the success rates of the attack decreased from 83%–100% for malicious and benign binaries against the three DNNs to 0%–20%, while the accuracy on the original samples was only slightly affected (e.g., it became 94% for *Endgame*). Masking less than 25% of the instructions' bytes was not as effective at mitigating attacks, while masking more than 25% led to a significant decrease in accuracy on the original samples.

5.3 Detecting Adversarial Examples

To prevent binaries transformed with our attacks (i.e., adversarial examples) from fooling malware detection, defenders may attempt to deploy methods to detect them. In cases of positive detections of adversarial examples, defenders may immediately classify them as malicious (regardless of whether they were originally malicious or benign). For example, because *Disp*-based attacks increase binaries' sizes and introduce additional `jmp` instructions, defenders may train statistical ML models that use features such as binaries' sizes and the ratio between `jmp` instructions and other instructions to detect adversarial examples. While training relatively accurate detection models may be feasible, we expect this task to be difficult, as the attacks increase binaries' sizes only slightly (1%–5%), and do not introduce many `jmp` instructions (7% median increase for binaries transformed via *Disp*-5). Furthermore, approaches for detecting adversarial examples are likely to be susceptible to evasion attacks (e.g., by introducing instructions after opaque predicates to decrease the ratio between `jmp` instructions and others). Last, another risk that defenders should take into account is that the defense should be able to precisely distinguish between adversarial examples and non-adversarial benign binaries that are transformed by similar methods to mitigate code-reuse attacks [52, 71].

5.4 Takeaways

While masking a subset of the bytes that pertain to instructions led to better performance on adversarial examples, it was still unable to prevent all evasion attempts. Although the defense may raise the bar for attackers and make attacks even more difficult if combined with a method to detect adversarial examples, these defenses do not provide formal guarantees and so attackers may be able to adapt to undermine them. For example, attackers may build on techniques for optimization over expectations to generate binaries that would mislead the DNNs even when masking a large number of instructions, in a similar manner to how attackers can evade image-classification DNNs under varying lighting conditions and camera angles [7, 30, 85, 86]. In fact, prior work has already demonstrated how defenses without formal guarantees are often vulnerable to adaptive, more sophisticated, attacks [6]. Thus, since there is no clear defense to prevent attacks against the DNNs that we studied in this work, or even general methods to prevent attackers from fooling ML models via arbitrary perturbations, we advocate for augmenting malware-detection systems with methods that are not based on ML (e.g., ones using templates to reason about the semantics of programs [17]), and against the use of ML-only detection methods, as has become recently popular [25].

6 Conclusion

We develop techniques to defend systems through deception by studying the proxy problem of malware detection; in particular, we develop evasion attacks on DNNs for malware detection. Differently from prior work, the attacks do not merely insert adversarially crafted bytes to mislead detection. Instead, guided by optimization processes, our attacks transform the instructions of binaries to fool malware detection while keeping functionality of the binaries intact. As a result, these attacks are challenging to defend against. We conservatively evaluated different variants of our attack against three DNNs under white-box and black-box settings and found the attacks successful as often as 100% of the time. Moreover, we found that the attacks pose a security risk to commercial anti-viruses, particularly ones using ML, achieving evasion success rates of up to 85%. We explored several potential defenses and found some to be promising. Nevertheless, adaptive adversaries remain a risk, and we recommend the deployment of multiple detection algorithms, including ones not based on ML, to raise the bar against such adversaries.

Acknowledgments We would like to thank Leyla Bilge, Sandeep Bhatkar, Yufei Han, and Kevin Roundy for helpful discussions. This work was supported in part by the Multidisciplinary University Research Initiative (MURI) Cyber Deception grant under ARO award W911NF-17-1-0370; by NSF grants 1801391 and 2113345; by the National Security Agency under award H9823018D0008; by gifts from Google and Nvidia, and from Lockheed Martin and NATO through Carnegie Mellon CyLab; by a CyLab Presidential Fellowship and a NortonLifeLock Research Group Fellowship; and by a DoD National Defense Science and Engineering Graduate fellowship.

Appendix 1: Comparison to Kreuk et al. and Success After Sanitization

While Kreuk et al.'s attack achieved success rates comparable to ours, their attack is easier to defend against. As a proof of concept, we implemented a sanitization method to defend against the attack using our alternate dataset described in Sect. 4.6. The method finds all the sections in a binary that do not contain instructions (using the IDAPro disassembler [40]) and masks the sections' content with zeros. As Kreuk et al.'s attack does not add functional instructions to the binaries, the defense masks the adversarial bytes that the attack introduces. Consequently, the evasion success rates of the attack drop significantly. In fact, except for when attempting to mislead the *Endgame* DNN with malicious binaries, the success rates of the *Kreuk* attacks dropped below 15%. This defense had little-to-no effect on our attacks, however: e.g., *Disp*-5 still achieved 92% and 100% success rates against *MalConv* for malicious and benign binaries, respectively. Moreover, the classification accuracy remained high both for malicious (99%) and benign (93%) binaries after the defense. Figure 7 in Appendix 1 presents the full results of the impact of sanitization on attacks' success on the *Kaggle* dataset.

Figure 7 shows the success rates of attacks when sanitizing bytes in sections that do not include instructions. In particular, we replaced byte values in such sections with zeros, as described in Sect. 4.4. Our attacks maintained high success rates after sanitization (e.g., >90% for *Disp*-5), whereas the success rates of the *Kreuk* attacks dropped below 15% in most cases.

Appendix 2: Our Attacks' Transferability to Commercial Anti-Viruses

Table 3 summarizes the effect of different attack variants on the number of positive detections (i.e., classification of binaries as malicious) by the anti-viruses featured on VirusTotal. Sect. 4.6 describes the experiment and explains the results.

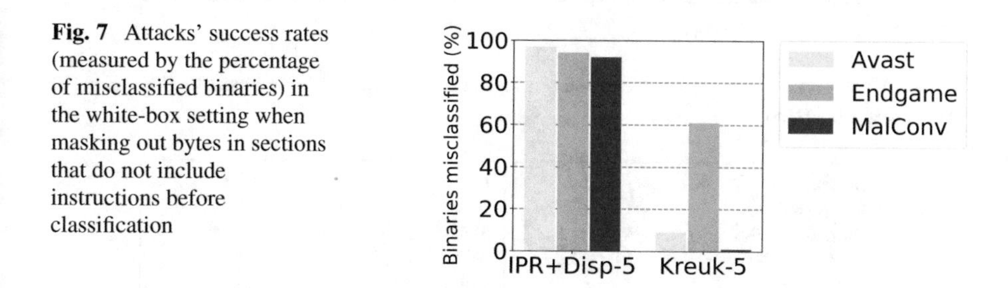

Fig. 7 Attacks' success rates (measured by the percentage of misclassified binaries) in the white-box setting when masking out bytes in sections that do not include instructions before classification

Table 3 The median number of VirusTotal anti-viruses that positively detected (i.e., as malicious) malicious binaries that were transformed by our white-box attacks (columns) to mislead the different DNNs (rows). The median number of anti-viruses that positively detected for the original malicious binaries is 55. Cases in which the change in the number of detections is statistically significant are in bold

DNN	*IPR*	*Disp*-1	*Disp*-3	*Disp*-5	*IPR+Disp*-1	*IPR+Disp*-3	*IPR+Disp*-5
AvastNet	–	**36**	**35**	**36**	**36**	**35**	**36**
Endgame	**33**	**35**	**36**	**35**	**35**	**36**	**35**
MalConv	–	**36**	**35**	**36**	**36**	**35**	**36**

Appendix 3: In-Place Normalization

In this section, we present a normalization process to map binaries to a standard form and undo the effect of the *IPR* transformations on classification. Specifically, the normalization process maps binaries to the functionally equivalent variant with the lowest lexicographic presentation that is achievable via the *IPR* transformation types. For each transformation type, we devise an operation that would decrease a binary's lexicographic representation when applied: *(1)* instructions would be replaced with equivalent ones only if the new instructions are lexicographically lower (*Eqv*); *(2)* registers in functions would be reassigned only if the byte representation of the first impacted instruction would decrease (*Regs*); *(3)* instructions would be reordered such that each time we would extract the instruction from the dependence graph with the lowest byte representation that does not depend on any of the remaining instructions in the graph (*Ord1*); and *(4)* `push` and `pop` instructions that save register values across function calls would be reordered to decrease the lexicographic representation while maintaining the last-in-first-out order (*Ord2*). Figure 8 depicts an example of replacing one instruction with an equivalent one via *Eqv* to decrease the lexicographic order of code.

```
sub eax, -0x20      (83e8e0)
test ebx, ebx       (85db)
```

(a)

```
add eax, 0x20       (83c020)
or ebx, ebx         (09db)
```

(b)

Fig. 8 An example of normalizing code via *Eqv*. The original code (**a**) is transformed via *Eqv* (**b**) to decrease the lexicographic order

```
push edx            (52)
push ebx            (53)
mov dh, 0x4         (b604)
mov bh, 0x3         (b703)
pop ebx             (5b)
pop edx             (5a)
```

(a)

```
push edx            (52)
push ebx            (53)
mov bh, 0x4         (b704)
mov dh, 0x3         (b603)
pop ebx             (5b)
pop edx             (5a)
```

(b)

```
push edx            (52)
push ebx            (53)
mov bh, 0x3         (b703)
mov dh, 0x4         (b604)
pop ebx             (5b)
pop edx             (5a)
```

(c)

```
push edx            (52)
push ebx            (53)
mov dh, 0x3         (b603)
mov bh, 0x4         (b704)
pop ebx             (5b)
pop edx             (5a)
```

(d)

Fig. 9 The normalization process can get stuck in a local minima. The lexicographic order of the original code (**a**) increases when reassigning registers (**b**) or reordering instructions (**c**). However, composing the two transformation (**d**) decreases the lexicographic order

Unfortunately, as shown in Fig. 9, when the different types of transformation types are composed, applying individual normalization operations does not necessarily lead to the binary's variant with the minimal lexicographic representation, as the procedure may be stuck in a local minima. To this end, we propose a stochastic algorithm that is guaranteed to converge to binaries' normalized variants if executed for a sufficiently large number of iterations.

The algorithm receives a binary x and the number of iterations *niters* as inputs. It begins by drawing a random variant of x, by applying all the transformation types to each function at random. The algorithm then proceeds to apply each of the individual normalization operations to decrease the lexicographic representation of the binary while self-supervising the normalization process. Specifically, the algorithm keeps track of the last iteration an operation decreased the binary's representation. If none of the four operations affects any of the functions, we deduce that the normalization process is stuck in a (global or local) minima, and a random binary is drawn again by randomizing all functions and the normalization process restarts.

When $niters \rightarrow \infty$ (i.e., the number of iterations is large enough). This algorithm would eventually converge to a global minima. Namely, it would find the variant of x with the minimal lexicographic representation. In fact, we are guaranteed to find $norm(x)$ even if we simply apply the transformation types at random x for $niters \rightarrow \infty$ iterations. When testing the algorithm with two binaries of moderate size, we found that *niters*=2000 was sufficient to converge for the same respective variants after every run. These variants are likely to be the global minima. However, executing the algorithm for 2000 iterations is computationally expensive, and impractical within the context of a widely deployed malware-detection system. Hence, for the purpose of our experiments, we set *niters*=10, which we found to be sufficient to successfully mitigate the majority of attacks.

References

1. Anderson, H.S., Kharkar, A., Filar, B., Roth, P.: Evading machine learning malware detection. Black Hat (2017)
2. Anderson, H.S., Roth, P.: Ember: An open dataset for training static PE malware machine learning models. Preprint (2018). arXiv:1804.04637
3. Armoun, S.E., Hashemi, S.: A general paradigm for normalizing metamorphic malwares. In: Proc. FIT (2012)
4. Arp, D., Spreitzenbarth, M., Hubner, M., Gascon, H., Rieck, K., Siemens, C.: Drebin: Effective and explainable detection of android malware in your pocket. In: Proc. NDSS (2014)
5. Athalye, A., Carlini, N.: On the robustness of the CVPR 2018 white-box adversarial example defenses. arXiv:1804.03286 (2018)
6. Athalye, A., Carlini, N., Wagner, D.: Obfuscated gradients give a false sense of security: Circumventing defenses to adversarial examples. In: Proc. ICML (2018)
7. Athalye, A., Engstrom, L., Ilyas, A., Kwok, K.: Synthesizing robust adversarial examples. In: Proc. ICML (2018)

8. Avast Software: Avast malware detection and blocking. https://www.avast.com/en-us/technology/malware-detection-and-blocking (2020). Accessed 12/09/2020
9. Baluja, S., Fischer, I.: Adversarial transformation networks: Learning to generate adversarial examples. In: Proc. AAAI (2018)
10. Barak, B., Bitansky, N., Canetti, R., Kalai, Y.T., Paneth, O., Sahai, A.: Obfuscation for evasive functions. In: Proc. TCC (2014)
11. Biggio, B., Corona, I., Maiorca, D., Nelson, B., Šrndić, N., Laskov, P., Giacinto, G., Roli, F.: Evasion attacks against machine learning at test time. In: Proc. ECML PKDD (2013)
12. Biondi, F., Enescu, M., Given-Wilson, T., Legay, A., Noureddine, L., Verma, V.: Effective, efficient, and robust packing detection and classification. Computers and Security (2018)
13. Carlini, N., Wagner, D.: Adversarial examples are not easily detected: Bypassing ten detection methods. In: Proc. AISec (2017)
14. Carlini, N., Wagner, D.: Towards evaluating the robustness of neural networks. In: Proc. IEEE S&P (2017)
15. Cheng, B., Ming, J., Fu, J., Peng, G., Chen, T., Zhang, X., Marion, J.Y.: Towards paving the way for large-scale windows malware analysis: Generic binary unpacking with orders-of-magnitude performance boost. In: Proc. CCS (2018)
16. Christodorescu, M., Jha, S.: Testing malware detectors. In: Proc. ISSTA (2004)
17. Christodorescu, M., Jha, S., Seshia, S.A., Song, D., Bryant, R.E.: Semantics-aware malware detection. In: Proc. IEEE S&P (2005)
18. Christodorescu, M., Kinder, J., Jha, S., Katzenbeisser, S., Veith, H.: Malware normalization. Tech. rep., U. Wisconsin-Madison (2005)
19. Chronicle: Virustotal. https://www.virustotal.com/ (2004–). Online; accessed 17 June 2019
20. Cisco: Clamav: Creating signature for clamav. https://www.clamav.net/documents/creating-signatures-for-clamav (2020). Accessed 12/10/2020
21. Cisse, M., Adi, Y., Neverova, N., Keshet, J.: Houdini: Fooling deep structured prediction models. In: Proc. NIPS (2017)
22. Cohen, J.M., Rosenfeld, E., Kolter, J.Z.: Certified adversarial robustness via randomized smoothing. Preprint (2019). arXiv:1902.02918
23. Collberg, C., Thomborson, C., Low, D.: A taxonomy of obfuscating transformations. Tech. rep., The University of Auckland (1997)
24. Coull, S., Gardner, C.: What are deep neural networks learning about malware? https://www.fireeye.com/blog/threat-research/2018/12/what-are-deep-neural-networks-learning-about-malware.html (2018). Online; accessed 1 July 2019
25. Cylance Inc.: Cylance: Artificial intelligence based advanced threat prevention. https://www.blackberry.com/us/en/cylance (2019). Accessed 6/17/2019
26. Dang, H., Huang, Y., Chang, E.C.: Evading classifiers by morphing in the dark. In: Proc. CCS (2017)
27. Demetrio, L., Biggio, B., Lagorio, G., Roli, F., Armando, A.: Explaining vulnerabilities of deep learning to adversarial malware binaries. Preprint (2019). arXiv:1901.03583
28. Demontis, A., Melis, M., Biggio, B., Maiorca, D., Arp, D., Rieck, K., Corona, I., Giacinto, G., Roli, F.: Yes, machine learning can be more secure! A case study on android malware detection. IEEE Transactions on Dependable and Secure Computing (2017)
29. Engstrom, L., Tsipras, D., Schmidt, L., Madry, A.: A rotation and a translation suffice: Fooling CNNs with simple transformations. In: Proc. NeurIPSW (2017)
30. Evtimov, I., Eykholt, K., Fernandes, E., Kohno, T., Li, B., Prakash, A., Rahmati, A., Song, D.: Robust physical-world attacks on machine learning models. In: Proc. CVPR (2018)
31. Feinman, R., Curtin, R.R., Shintre, S., Gardner, A.B.: Detecting adversarial samples from artifacts. arXiv:1703.00410 (2017)
32. Fleshman, W., Raff, E., Sylvester, J., Forsyth, S., McLean, M.: Non-negative networks against adversarial attacks. Preprint (2018). arXiv:1806.06108
33. Goodfellow, I.J., Shlens, J., Szegedy, C.: Explaining and harnessing adversarial examples. In: Proc. ICLR (2015)

34. Grosse, K., Manoharan, P., Papernot, N., Backes, M., McDaniel, P.: On the (statistical) detection of adversarial examples. Preprint (2017). arXiv:1702.06280
35. Grosse, K., Papernot, N., Manoharan, P., Backes, M., McDaniel, P.: Adversarial examples for malware detection. In: Proc. ESORICS (2017)
36. Guarnieri, C., Tanasi, A., Bremer, J., Schloesser, M.: The Cuckoo Sandbox. https://cuckoosandbox.org/ (2012). Accessed 6/21/2019
37. Guo, C., Rana, M., Cisse, M., van der Maaten, L.: Countering adversarial images using input transformations (2018)
38. Harris, L.C., Miller, B.P.: Practical analysis of stripped binary code. ACM SIGARCH Comput. Architect. News **33**(5), 63–68 (2005)
39. Herley, C.: So long, and no thanks for the externalities: the rational rejection of security advice by users. In: Proc. NSPW (2009)
40. Hex-Rays: IDA: About. https://www.hex-rays.com/products/ida/. Online; accessed 13 September 2019
41. Hu, W., Tan, Y.: Generating adversarial malware examples for black-box attacks based on GAN. Preprint (2017). arXiv:1702.05983
42. Ilyas, A., Engstrom, L., Madry, A.: Prior convictions: Black-box adversarial attacks with bandits and priors. In: Proc. ICLR (2019)
43. Incer, I., Theodorides, M., Afroz, S., Wagner, D.: Adversarially robust malware detection using monotonic classification. In: Proc. IWSPA (2018)
44. Jindal, C., Salls, C., Aghakhani, H., Long, K., Kruegel, C., Vigna, G.: Neurlux: Dynamic malware analysis without feature engineering. In: Proc. ACSAC (2019)
45. Junod, P., Rinaldini, J., Wehrli, J., Michielin, J.: Obfuscator-llvm–software protection for the masses. In: Proc. IWSP (2015)
46. Kannan, H., Kurakin, A., Goodfellow, I.: Adversarial logit pairing. Preprint (2018). arXiv:1803.06373
47. Kantchelian, A., Tygar, J., Joseph, A.D.: Evasion and hardening of tree ensemble classifiers. In: Proc. ICML (2016)
48. Kennedy, J., Batchelor, D., Robertson, C., Satran, M., LeBLanc, M.: PE format. https://docs.microsoft.com/en-us/windows/desktop/debug/pe-format (2019). Accessed on 06-03-2019
49. Kolosnjaji, B., Demontis, A., Biggio, B., Maiorca, D., Giacinto, G., Eckert, C., Roli, F.: Adversarial malware binaries: Evading deep learning for malware detection in executables. In: Proc. EUSIPCO (2018)
50. Kolter, J.Z., Maloof, M.A.: Learning to detect and classify malicious executables in the wild. Journal of Machine Learning Research (2006)
51. Kolter, J.Z., Wong, E.: Provable defenses against adversarial examples via the convex outer adversarial polytope. In: Proc. ICML (2018)
52. Koo, H., Polychronakis, M.: Juggling the gadgets: Binary-level code randomization using instruction displacement. In: Proc. AsiaCCS (2016)
53. Koo, H., Chen, Y., Lu, L., Kemerlis, V.P., Polychronakis, M.: Compiler-assisted code randomization. In: Proc. IEEE S&P (2018)
54. Krčál, M., Švec, O., Bálek, M., Jašek, O.: Deep convolutional malware classifiers can learn from raw executables and labels only. In: Proc. ICLRW (2018)
55. Kreuk, F., Barak, A., Aviv-Reuven, S., Baruch, M., Pinkas, B., Keshet, J.: Adversarial examples on discrete sequences for beating whole-binary malware detection. In: Proc. NeurIPSW (2018)
56. Kruegel, C., Robertson, W., Valeur, F., Vigna, G.: Static disassembly of obfuscated binaries. In: Proc. USENIX Security (2004)
57. Kurakin, A., Goodfellow, I., Bengio, S.: Adversarial machine learning at scale. In: Proc. ICLR (2017)
58. Larsen, P., Homescu, A., Brunthaler, S., Franz, M.: SoK: Automated software diversity. In: Proc. IEEE S&P (2014)
59. Le, Q., Mikolov, T.: Distributed representations of sentences and documents. In: Proc. ICML (2014)

60. Lecuyer, M., Atlidakis, V., Geambasu, R., Hsu, D., Jana, S.: Certified robustness to adversarial examples with differential privacy. In: Proc. IEEE S&P (2019)
61. Liao, F., Liang, M., Dong, Y., Pang, T., Zhu, J., Hu, X.: Defense against adversarial attacks using high-level representation guided denoiser. In: Proc. CVPR (2018)
62. Madry, A., Makelov, A., Schmidt, L., Tsipras, D., Vladu, A.: Towards deep learning models resistant to adversarial attacks. In: Proc. ICLR (2018)
63. Massalin, H.: Superoptimizer: A look at the smallest program. ACM SIGARCH Computer Architecture News **15**(5), 122–126 (1987)
64. Meng, D., Chen, H.: Magnet: A two-pronged defense against adversarial examples. In: Proc. CCS (2017)
65. Meng, X., Miller, B.P., Jha, S.: Adversarial binaries for authorship identification. Preprint (2018). arXiv:1809.08316
66. Metzen, J.H., Genewein, T., Fischer, V., Bischoff, B.: On detecting adversarial perturbations. In: Proc. ICLR (2017)
67. Mirman, M., Gehr, T., Vechev, M.: Differentiable abstract interpretation for provably robust neural networks. In: Proc. ICML (2018)
68. Moser, A., Kruegel, C., Kirda, E.: Limits of static analysis for malware detection. In: Proc. ACSAC (2007)
69. Oberhumer, M., Molnar, L., Reiser, J.: UPX: The ultimate packer for executables. https://upx.github.io/. Online; accessed 1/13/2020
70. Papernot, N., McDaniel, P., Jha, S., Fredrikson, M., Celik, Z.B., Swami, A.: The limitations of deep learning in adversarial settings. In: Proc. IEEE Euro S&P (2016)
71. Pappas, V., Polychronakis, M., Keromytis, A.D.: Smashing the gadgets: Hindering return-oriented programming using in-place code randomization. In: Proc. IEEE S&P (2012)
72. Park, D., Khan, H., Yener, B.: Generation evaluation of adversarial examples for malware obfuscation. In: Proc. ICMLA, pp. 1283–1290 (2019)
73. Pierazzi, F., Pendlebury, F., Cortellazzi, J., Cavallaro, L.: Intriguing properties of adversarial ml attacks in the problem space. In: Proc. IEEE S&P (2020)
74. Qin, Y., Carlini, N., Goodfellow, I., Cottrell, G., Raffel, C.: Imperceptible, robust, and targeted adversarial examples for automatic speech recognition. In: Proc. ICML (2019)
75. Quiring, E., Maier, A., Rieck, K.: Misleading authorship attribution of source code using adversarial learning. In: Proc. USENIX Security (2019)
76. Raff, E., Barker, J., Sylvester, J., Brandon, R., Catanzaro, B., Nicholas, C.K.: Malware detection by eating a whole exe. In: Proc. AAAIW (2018)
77. Roberts, M.: Virusshare. https://virusshare.com/ (2012). Online; accessed 18 June 2019
78. Ronen, R., Radu, M., Feuerstein, C., Yom-Tov, E., Ahmadi, M.: Microsoft malware classification challenge. Preprint (2018). arXiv:1802.10135
79. Rosenberg, I., Shabtai, A., Rokach, L., Elovici, Y.: Generic black-box end-to-end attack against state of the art API call based malware classifiers. In: Proc. RAID (2018)
80. Roundy, K.A., Miller, B.P.: Binary-code obfuscations in prevalent packer tools. ACM Computing Surveys (CSUR) **46**(1), 4 (2013)
81. Samangouei, P., Kabkab, M., Chellappa, R.: Defense-GAN: Protecting classifiers against adversarial attacks using generative models. In: Proc. ICLR (2018)
82. Schkufza, E., Sharma, R., Aiken, A.: Stochastic superoptimization. In: Proc. ASPLOS (2013)
83. Schönherr, L., Kohls, K., Zeiler, S., Holz, T., Kolossa, D.: Adversarial attacks against automatic speech recognition systems via psychoacoustic hiding. In: Proc. NDSS (2019)
84. Sconzo, M.: Packer yara ruleset. https://github.com/sooshie/packerid (2014). Online; accessed 18 June 2019
85. Sharif, M., Bhagavatula, S., Bauer, L., Reiter, M.K.: Accessorize to a crime: Real and stealthy attacks on state-of-the-art face recognition. In: Proc. CCS (2016)
86. Sharif, M., Bhagavatula, S., Bauer, L., Reiter, M.K.: Adversarial generative nets: Neural network attacks on state-of-the-art face recognition. Preprint (2017). arXiv:1801.00349
87. Srinivasan, V., Marban, A., Müller, K.R., Samek, W., Nakajima, S.: Counterstrike: Defending deep learning architectures against adversarial samples by langevin dynamics with supervised denoising autoencoder. Preprint (2018). arXiv:1805.12017

88. Srndic, N., Laskov, P.: Practical evasion of a learning-based classifier: A case study. In: Proc. IEEE S&P (2014)
89. Suciu, O., Coull, S.E., Johns, J.: Exploring adversarial examples in malware detection. In: Proc. AAAIW (2018)
90. Symantec: How does Symantec Endpoint Protection use advanced machine learning? https://techdocs.broadcom.com/us/en/symantec-security-software/endpoint-security-and-management/endpoint-protection/all/Using-policies-to-manage-security/preventing-and-handling-virus-and-spyware-attacks-v40739565-d49e172/how-does-use-advanced-machine-learning-v120625733-d47e275.html (2019). Accessed on 01-12-2020
91. Szegedy, C., Zaremba, W., Sutskever, I., Bruna, J., Erhan, D., Goodfellow, I.J., Fergus, R.: Intriguing properties of neural networks. In: Proc. ICLR (2014)
92. Szor, P.: The Art of Computer Virus Research and Defense. Pearson Education (2005)
93. TrendMicro: Trendmicro machine learning. https://www.trendmicro.com/vinfo/us/security/definition/machine-learning (2020). Accessed 12/09/2020
94. Ugarte-Pedrero, X., Balzarotti, D., Santos, I., Bringas, P.G.: SoK: Deep packer inspection: A longitudinal study of the complexity of run-time packers. In: Proc. IEEE S&P (2015)
95. Vipre: Vipre android security. https://www.vipre.com/vipre-android-security/ (2020). Accessed 12/10/2020
96. VirusTotal: Packer yara ruleset. https://github.com/Yara-Rules/rules/tree/master/Packers (2016). Online; accessed 18 June 2019
97. Wagner, D., Soto, P.: Mimicry attacks on host-based intrusion detection systems. In: Proc. CCS (2002)
98. Walenstein, A., Mathur, R., Chouchane, M.R., Lakhotia, A.: Normalizing metamorphic malware using term rewriting. In: Proc. SCAM (2006)
99. Wang, S., Wang, P., Wu, D.: Uroboros: Instrumenting stripped binaries with static reassembling. In: Proc. SANER (2016)
100. Xie, C., Wu, Y., van der Maaten, L., Yuille, A., He, K.: Feature denoising for improving adversarial robustness. Preprint (2018). arXiv:1812.03411
101. Xu, W., Evans, D., Qi, Y.: Feature squeezing: Detecting adversarial examples in deep neural networks. In: Proc. NDSS (2018)
102. Xu, W., Qi, Y., Evans, D.: Automatically evading classifiers. In: Proc. NDSS (2016)
103. Zhang, H., Yu, Y., Jiao, J., Xing, E.P., Ghaoui, L.E., Jordan, M.I.: Theoretically principled trade-off between robustness and accuracy. Preprint (2019). arXiv:1901.08573